Pitman Research Notes in Mathematics Series

Main Editors
H. Brezis, Université de Paris
R.G. Douglas, State University of New York at Stony Brook
A. Jeffrey, University of Newcastle upon Tyne *(Founding Editor)*

Editorial Board
H. Amann, University of Zürich
R. Aris, University of Minnesota
G.I. Barenblatt, University of Cambridge
A. Bensoussan, INRIA, France
P. Bullen, University of British Columbia
S. Donaldson, University of Oxford
R.J. Elliott, University of Alberta
R.P. Gilbert, University of Delaware
D. Jerison, Massachusetts Institute of Technology
K. Kirchgässner, Universität Stuttgart
B. Lawson, State University of New York at Stony Brook
B. Moodie, University of Alberta
S. Mori, Kyoto University
L.E. Payne, Cornell University
G.F. Roach, University of Strathclyde
I. Stakgold, University of Delaware
W.A. Strauss, Brown University
S.J. Taylor, University of Virginia

Submission of proposals for consideration
Suggestions for publication, in the form of outlines and representative samples, are invited by the Editorial Board for assessment. Intending authors should approach one of the main editors or another member of the Editorial Board, citing the relevant AMS subject classifications. Alternatively, outlines may be sent directly to the publisher's offices. Refereeing is by members of the board and other mathematical authorities in the topic concerned, throughout the world.

Preparation of accepted manuscripts
On acceptance of a proposal, the publisher will supply full instructions for the preparation of manuscripts in a form suitable for direct photo-lithographic reproduction. Specially printed grid sheets can be provided and a contribution is offered by the publisher towards the cost of typing. Word processor output, subject to the publisher's approval, is also acceptable.

Illustrations should be prepared by the authors, ready for direct reproduction without further improvement. The use of hand-drawn symbols should be avoided wherever possible, in order to maintain maximum clarity of the text.

The publisher will be pleased to give any guidance necessary during the preparation of a typescript, and will be happy to answer any queries.

Important note
In order to avoid later retyping, intending authors are strongly urged not to begin final preparation of a typescript before receiving the publisher's guidelines. In this way it is hoped to preserve the uniform appearance of the series.

Longman Group Ltd
Longman House
Burnt Mill
Harlow, Essex, CM20 2JE
UK
(Telephone (0) 1279 426721)

Titles in this series. A full list is available from the publisher on request.

301 Generalized fractional calculus and applications
  V Kiryakova
302 Nonlinear partial differential equations and their applications. Collège de France Seminar Volume XII
  H Brezis and J L Lions
303 Numerical analysis 1993
  D F Griffiths and G A Watson
304 Topics in abstract differential equations
  S Zaidman
305 Complex analysis and its applications
  C C Yang, G C Wen, K Y Li and Y M Chiang
306 Computational methods for fluid-structure interaction
  J M Crolet and R Ohayon
307 Random geometrically graph directed self-similar multifractals
  L Olsen
308 Progress in theoretical and computational fluid mechanics
  G P Galdi, J Málek and J Necas
309 Variational methods in Lorentzian geometry
  A Masiello
310 Stochastic analysis on infinite dimensional spaces
  H Kunita and H-H Kuo
311 Representations of Lie groups and quantum groups
  V Baldoni and M Picardello
312 Common zeros of polynomials in several variables and higher dimensional quadrature
  Y Xu
313 Extending modules
  N V Dung, D van Huynh, P F Smith and R Wisbauer
314 Progress in partial differential equations: the Metz surveys 3
  M Chipot, J Saint Jean Paulin and I Shafrir
315 Refined large deviation limit theorems
  V Vinogradov
316 Topological vector spaces, algebras and related areas
  A Lau and I Tweddle
317 Integral methods in science and engineering
  C Constanda
318 A method for computing unsteady flows in porous media
  R Raghavan and E Ozkan
319 Asymptotic theories for plates and shells
  R P Gilbert and K Hackl
320 Nonlinear variational problems and partial differential equations
  A Marino and M K V Murthy
321 Topics in abstract differential equations II
  S Zaidman
322 Diffraction by wedges
  B Budaev
323 Free boundary problems: theory and applications
  J I Diaz, M A Herrero, A Liñan and J L Vazquez
324 Recent developments in evolution equations
  A C McBride and G F Roach
325 Elliptic and parabolic problems: Pont-à-Mousson 1994
  C Bandle, J Bemelmans, M Chipot, J Saint Jean Paulin and I Shafrir
326 Calculus of variations, applications and computations: Pont-à-Mousson 1994
  C Bandle, J Bemelmans, M Chipot, J Saint Jean Paulin and I Shafrir
327 Conjugate gradient type methods for ill-posed problems
  M Hanke
328 A survey of preconditioned iterative methods
  A M Bruaset
329 A generalized Taylor's formula for functions of several variables and certain of its applications
  J-A Riestra
330 Semigroups of operators and spectral theory
  S Kantorovitz
331 Boundary-field equation methods for a class of nonlinear problems
  G N Gatica and G C Hsiao
332 Metrizable barrelled spaces
  J C Ferrando, M López Pellicer and L M Sánchez Ruiz
333 Real and complex singularities
  W L Marar
334 Hyperbolic sets, shadowing and persistence for noninvertible mappings in Banach spaces
  B Lani-Wayda
335 Nonlinear dynamics and pattern formation in the natural environment
  A Doelman and A van Harten
336 Developments in nonstandard mathematics
  N J Cutland, V Neves, F Oliveira and J Sousa-Pinto

Nigel J Cutland
University of Hull

Vítor Neves
Universidade da Beira Interior, Portugal

Franco Oliveira
Universidade de Lisboa, Portugal

and

José Sousa-Pinto
Universidade de Aveiro, Portugal

(Editors)

Developments in nonstandard mathematics

 LONGMAN

Copublished in the United States with
John Wiley & Sons Inc., New York.

Longman Group Limited
Longman House, Burnt Mill, Harlow
Essex CM20 2JE, England
and Associated Companies throughout the world.

*Copublished in the United States with
John Wiley & Sons Inc., 605 Third Avenue, New York, NY 10158*

© Longman Group Limited 1995

All rights reserved; no part of this publication may be reproduced, stored
in a retrieval system, or transmitted in any form or by any means,
electronic, mechanical, photocopying, recording, or otherwise, without
the prior written permission of the Publishers, or a licence permitting
restricted copying in the United Kingdom issued by the Copyright
Licensing Agency Ltd, 90 Tottenham Court Road, London, W1P 9HE

First published 1995

AMS Subject Classifications: (Main) 03H05, 26E35, 28E05
(Subsidiary) 03H15, 30G06, 46S20

ISSN 0269-3674

ISBN 0 582 27970 4

British Library Cataloguing in Publication Data

A catalogue record for this book is
available from the British Library

Library of Congress Cataloging-in-Publication Data

Developments in nonstandard mathematics / N.J. Cutland ... [et al.]
   (editors).
    p. cm. -- (Pitman research notes in mathematics series, ISSN
0269-3674 ; ??)
    Proceedings of the International Colloquium on Nonstandard
Mathematics, held at the University of Aveiro, Portugal, July 18-22,
1994.
    1. Nonstandard mathematical analysis--Congresses. I. Cutland,
Nigel. II. International Colloquium on Nonstandard Mathematics
(1994 : University of Aveiro) III. Series.
QA299.82.D48 1995
510--dc20   95-21998
                                                                            CIP

Printed and bound in Great Britain
by Biddles Ltd, Guildford and King's Lynn

in memoriam

ABRAHAM ROBINSON

ABRAHAM ROBINSON (1918-1974)

As published previously in Joseph W. Dauben, *Abraham Robinson. The Creation of Nonstandard Analysis, A Personal and Mathematical Odyssey* (Princeton University Press, 1995); reproduced here from the collection of Mrs. Abraham Robinson with her kind permission.

Abraham Robinson
a biographical note

Abraham Robinson was born on October 6, 1918 in Waldenburg in Lower Silesia a few months before his father Abraham Robinsohn, a philosopher, passed on. He received his education at a private school in Breslau.

After the rise of nazism in Germany, the Robinsohn family fled in 1933 to Palestine. In 1936, Abraham Robinson enrolled as a student in philosophy and mathematics at the Hebrew University in Jerusalem. His teacher, Abraham Fraenkel, inspired Robinson to concentrate his studies in mathematical logic and set theory. Robinson's first research paper on the independence of the axiom of definiteness (extensionality) in set theory appeared in 1939 in the Journal of Symbolic Logic. During that eventful year Robinson was awarded a scholarship to continue his studies of mathematics at the Sorbonne. Shortly before the start of the second world war Robinson arrived in Paris without having received yet a degree from the Hebrew University. The following year in June, when the German Army entered Paris, Robinson barely escaped to England.

In January of 1941, after a brief service in the Free French Army under the Gaulle, Robinson joined the British Army as a Scientific Officer in the Ministry of Aircraft Production of the Royal Aircraft Establishment in Farnborough, where he was assigned the task of analyzing the design of supersonic airfoils. By the end of the war, Robinson was recognized as one of the leading experts on wing theory. A seminal work on this subject with his former student J. A. Laurmann appeared in 1956.

After the war Robinson returned briefly to Jerusalem to receive the M. Sc. degree of the Hebrew University. Thereafter, he returned to his work in mathematical logic that soon began to overshadow his research in aeronautics. In 1949, Robinson received his Ph.D. in mathematics under the direction of Professor P. Dienes of the University of London. His thesis entitled "The Metamathematics of Algebraic Systems" attracted a great deal of attention by the specialists in the field and earned Robinson an invitation to speak about his work at the first Internal Congress of Mathematicians held after the war in 1950 in Cambridge, Massachusetts.

During those early years after the war in England, Robinson was a Faculty member of the newly established College of Aeronautics in Cranfield. In 1951, because of his world-wide reputation as an applied mathematician, Robinson was enticed by the University of Toronto to join its Department of Applied Mathematics and become its head a few years later.

When in 1957 his former teacher Fraenkel retired, Robinson became Fraenkel's successor and Chairman of the Department of Mathematics of the Hebrew University. Back in Jerusalem Robinson returned full time to his research in logic and model theory.

In the fall of 1960 while visiting the Institute for Advanced Study in Princeton, N.J., Robinson inspired by Skolem's earlier work on models of arithmetic, hit upon the idea to apply Skolem's methodology to the system of axioms for the real numbers and nonstandard analysis was born. This discovery was published in a paper entitled "Non-standard Analysis" and appeared in 1961 in the April issue of the Proceedings of the Royal Academy of Amsterdam. In this remarkable paper Robinson showed how model theoretic methods could be used to provide a complete solution to the centuries old problem of G. Leibniz to develop a rigorous theory of infinitesimals as a foundation for the infinitesimal calculus.

From 1962 till 1967 Robinson served as a Professor of Mathematics and Philosophy at the University of California in Los Angeles. At UCLA Robinson wrote his celebrated work "Non-standard Analysis" which appeared in 1966. In this ground breaking work Robinson clearly and convincingly showed how model theory can be used to unite mathematical logic with modern mathematics. We may also mention here that in 1967 the first Internal Symposium on nonstandard mathematics was held at the California Institute of Technology under the direction of W. A. J. Luxemburg. Thereafter in 1967 Robinson joined the Faculty of Yale University where he continued vigorously his work in model theory and nonstandard mathematics until his untimely death in April of 1974.

For more information of the life and work of Abraham Robinson we refer the interested reader to a recent biography of Robinson authored by Joseph Warren Dauben with a foreword by Benoit Mandelbrot entitled "Abraham Robinson. The Creation of Nonstandard Analysis, A Personal and Mathematical Odyssey", which appeared in January of 1995 as a publication of the Princeton University Press.

The editors

Contents

Preface

K. D. STROYAN.
The infinitesimal rule of three ... 1

MARK MCKINZIE AND CURTIS TUCKEY.
Nonstandard methods in the precalculus curriculum ... 23

VÍTOR NEVES.
Difference quotients and smoothness ... 35

HERMANN RENDER.
Continuous maps with special properties ... 42

STEVEN C LETH.
Some nonstandard methods in geometric topology ... 50

THOMAS WALTER.
Delayed bifurcations in perturbed systems analysis of slow passage of Suhl-threshold ... 61

ALAIN ROBERT.
Functional analysis and NSA ... 73

W. A. J. LUXEMBURG.
Near-Standard compact internal linear operators ... 91

YVETTE FENEYROL-PERRIN.
Discrete Fredholm's equations ... 99

R. F. HOSKINS.
Nonstandard theory of generalized functions ... 108

CHRIS IMPENS.
Representing distributions by nonstandard polynomials ... 119

MICHAEL OBERGUGGENBERGER.
Contributions of nonstandard analysis to partial differential equations ... 130

NIGEL CUTLAND.
Loeb measure theory   151

DAVID A. ROSS.
Unions of Loeb nullsets: the context   178

SIU-AH NG.
Gradient lines and distributions of functionals in infinite dimensional
Euclidean spaces   186

SERGIO ALBEVERIO AND JIANG-LUN WU.
Nonstandard flat integral representation of the free Euclidean field
and a large deviation bound for the exponential interaction   198

MICHAEL BENEDIKT.
Nonstandard analysis in selective universes   211

J. M. ALDAZ.
Lattices and monads   224

H. JEROME KEISLER.
A neometric survey   233

SERGIO FAJARDO AND H. JEROME KEISLER.
Long sequences and neocompact sets   251

Preface

This book presents a selection of contributions to the **Colóquio Internacional de Matemática Não Standard 1994** (CIMNS94), ranging from the teaching of pre-calculus to new areas for mathematical research.

The CIMNS94 was held at the University of Aveiro, Portugal, in July, *in memoriam* of Abraham Robinson, the creator of nonstandard mathematics, on the twentieth anniversary of his death. W. A. J. Luxemburg, who worked with Robinson in the early development of this new field, enthusiastically offered to open the conference with a talk which is summarized in the foregoing biographical note.

One may distinguish five parts in the whole volume: the first, headed by Stroyan's discussion of the axiomatic and superstructure approaches to "mathematics with infinitesimals", ends with an application to ordinary differential equations and systems analysis; the second begins with Robert's course on functional analysis and ends with Fredholm equations; Hoskin's article on generalized functions and Oberguggenberger's survey on partial differential equations delimit the third; Cutland's course on measure theory initiates the fourth, which also includes overviews, "hard" analysis and proposals of new paths for investigation; the last consists of Keisler's survey of a "new method for existence proofs" with an immediate application on the very last article.

In general the more technical articles were placed at the end of each part, sometimes as a means of transition from one subject to another.

We expect that the survey articles and papers based on courses presented at the conference attract the curiosity of newcomers to the field of nonstandard mathematics — which has revealed itself as a powerful tool of discovery.

There would never have been a CIMNS94 without the support of institutions and friends.

We thank the Fundação Luso-Americana para o Desenvolvimento, the Junta Nacional de Investigação Científica e Tecnológica, the British Council, the Departamento de Matemática da Faculdade de Ciências de Lisboa, the Departamento de Matemática da Universidade de Aveiro, the Departamento de Matemática/Informática da Universidade da Beira Interior and the Centro de Matemática e Aplicações Fundamentais for their sponsorship.

We are obliged to Arala Chaves (Faculdade de Ciências da Universidade do Porto), Marques de Sá and Graciano de Oliveira (Faculdade de Ciências e Tecnologia da Universidade of Coimbra), António Bivar Weinholtz (Faculdade de Ciências da Universidade de Lisboa), J. Campos Ferreira (Instituto Superior Técnico-Universidade Técnica de Lisboa), for having shown their support by presiding over sessions or taking part in discussions, as well as to Tom Lindstrøm for his talk and other supporting activities.

We are indebted to Agostinho Flor, Mário Chuva, Paula Prata, João Riço Nunes, João Muranho, José Rogeiro and Ana H. Roque for their help, at both universities of Aveiro and of Beira Interior.

Special recognition is due to Batel Anjo, at Aveiro, and António Tomé, at Beira Interior, for their unfailing aid.

A word of thanks to the staff at Longman, for their understanding of our editing troubles.

Praise to our wives for their help in many ways during the preparation and completion of the different parts of the whole act — both the meeting and this volume.

August 1995
The editors

K. D. STROYAN
The infinitesimal rule of three

0. Introduction

This article is a beginner's introduction to modern Infinitesimal Analysis. There are three main approaches to the foundations: Keisler's "elementary" axioms, Nelson's axioms, and Robinson's 'superstructures.' This article will use all three and allow us to compare and contrast some of their strengths and weaknesses. To set a familiar context, all three will be used to show how to rigorously support intuitive approximations in calculus.

Intuitively, a function with a positive derivative is increasing, yet the pointwise definition of derivative does not give this result. We defend the "modest proposal" that the definition you learned for the derivative is overgeneralized and that intuition is right. We use infinitesimals to formulate a stronger notion of derivative that agrees with intuition and is compatible with classical calculus. This approach yields straightforward proofs of all the fundamental classical results with none of the pointwise pathologies. These results provide us with a non-standard introduction to the theory of infinitesimals; one which gives new insight to an old subject as well.

The infinitesimal formulation of strong derivative is simple, in fact, every bit as elementary as the pointwise derivative in the traditional non-infinitesimal approach. This is an important part of our "modest proposal." However, there are differences in the three basic foundations for Infinitesimal Analysis, even in formulating proofs of intuitive results of calculus. Keisler's axioms have the least logical "overhead"; in two typeset pages, you can show what infinitesimals are and how they are logically equivalent to epsilons and deltas. Uniformity of approximations is clearest in Nelson's approach. Fundamental results of calculus give us comparisons of the three foundations for Infinitesimal Analysis.

1. Intuitive Approximations

The two main ideas of calculus are that small changes in smooth functions are nearly linear and that integrals are approximately sums of 'slices.' These two things are connected by the following 'intuitive' argument. Let $F[x]$ be a real valued function of a real variable (defined on the whole real line, for simplicity). We will use the notation $a \sim b$ to intuitively mean a is approximately equal to b. The intuitive limit

$$\lim_{\Delta x \to 0} \frac{F[x + \Delta x] - F[x]}{\Delta x} = f(x)$$

can also be written
$$\frac{F[x+\Delta x] - F[x]}{\Delta x} \sim f(x), \quad \text{when } \Delta x \sim 0$$
or, expressing the error explicitly as ε, by
$$\frac{F[x+\Delta x] - F[x]}{\Delta x} = f(x) + \varepsilon, \quad \text{with } \varepsilon \sim 0 \quad \text{when } \Delta x \sim 0$$
This form can be used to express the change in the function more directly as
$$F[x+\Delta x] - F[x] = f(x) \cdot \Delta x + \varepsilon \cdot \Delta x, \quad \text{with } \varepsilon \sim 0 \quad \text{when } \Delta x \sim 0 \quad \text{(LL)}$$
We will temporarily take the formula (LL) to be the intuitive meaning of "local linearity," with $f(x)$, the intuitive derivative of $F[x]$, yielding the slope of the linear function of the change, $L(\Delta x) = f(x) \cdot \Delta x$

The intuitive idea of the integral, $\int_a^b f(x)\, dx$ is the approximation
$$\int_a^b f(x)\, dx \sim \text{Sum}[f(x)\Delta x \ : \ a \leq x \leq b, x \text{ in steps of size } \Delta x] \quad \text{(IS)}$$
where the sum is taken in steps of size Δx.

Now suppose we want to compute the integral $\int_a^b f(x)\, dx$ and know a function $F[x]$ satisfying formula (LL) for the desired integrand $f(x)$. We first re-write (LL) as
$$f(x) \cdot \Delta x = F[x+\Delta x] - F[x] - \varepsilon \cdot \Delta x, \quad \text{with } \varepsilon \sim 0 \quad \text{when } \Delta x \sim 0 \quad \text{(dL)}$$
and substitute (dL) into the integral approximation (IS)
$$\int_a^b f(x)\, dx \sim \text{Sum}[f(x)\Delta x \ : \ a \leq x \leq b, \text{step}\Delta x]$$
$$\sim \text{Sum}[F[x+\Delta x] - F[x] - \varepsilon \cdot \Delta x \ : \ a \leq x \leq b, \text{step}\Delta x]$$
$$\sim \text{Sum}[F[x+\Delta x] - F[x] \ : \ a \leq x \leq b, \text{step}\Delta x]$$
$$- \text{Sum}[\varepsilon \cdot \Delta x \ : \ a \leq x \leq b, \text{step}\Delta x]$$

The telescoping sum can be computed exactly:
$$\text{Sum}[F[x+\Delta x] - F[x] \ : \ a \leq x \leq b, \text{step}\Delta x]$$
$$= (F[a+\Delta x] - F[a]) + \ldots + (F[b] - F[b-\Delta x])$$
$$= F[b] - F[a]$$

The remaining sum can be estimated:
$$|\text{Sum}[\varepsilon \cdot \Delta x \ : \ a \leq x \leq b, \text{step}\Delta x]|$$
$$\leq \text{Max}[|\varepsilon| \ : \ a \leq x \leq b, \text{step}\Delta x] \cdot \text{Sum}[\Delta x \ : \ a \leq x \leq b, \text{step}\Delta x]$$
$$\leq \text{Max}[|\varepsilon|] \cdot (b-a)$$
$$\sim 0$$
since each $\varepsilon \sim 0$ is small and $(b-a)$ is fixed.

Combining these intuitive computations we see that the integral is approximately $F[b] - F[a]$. Since both quantities are fixed, we have shown

The Intuitive Fundamental Theorem of Integral Calculus
If $F[x]$ and $f(x)$ satisfy (LL), then
$$\int_a^b f(x)\ dx = F[b] - F[a]$$

This is a simple argument, but the result is false if we take the traditional definition of the derivative, so the technical details really do matter. There are everywhere pointwise differentiable functions $F[x]$ whose derivative $f(x)$ is not Riemann integrable, but the telescoping sum argument is more compelling than the counterexample. We offer it as our first justification to the claim that the traditional definition of the derivative is what's wrong. The question we wish to pose is: What is needed to make this argument correct? After all, we only need a good formulation of formula (LL).

At the most basic level, we ask: What are the properties of our notion of approximate equality $\sim$ used in the proof? At the last step of estimating the sum of errors, we needed to know that the error $\text{Max}[\varepsilon]$ was small and that small times an ordinary amount $(b-a)$ was also small. These two steps come from Keisler's two axioms.

If we restrict our approach to the ordinary "real" numbers, it is clear that no fixed notion of "small" will do, because once we have a small nonzero real number, we can then change $(b-a)$ to make the integral error large. We have collectively known since Weierstrass that it is possible to take inverse estimates with real numbers, but we choose here to take an extension of the real numbers in order to make our notion of approximate equality precise. In other words, we want to have small, medium and large numbers in a technically correct sense. These numbers will satisfy 'small times medium is small' (Rule 3 below). Keisler's function extension axiom is needed to show that $\text{Max}[\varepsilon]$ is small.

2. Hyperreal Numbers as an Ordered Field Extension

Any ordered field extension of the reals is automatically non-archimedean. This is a basic consequence of algebra and simply means that there are numbers $\delta \neq 0$ such that
$$|\delta| < \frac{1}{m} \quad \text{for every natural number} \quad m = 1, 2, 3, \ldots \tag{Infsml}$$
Numbers satisfying (Infsml) are called "infinitesimal" and zero is the only real infinitesimal by the archimedean property of the reals. The extensions of the reals used in modern Infinitesimal Analysis are called "hyperreal" numbers and we will see that we need the hyperreal numbers to be more than an algebraic extension of the ordered field of reals. We write $a \approx b$ for two hyperreals, if $b - a$ satisfies condition (Infsml), and say "a is infinitely close to b."

The fact that hyperreals form an ordered field means that there are many infinitesimals: If $\delta > 0$ is infinitesimal, then positive and negative multiples are different infinitesimals,
$$\cdots < -2\delta < -\delta < 0 < \delta < 2\delta < \cdots$$

There are also many points near each number a
$$\cdots < a - 2\delta < a - \delta < a < a + \delta < a + 2\delta < \cdots$$
and
$$a - 2\delta \approx a - \delta \approx a \approx a + \delta \approx a + 2\delta$$

Moreover, reciprocals of nonzero numbers must exist (by the field axioms) and if $0 < \delta \approx 0$, ordered field properties show that $1/\delta$ is "infinite" or "unlimited" in the sense that
$$\frac{1}{\delta} > m \quad \text{for every natural number } m = 1, 2, 3, \ldots$$

A hyperreal number b is said to be "finite" or "limited" if there is an ordinary real number r such that
$$|b| \leq r$$
(The term "limited" is inclusive, so infinitesimals and ordinary reals are finite.)

Proof of the following basic rules of small, medium and large numbers is a simple exercise on using the definitions above. We leave the proof to the reader.

Theorem on Rules of Small, Medium and Large Hyperreal Numbers
The following rules are satisfied by hyperreal numbers:
(1) If a and b are finite, so are $a + b$ and $a \cdot b$.
(2) If $\varepsilon \approx 0$ and $\delta \approx 0$ are infinitesimal, so is $\varepsilon + \delta$
(3) If $\delta \approx 0$ is infinitesimal and b is finite, then $b \cdot \delta \approx 0$ is infinitesimal.
(4) The ratio $\dfrac{1}{0}$ is still undefined and $\dfrac{1}{x}$ is infinite only when $x \approx 0$ is infinitesimal.

Rule (3) in the preceding theorem says 'small times medium is small' in a technically correct way and this will help make the intuitive estimate
$$\text{Max}[\varepsilon] \cdot (b - a) \approx 0$$
valid in the last step of the proof of the Fundamental Theorem.

If we wish to replace the intuitive $\sim$ of our proof above with the technical $\approx$ of a hyperreal field, we need to be able to write the technical form of (LL) with Δx infinitesimal. This is only a triviality for polynomials, but the deep results of classical calculus require that we can at least write (LL) for the classical transcendental functions. A simple description of how functions extend to the hyperreals is the most elementary foundation for Infinitesimal Analysis given by Keisler as outlined in the next section. Keisler's approach also covers the $\text{Max}[\varepsilon]$ function.

Exercise
(1) Let $F[x] = x^3$ and use field properties to show that
$$F[x + \delta x] - F[x] = f(x) \cdot \delta x + \varepsilon \cdot \delta x$$
where $f(x) = 3 \cdot x^2$ and $\varepsilon = \delta x(3x + \delta x)$.
(2) If x is finite and $\delta x \approx 0$ is infinitesimal, show how the rules of small, medium and large numbers in the Theorem above make $\varepsilon \approx 0$.

(3) Given a nonzero $\delta x \approx 0$, use ordered field properties to show that there are hyperreal values of x so that ε is not infinitesimal.

One final important connection between finite hyperreal numbers and ordinary real numbers is the following result.

The Standard Part Theorem
Every finite hyperreal number b is infinitely near some ordinary real r called its "standard part," $r = st(b)$ and $b \approx r$.

Proof: Define a cut $r = (\{x | x \leq b, x \text{ real}\}, \{x | x > b, x \text{ real}\})$. Both sets are nonempty since b is limited. The cut is the standard part $r = st(b)$, because $|r - b| < 1/m$ for each natural number, $r \approx b$.

3. Keisler's Axioms: Extension of Real Functions

Keisler's elementary approach to Infinitesimal Analysis just requires the following two axioms (cf. [K] or [S]).

Keisler's Ordered Field Axiom
Robinson's "hyperreal" numbers are an ordered field extension of Dedekind's "real" numbers.

The hyperreal numbers are not just an algebraic extension, rather, every real function of finitely many real variables has a "natural" extension to hyperreal numbers. These natural extensions satisfy

Keisler's Function Extension Axiom
Every logical real statement about real functions that holds for all real numbers also holds for all hyperreal numbers when the functions in the statement are replaced by their natural extensions.

For example, the identities
$$\sin[\theta + \delta] = \sin[\theta]\cos[\delta] + \sin[\delta]\cos[\theta]$$
$$\sin^2[\theta] + \cos^2[\theta] = 1$$
both hold for all hyperreal θ and δ.

The precise meaning of "logical real statement" is an implication
$$S \Rightarrow T$$
where S and T are finite sets of real logical formulas.

In the case of the pair of hyperreal trig identities above, $S = \{\theta = \theta, \delta = \delta\}$ and T is the set consisting of the two identities. We know that for all reals $S \Rightarrow T$, so the axiom gives $S \Rightarrow T$ for hyperreals.

"Real logical formulas" consist of:

(1) *An equation, $E_1 = E_2$*
(2) *An inequality $E_1 < E_2, E_1 \leq E_2, E_1 \neq E_2$.*
(3) *A statement of the form "E_1 is defined" or "E_2 is undefined"*

where E_i are "real logical expressions."
"Real logical expressions" are built using the following rules:

(a) *A real number is a logical real expression.*
(b) *A variable standing alone is a logical real expression.*
(c) *If $E_1, \ldots, E_n$ are logical real expressions and $f(R_1, \ldots, R_n)$ is a real function, then $f(E_1, \ldots, E_n)$ is a real logical expression.*

Notice that a "real logical formula" can not directly refer to the technical notion of approximation $\approx$, that is, "$a \approx b$" is NOT a REAL logical formula. In particular, the next definition is not written as a real logical formula. Notice that the definition refers to two ordinary real functions, but that the condition making one the derivative of the other is written in terms of the natural extensions of the functions, since both x and $x + \delta x$ need not be ordinary real numbers (though one of them they may be, too). The definition is just a technical form of the intuitive approximation (LL).

Definition of Derivable Function
A real function $F[x]$ defined on the open interval (a,b) is said to be derivable on the interval if there is a real function $f(x)$ such that whenever x is a hyperreal number NOT infinitely close to either a or b and whenever $\delta x \approx 0$, the change in $F[x]$ is approximately linear in the sense that

$$F[x + \delta x] - F[x] = f(x) \cdot \delta x + \varepsilon \cdot \delta x, \text{ with } \varepsilon \approx 0.$$

For example, the sine function is derivable with derivative cosine, so we have both of the following equations, one exact (a consequence of Axiom 2), one approximate:

$$\sin[\theta + \delta\theta] = \sin[\theta]\cos[\delta\theta] + \sin[\delta\theta]\cos[\theta]$$
$$\sin[\theta + \delta\theta] = \sin[\theta] \quad + \quad \delta\theta \cdot \cos[\theta] \quad + \quad \varepsilon \cdot \delta\theta$$

with $\varepsilon \approx 0$, provided $\delta\theta \approx 0$. (This is just the statement that sine is derivable; we do not offer the proof here.)

Before we discuss the traditional "epsilon - delta" version of the definition of derivable function, let us apply it to proving a rigorous form of

The First Fundamental Theorem of Integral Calculus
If $F[x]$ is derivable with derivative $f(x)$ on an open interval containing $[a,b]$, then

$$\int_a^b f(x)\ dx = F[b] - F[a]$$

This theorem only refers to real functions and real intervals, but the proof will use hyperreals and extended functions. We need to fix a definition of the integral in order to make the statement precise. In order to do this, we need to extend summation to infinitesimal steps. Keisler's approach to the foundations of infinitesimal analysis uses function extensions to do this. If $f(x)$ is a real function defined on $[a,b]$, then the following defines another real function of the variable Δx:

$$s(\Delta x) = \text{Sum}[f(x)\ \Delta x\ :\ a \leq x \leq b, x \text{ in steps of size } \Delta x]$$

The notation, $\text{Sum}[f(x)\,\delta x\ :\ a \le x \le b, x \text{ in steps of size } \delta x]$ when $\delta x \approx 0$ simply means the natural extension of the function s,

$$s(\delta x) = \text{Sum}[f(x)\,\delta x\ :\ a \le x \le b, x \text{ in steps of size } \delta x]$$

We will use this extension of summation to give a precise definition of the integral. (Our definition is satisfactory for continuous functions and Lebesgue integrals are discussed in another article in this volume.)

Definition of Integrable Function
A real function $f(x)$ defined on the interval $[a,b]$ has the integral $I = \int_a^b f(x)\,dx$, if there is a single real number I such that whenever $\delta x \approx 0$,

$$I \approx \text{Sum}[f(x)\,\delta x\ :\ a \le x \le b, x \text{ in steps of size } \delta x]$$

Proof of the Fundamental Theorem proceeds by the intuitive computations given above, replacing the intuitive $\sim$ with the precise $\approx$. The computations show that for any infinitesimal δx,

$$\text{Sum}[f(x)\delta x\ :\ a \le x \le b, \text{step}\delta x] = F[b'] - F[a] - \text{Sum}[\,\varepsilon \cdot \delta x\ :\ a \le x \le b, \text{step}\delta x]$$

(where b' is the last value of x in the sum by steps of size δx.) The equality

$$\text{Sum}[F[x+\delta x] - F[x]\ :\ a \le x \le b, \text{step}\delta x] = F[b'] - F[a]$$

is a consequence of Keisler's Function Extension Axiom. The ordinary real function of Δx given by

$$s(\Delta x) = \text{Sum}[F[x+\Delta x] - F[x]\ :\ a \le x \le b, \text{step}\Delta x]$$

satisfies

$$\Delta x > 0 \Rightarrow \{s(\Delta x) = F[b'] - F[a]\ \&\ b' \le b\ \&\ b - b' \le \Delta x\},$$

hence so does $s(\delta x)$ (where $b' = b'(\Delta x)$ is actually another real function).

It is easy to show that $F[b'] \approx F[b]$, since we know $|b - b'| \le \delta x$. Our definition of derivative gives

$$F[b'] - F[b] = (f(b) + \varepsilon) \cdot (b' - b)$$

which is of the form, 'a real plus an infinitesimal times an infinitesimal,' hence is infinitesimal. Thus,

$$\text{Sum}[F[x+\delta x] - F[x]\ :\ a \le x \le b, \text{step}\delta x] \approx F[b] - F[a]$$

whenever $\delta x \approx 0$ is a positive infinitesimal.

Estimating $\text{Sum}[\,\varepsilon \cdot \delta x\ :\ a \le x \le b, \text{step}\delta x]$ is the only difficulty in the proof. Keisler's approach requires us to concoct real functions in order to make proper intuitive statements. The concoctions (given next) are "elementary", but a little less intuitive than methods based on Nelson's axioms or on superstructures. (Their advantage is that they require minimal formal logic.)

The real sum $\text{Sum}[\,\varepsilon \cdot \Delta x\ :\ a \le x \le b, \text{step}\Delta x]$, when Δx is real is a sum of finitely many terms. This means that the maximum of the terms is attained,

$$\text{Max}[\,\varepsilon\ :\ a \le x \le b, \text{step}\Delta x] = \varepsilon_M$$

We express this by defining a function ξ that gives the location of the maximal ε. Recall that ε satisfies:
$$\varepsilon \cdot \Delta x = \varepsilon(x, \Delta x) \cdot \Delta x = F[x + \Delta x] - F[x] - f(x) \cdot \Delta x \tag{Er}$$
The real function $\xi(\Delta x)$ is simply the x-location of the maximal one, so for all positive real Δx, the real function
$$s(\Delta x) = \text{Sum}[\ \varepsilon \cdot \Delta x\ :\ a \leq x \leq b, \text{step}\Delta x]$$
satisfies
$$|s(\Delta x)| \leq |\varepsilon(\xi(\Delta x), \Delta x)| \cdot (b - a)$$
so by Keisler's Function Extension Axiom, when $\delta x \approx 0$ is a positive infinitesimal,
$$|s(\delta x)| \leq |\varepsilon(\xi(\delta x), \delta x)| \cdot (b - a)$$
or in more intuitive terms,
$$|\text{Sum}[\varepsilon \cdot \Delta x\ :\ a \leq x \leq b, \text{step}\Delta x]|$$
$$\leq \text{Max}[|\varepsilon|\ :\ a \leq x \leq b, \text{step}\Delta x] \cdot \text{Sum}[\Delta x\ :\ a \leq x \leq b, \text{step}\Delta x]$$
$$\leq \text{Max}[|\varepsilon|] \cdot (b - a)$$

We also know from the formula (Er) above that
$$\varepsilon(\xi, \delta x) \cdot \delta x = F[x + \delta x] - F[x] - f(x) \cdot \delta x$$
so by our definition of derivative, $\varepsilon(\xi, \delta x) \approx 0$. The hyperreal rule that infinitesimal times finite is infinitesimal now shows that for every positive infinitesimal δx,
$$\text{Sum}[f(x)\ \delta x\ :\ a \leq x \leq b, \text{step}\delta x] \approx F[b] - F[a]$$

This completes the proof, but it is simple yet a strange proof from a non-infinitesimal point of view. Not only have we shown what the value of the integral is, we have also shown that the integral exists (according to our definition of integrable). This would not work for pointwise derivatives (where the integral need not exist). The next exercise shows why we don't encounter the pointwise pathology.

As part of the proof of the Fundamental Theorem, we showed that $F[b'] \approx F[b]$. The idea generalizes easily to show 'derivable implies continuous,' $F[R_1] \approx F[R_2]$, whenever $R_1 \approx R_2$ and neither lies near an endpoint of the interval of derivability. The derivative of a derivable function is also automatically continuous. You should verify these simple facts before we unravel the mystery from the point of view of non-infinitesimal limits.

Exercise: Strong Derivatives are Continuous
Let $F[x]$ be derivable on a real open interval with derivative $f(x)$. Show that the derivative $f(x)$ is continuous by the following steps.
(1) Let $x_1 \approx x_2$ be different points in the interval, not near the endpoints. Write the local linearity approximation first with $x = x_1$ and $\delta x = x_2 - x_1$ to obtain
$$F[x_2] - F[x_1] = f(x_1) \cdot (x_2 - x_1) + \varepsilon_1 \cdot (x_2 - x_1)$$

(2) Next write the local linearity with $x = x_2$ and $\delta x = x_1 - x_2$ to obtain
$$F[x_2] - F[x_1] = f(x_2) \cdot (x_2 - x_1) + \varepsilon_2 \cdot (x_2 - x_1)$$
(Hint: check signs.)
(3) Subtract these two equations and divide by $(x_2 - x_1)$ to show that
$$f(x_2) - f(x_1) = \varepsilon_1 - \varepsilon_2$$
(4) Conclude that the DERIVATIVE $f(x)$ is continuous,
$$f(x_2) \approx f(x_1)$$

This exercise and the proof of the Fundamental Theorem show that the definition of derivable is not just a simple-minded translation of the pointwise derivative given at the beginning of the article. The definition IS stronger. At the same time, we can show that all the classical functions are derivable (on intervals where both the function and derivative are defined). This can be shown directly - and simply, so we could just proceed with calculus. From a non-infinitesimal point of view, the foundational question remains: What is the 'epsilon-delta' equivalent of the definition of derivable? We will use this question to illustrate the use of Nelson's axioms for Infinitesimal Analysis in the section 6.

It turns out that a function is derivable if and only if it is continuously differentiable in the traditional weak pointwise sense. (In other words, the 'converse' of the previous exercise holds.) This equivalence shows our reader (who already knows calculus) why the pointwise pathologies do not arise with derivable functions. The results are true and the direct approximation given in the definition of derivable shortens, simplifies and clarifies a number of proofs of basic results of calculus.

4. Nelson's Axioms, Superstructures, and Higher Analysis

Keisler's Axioms provide an elegant solution to Leibniz' 300 year old question of how one might rigorously develop calculus using infinitesimals. Robinson's original paper solved this question at about the same level as Keisler's axioms, but with somewhat more cumbersome formal logic. Keisler's axioms require us to concoct functions in order to say things like, 'There is an x where the error is maximal ...' It's time to bite the bullet and introduce logic with quantifiers (so we can rigorously say 'there is an x...') and we wish to solve another problem at the same time.

After Robinson's invention of rigorous Infinitesimal Analysis, it soon became clear that his methods could be extended to a spectrum of problems in analysis, as shown already in 1966 [R-B]. More recent developments in Infinitesimal Analysis can be found in [C], but Robinson's book truly is a classic. (You should at least read the final Chapter X.) Modern analysis studies spaces of functions and the like, so Keisler's axiom extending only real functions does not cover modern analysis so well. Robinson's solution to this difficulty evolved into the method of 'superstructures.'

A superstructure is a set that 'contains everything in analysis' and one constructs an extended 'non-standard model' of the formal logic of this set. The new sets that are 'known to the logic' are called "internal," while the others are called "external." The basic technical difficulty in learning the 'superstructure' foundations for Infinitesimal Analysis is in learning to recognize internal and external sets. (Analogous to measurable and non-measurable sets in analysis.) We will get to the details, but the point is that it is possible to apply Infinitesimal Analysis to 'all of analysis' and the first three decades of development saw a host of applications. These developments will continue into the future, despite confusion in the general public about whether or not infinitesimals are 'necessary,' compounded by prejudice about whether they are desirable.

In 1973, Robinson [R-M] posed a number of metamathematical questions, including: 'Is there a methodical way to translate infinitesimal proofs into traditional proofs?' Nelson [N-S] answered this by giving a formalization of Infinitesimal Analysis (that covers many proofs) together with a syntactical procedure to convert proofs using infinitesimals into proofs that only refer to traditional set theory. Nelson's original paper [N-I] is meticulously well-written, but nevertheless caused considerable confusion because of misunderstandings over his precise use of set formation which made all official sets internal. Nelson's theory certainly can discuss external objects and it can do so formally and systematically. (They just aren't "sets," but are like "classes".)

Superstructures (and saturated models) are in a precise sense more powerful than Nelson's approach (see [HK]), but the syntactical structure of Nelson's Axioms is useful even there. Since we want to develop all three foundations for Infinitesimal Analysis, we will introduce Nelson's approach in the context of superstructures. This is not quite what Nelson had in mind, but it will give us some insight into "Internal Set Theory" and allow us to compare all three.

5. The Definition of Superstructure

Let $\mathbb{R}$ denote Dedekind's "real" numbers as a set of atoms (or "points"), that is, so that if $r \in \mathbb{R}$ is a real number, then r has no elements in the sense that it is a 'point' (not the empty set). In other words, the formal statement $a \in r$ is not a legitimate formula. Let $\mathfrak{P}(X)$ denote the "power set" consisting of all subsets of another set X. Inductively define a sequence of sets $R_0 = \mathbb{R}$, and $R_{n+1} = \mathfrak{P}(\cup_{k=0}^{n} R_k)$ for $n \in \mathbb{N}$. Finally, the superstructure of 'all of classical analysis' is

$$\mathfrak{R} = \cup_{n \in \mathbb{N}} R_n$$

(This could be done over any set of atoms, R_0, but we only need the basic example.)

The superstructure $\mathfrak{R}$ should NOT be taken very seriously. It is only a way to 'code' objects from analysis so that we can rigorously define formal statements about familiar objects from analysis. All we need to understand is why $\mathfrak{R}$ has sufficient elements to code 'everything in analysis.' Officially, our formal statements about analysis can only refer to elements of $\mathfrak{R}$, but we usually do not write out all the details of such statements. Rather, we observe that an informal statement of analysis could be formalized and we take special notice of the bounds on the quantifiers of the statement. These quantifier bounds are technically the most important thing, as we shall see.

First, we need to see that $\mathfrak{R}$ contains all real functions of finitely many real variables as elements, $f \in \mathfrak{R}$. This will mean that the logic of $\mathfrak{R}$ can discuss all the things covered by Keisler's Axioms. The specific details of how the function is represented logically are not important, but a function can be thought of as a set of ordered pairs. Set theoretically, an ordered pair can be thought of as the set $(x,y) = \{\{x\},\{x,y\}\}$, so an ordered pair of real numbers lies in R_2, since $\{x\} \& \{x,y\} \in \mathfrak{P}(\mathbb{R}) = R_1$ and the set $(x,y) = \{\{x\},\{x,y\}\} \in \mathfrak{P}(R_0 \cup R_1) = R_2$. Any function of one real variable (represented as a particular set of ordered pairs) is thus an element of R_3.

An n-tuple of real numbers can be thought of as a function from the set $\{1, 2, \ldots, n\}$ into $\mathbb{R}$, so sets of n-tuples lie in R_3. A function of n real variables can be represented as a set of ordered pairs with first element an n-tuple and second element the function value. As a result, all real functions of n real variables are elements in R_6. The 6 doesn't matter (and the same function could also be represented in other R_n's), what matters is that each real function of n real variables, f, is an element of the formal model of analysis, $f \in \mathfrak{R}$.

Functions of a complex variable can be represented as functions between ordered pairs of real numbers (the real and imaginary parts of the input and output). Sets of ordered pairs of reals lie in R_3, so functions between them lie in R_6. The space of holomorphic functions defined on the unit disk is thus represented as a subset of R_6 and therefore is an element of R_7 and $Hol(D) \in \mathfrak{R}$. Seminorms on the space of holomorphic functions defined on the disk are functions from an element of R_7 into R_0, so these too are elements of $\mathfrak{R}$.

In short, all the specific functions, spaces and 'functionals' of classical analysis are representable as elements of the superstructure $\mathfrak{R}$.

Exercise: Functions and spaces of classical analysis are elements of $\mathfrak{R}$

(1) Let $A \subseteq R_m$ and $B \subseteq R_n$ be subsets of levels of the superstructure and suppose $f : A \to B$ is a function. Show that $f \in R_p$ for $p = Max[m,n] + 3$ so that $f \in \mathfrak{R}$.

(2) If F is any set of functions which all have the common domain $A \subseteq R_m$ and common range $B \subseteq R_n$, show that $F \in R_{p+1}$ and $F \in \mathfrak{R}$.

(3) If $\phi : F \to \mathbb{R}$ is a function from the space of functions of part (2) into the real numbers (such as a seminorm on a linear space of real functions), show that $\phi \in R_{p+4}$ and $\phi \in \mathfrak{R}$.

(4) Pick your favorite specific object from classical analysis, say a flow on the infinite dimensional manifold of all diffeomorphisms of a 5-sphere, and show that it can be represented set-theoretically as an element of $\mathfrak{R}$. (The 5-sphere can be represented in $\mathbb{R}^6 \in R_?$, a diffeomorphism on that is an element of $R_{?+??}$, the set of those is an element of $R_{?+??+1}$, a flow is a function from ...)

In summary, we have constructed a set $\mathfrak{R}$ that contains (representations) of every specific object in classical analysis (L^p–spaces, norms, manifolds, etc.). There is one thing that is a little confusing about the set $\mathfrak{R}$: the elements of $\mathfrak{R}$ are of many 'types,' that is, elements of $\mathfrak{R}$ include real numbers, real functions, spaces of functions (as elements of $\mathfrak{R}$), etc. When we define a map $*: \mathfrak{R} \to \mathfrak{H}$ it will extend specific numbers, the set of all numbers, the set of all functions, the set of all spaces of sets of functions with such and such property, etc...

6. Standard, Internal and External Sets and Functions

Infinitesimal Analysis is (or may be) based on an extension of $\mathfrak{R}$ given by a mapping that respects the levels of the superstructure,

$$*: \mathfrak{R} \to \mathfrak{H}$$

Where $\mathfrak{H}$ is a superstructure based on a new larger set of atoms, H_0, with $H_{n+1} = \mathfrak{P}(\cup_{k=0}^n H_k)$ for $n \in \mathbb{N}$, and,

$$\mathfrak{H} = \cup_{n \in \mathbb{N}} H_n$$

The mapping works like this:
(1) If $r \in \mathbb{R}$ is a real number then $*r = r \in H_0$, so we drop the $*$ on reals.
(2) Since $\mathbb{R} = R_0 \in \mathfrak{R}$, $*\mathbb{R} =* R_0$ is defined and in this case, $*\mathbb{R} = H_0$, the new ground set of atoms. The set of hyperreal numbers $H_0 =*\mathbb{R}$ is larger than the real numbers, of course,

$$\mathbb{R} \subset *\mathbb{R} \quad \text{(strictly.)}$$

(3) In general, $*$ respects set containment, if $a \in A$, with both in $\mathfrak{R}$, then $*a \in* A$, both in $\mathfrak{H}$. In particular, every ordinary real $r \in \mathbb{R}$ satisfies $r =*r \in *\mathbb{R}$, but not every $s \in *\mathbb{R}$ is of the form $s =*r$ for some $r \in \mathbb{R}$.

Exercise
Show that $*R_n \subseteq H_n$ for all $n \in \mathbb{N}$.

Every real function f is an element of $\mathfrak{R}$ at some level, so $*f \in \mathfrak{H}$. In fact, any function $f : A \to B$ where $A \in \mathfrak{R}$ and $B \in \mathfrak{R}$ satisfies $f \in \mathfrak{R}$ (when represented as a set of ordered pairs), so $*f \in \mathfrak{H}$. Since $*$ respects $\in$ (property (3) above), $*f$ is a function, $*f :* A \to* B$ and if $b = f(a)$, $*b =* f(*a)$.

Notation for Natural Extensions of Functions in a Superstructure
If $f : A \to B$ is a function in $\mathfrak{R}$, since $*f :* A \to* B$ is an extension, we drop the $*$ and write $f(x)$ for the new function (provided the context is clear).

The transfer property of $*$ or Nelson's "transfer axiom" (both given below) will show that extended functions are 'natural' in Keisler's sense. The full logic of a superstructure will allow us to extend the idea.

Property (3) of the $*$ map can be quite confusing, since all infinite sets are enlarged under $*$ and most of us are not used to working with different higher type sets at the same time. Here is an important example. The set of all subsets of the real numbers is part of formal real analysis, $\mathfrak{P}(\mathbb{R}) = R_1 \in \mathfrak{R}$, so $^*R_1 =\, ^*\mathfrak{P}(\mathbb{R})$ is defined. A fact of fundamental importance is the strict inclusion $^*\mathfrak{P}(\mathbb{R}) \subset \mathfrak{P}(^*\mathbb{R})$:

$$^*R_1 =\, ^*\mathfrak{P}(\mathbb{R}) \subset \mathfrak{P}(H_0) = \mathfrak{P}(^*\mathbb{R}) = H_1 \quad \text{but} \quad ^*R_1 =\, ^*\mathfrak{P}(\mathbb{R}) \neq \mathfrak{P}(^*\mathbb{R}) = H_1$$

This strict inclusion follows from the transfer and idealization properties of * which we give below. The inclusion means that extensions of sets of reals are sets of hyperreals, while the strictness of the inclusion means that not all sets of hyperreals arise this way.

Definitions of standard, internal and external sets & Notation

A set (or function viewed as a set) $A \in \mathfrak{H}$ is called (extended) standard if there is a set $B \in \mathfrak{R}$ such that $A =\, ^*B$.

A set (or function viewed as a set) $A \in \mathfrak{H}$ is called internal if there is a set $B \in \mathfrak{R}$ such that $A \in\, ^*B$. ("Internal" is inclusive, so extended standard sets are internal.)

A set $A \in \mathfrak{H}$ which is not internal is called external.

The set of standard elements of a set $A \in \mathfrak{H}$ is denoted

$$^\sigma A = \{a \in A \,:\, \text{there exists } b \in \mathfrak{R} \text{ with } ^*b = a\}$$

When $A =\, ^*B$, we denote

$$^\sigma B =\, ^\sigma(^*B) = \{a \in\, ^*B \,:\, \text{there exists } b \in \mathfrak{R} \text{ with } ^*b = a\}$$

The real part of a set $A \in \mathfrak{H}$ is the set

$$^\circ A = \{b \,:\, ^*b \in A\} \in \mathfrak{R}$$

The (nonstandard) notations $^\circ$ and $^\sigma$ are not defined at the ground level, because points are not sets. The notation is vacuous at the first level, if $B \subset \mathbb{R}$,

$$^{\circ*}\mathbb{R} = \mathbb{R} =\, ^\sigma\mathbb{R} \quad \& \quad ^{\circ*}B = B =\, ^\sigma B$$

(because $\mathbb{R} \subset H_0$) and thus the idea is often overlooked. (Perhaps it is an excess, anyway. Keisler calls it the notation that doesn't mean anything - with a smile.) However, at the level of H_1 and above, these sets must sometimes be distinguished. The notation does this. You can ignore it except when you need it if you wish.

If $A \subset \mathbb{R}$, then $^*A \in\, ^*\mathfrak{P}(\mathbb{R})$, but not every element of $^*\mathfrak{P}(\mathbb{R})$ arises this way, and moreover, there are subsets of the hyperreals that are not internal. This statement can be written as the strict inclusions

$$^\sigma\mathfrak{P}(\mathbb{R}) \subset\, ^*\mathfrak{P}(\mathbb{R}) \subset \mathfrak{P}(^*\mathbb{R})$$

These sets are notation respectively for the set of (extended) standard subsets of the hyperreal numbers, the set of internal sets of hyperreal numbers and the set of all subsets of hyperreal numbers. That is:

If $A \in\, ^\sigma \mathfrak{P}(\mathbb{R})$, then $A =\, ^*B$ is standard, while the set $^\sigma\mathfrak{P}(\mathbb{R})$ is external.

If $A \in\, ^*\mathfrak{P}(\mathbb{R})$, then A is internal, while the set $^*\mathfrak{P}(\mathbb{R})$ is standard.

If $A \in \mathfrak{P}(^*\mathbb{R})$, then A could be external, while the set $\mathfrak{P}(^*\mathbb{R})$ is external.

(Below, we will use logic to show that when $^\sigma A$ is infinite, it is external. This in turn shows that $\mathfrak{P}(^*\mathbb{R})$ is external, because it contains external elements.)

The set $°A$ is part of 'classical real analysis', $°A \in \mathfrak{R}$, whereas $^\sigma A \in \mathfrak{H}$ is (an external) part of hyperreal analysis. The set $^*(°A)$ is a standard set in $\mathfrak{H}$ and usually $^\sigma A \neq {^*}(°A) \neq A$. This goofy construction is related to Nelson's Axiom (S), as we shall see. Perhaps Nelson's original formulation is more elegant than my excessive notation, perhaps it is just as confusing. You decide.

7. The Formal Logic of Infinitesimal Analysis

The formal language of infinitesimal analysis has the following:

- constants: one for each hyperreal number and each internal set
- variables: $x_1, x_2, x_3, \ldots$
- relations: $=$ and $\in$
- parentheses:) and (
- connectives: $\wedge$ (and), $\vee$ (or) and $\neg$ (not) $\Rightarrow$ (implies), $\Leftrightarrow$ (if and only if)
- internal quantifiers: $\forall$ (for all) and $\exists$ (there exists)
- external quantifiers: $\forall^{st}$ (for all standard) and $\exists^{st}$ (there exists standard)

The formulas of the language are defined inductively:

- If x and y are variables and a and b are constants, $(x = y)$, $(x \in y)$, $(a = x)$, $(a \in x)$, $(x \in a)$, $(a = b)$, and $(a \in b)$ are formulas,
- if ϕ and ψ are formulas, $(\phi \wedge \psi)$, $(\phi \vee \psi)$, $(\phi \Rightarrow \psi)$, $(\phi \Leftrightarrow \psi)$ and $(\neg \phi)$ are formulas,
- if ϕ is a formula (that does not contain a bound occurrence of the variable x), a is a standard set constant and Q is one of the four quantifiers, then $(Qx \in a\ \phi)$ is a formula.

Notice that all quantifiers are bounded by a standard set, $a =^* b$, and thus have the forms:

(1) $\quad \forall x \in a \equiv \forall x \in {^*}b, \qquad \exists x \in a \equiv \exists x \in {^*}b$

(2) $\quad \forall^{st} x \in a \equiv \forall x \in^\sigma a, \qquad \exists^{st} x \in a \equiv \exists x \in^\sigma a$

Definitions

A logical formula ϕ is said to be *internal* if it does not contain the external quantifiers.

A logical formula ϕ is said to be *standard* if it does not contain the external quantifiers and only contains standard constants.

A logical formula ϕ is said to be a *sentence* if every variable is bound by a quantifier (that is, occurs in one and perhaps other places.)

A standard logical sentence has a truth value when interpreted as a statement about the elements of $\mathfrak{H}$, since it has no free variables and reduces to statements about $=$

and $\in$ on variables and standard constants. The sentence could be either true or false, for example,

$$(\forall x \in {}^*\mathbb{R} \ (\forall y \in {}^*\mathbb{R} \ (\sin[x+y] = \sin[x] + \sin[y])))$$
$$(\forall x \in {}^*\mathbb{R} \ (\forall y \in {}^*\mathbb{R} \ (\sin[x+y] = \sin[x]\cos[y] + \cos[x]\sin[y])))$$

However, a standard sentence can also be interpreted in $\mathfrak{R}$, simply by erasing stars:

$$(\forall x \in \mathbb{R} \ (\forall y \in \mathbb{R} \ (\sin[x+y] = \sin[x]\cos[y] + \cos[x]\sin[y])))$$

We are using the short-hand of writing $\sin[x+y]$ and $\sin[x]\cos[y] + \cos[x]\sin[y]$ for the standard functions of two variables that they define. Moreover, if we really view functions as sets of ordered pairs, we need to write '$(x, y) \in \sin$' rather than $y = \sin[x]$, etc. This is not an interesting detail of the formality, since it is clear that the language has the expressive power. The quantifier bounds are important.

Consider Archimedes' Axiom: 'There are no infinitesimals':

$$(\forall \delta \in \mathbb{R}^+ \ (\exists n \in \mathbb{N} \ (\delta > \frac{1}{n}))) \tag{AA}$$

This statement is true in $\mathfrak{R}$, but it is not a statement of our language, because we can only use extended standard constants. Formally we must write:

$$(\forall \delta \in {}^*\mathbb{R}^+ \ (\exists n \in {}^*\mathbb{N} \ (\delta > \frac{1}{n}))) \tag{*AA}$$

Our formal language automatically refers to $\mathfrak{H}$, but if we wrote a formal sentence about elements of $\mathfrak{R}$ and simply put a star on every constant, then we would have a legal standard formal sentence of our language. However we approach infinitesimals, we need to arrange for the following:

The Star Transfer Principle
*A standard formal sentence is true in $\mathfrak{H}$ if and only if it is true in $\mathfrak{R}$ (when interpreted without * 's).*

In Robinson's approach, we make a set-theoretical construction of the map * and prove the transfer principle as a theorem (of ordinary set theory - there's nothing "non-standard" about it.) We will see that we can formulate transfer completely within the formal language of infinitesimal analysis, that is, we can use external quantifiers instead of erasing stars. This is Nelson's Axiom (T). In Nelson's full approach, one simply dispenses with the classical model (represented by $\mathfrak{R}$ here) and gives three additional axioms for set theory with infinitesimals.

The * form of Archimedes' Axiom (*AA) is true in $\mathfrak{H}$ (by transfer), but does NOT say that there are no infinitesimals in *$\mathbb{R}$, because the set *$\mathbb{N}$ is larger than $\mathbb{N}$. All formal statements about real analysis remain true in hyperreal analysis (that is, $\mathfrak{H}$ internally forms a "non-standard model of the language of $\mathfrak{R}$"), but the quantifier bounds make the formal meaning weaker than the original meaning in $\mathfrak{R}$.

The external formal sentence asserting the existence of infinitesimals is also true in $\mathfrak{H}$ (although we have not shown this yet). Officially, we can not write "$\delta \approx 0$" in our formal language (because the relation $\approx$ is not internal), but the external quantifier allows us to express the idea.

(Infsml)
$$\exists \delta \in {}^*\mathbb{R}^+ \ (\delta \approx 0) \Leftrightarrow (\exists \delta \in {}^*\mathbb{R}^+ \ (\forall^{st} n \in {}^*\mathbb{N} \ (\delta < \frac{1}{n})))$$
$$\Leftrightarrow (\exists \delta \in {}^*\mathbb{R}^+ \ (\forall n \in \mathbb{N} \ (\delta < \frac{1}{n})))$$

Up to this point we have not shown that * actually makes infinite sets larger, in particular that $\mathbb{N} \subset {}^*\mathbb{N}$ (strictly). Both transfer and extension are essential to Infinitesimal Analysis and we will approach extension first with Nelson's idealization axiom. Once we know this, both (*AA) and (Infsml) must hold.

8. Nelson's Axioms: (T), (I), & (S)

The transfer principle (given above in terms of both $\mathfrak{R}$ and $\mathfrak{H}$) can be expressed completely within our language by:

The Formal Transfer Principle
If $\phi(x, b)$ is a standard formula with only the free variables $x = (x_1, \ldots, x_m)$ and only the standard constants $b = (b_1, \ldots, b_n)$, and if a is also a standard constant,
$$(\forall x \in^\sigma a \ \phi(x, b)) \Leftrightarrow (\forall x \in a \ \phi(x, b))$$

abbreviated
$$(\forall^{st} x \ \phi) \Leftrightarrow (\forall x \ \phi) \hspace{3cm} \text{(Nelson's Axiom T)}$$

This is the same transfer principle described above, because if we know $(\forall x \in^\sigma a \ \phi(x, b))$, then we know that for each $y \in c$ we have $\phi({}^*y, {}^*d)$ and by unstar transfer, $\phi(y, d)$, so $(\forall y \in c \ \phi(y, d))$. Using star transfer we get, $(\forall x \in {}^*c \ \phi(x, {}^*d))$ this is the desired conclusion, $(\forall x \in {}^*c \ \phi(x, {}^*d)) \equiv (\forall x \in a \ \phi(x, b))$.

If we construct the * map set-theoretically, we need to prove the transfer principle, but we may also take it as one of three axioms for Infinitesimal Analysis. Just for the record, we will assume Nelson's axioms now and notice that Axiom (T) implies (*AA) above. The transfer axiom would be satisfied for an exact copy of $\mathfrak{R}$, so we need an axiom for extension to have infinitesimals.

If $B = {}^* C$ is a standard set in $\mathfrak{H}$, we will denote the set of finite subsets of C by $\mathfrak{F}(C)$. Every element $Y \in \mathfrak{F}(C)$ satisfies (the abbreviated sentence)
$$(\exists m \in \mathbb{N} \ (\exists f : Y \to \{1, \ldots, m\} \ (f \text{ is one-to-one and onto})))$$
which says that Y has m elements. If C is also an infinite set, we know that
$$(\forall m \in \mathbb{N} \ (\exists Y \in \mathfrak{F}(C) \ (Y \text{ has } m \text{ elements})))$$
We want to explore the difference between
$${}^\sigma \mathfrak{F}(B) \equiv^\sigma ({}^*(\mathfrak{F}(C))) \subset {}^* \mathfrak{F}(C) \supset \mathfrak{F}({}^*C) \equiv \mathfrak{F}(B)$$

Transfer shows that a standard finite subset $Y =^* Z$ of B has the form $Y = \{^*c_1, \ldots, ^*c_m\} = \{b_1, \ldots, b_m\}$ for standard constants, because a formal sentence can name every element:

$$(\forall x \in {}^*C \ (x \in {}^*Z \Leftrightarrow (x =^* c_1 \vee \ldots \vee x =^* c_m)))$$

Finite sets of B need not be finite sets of standard elements (once we know there are internal nonstandard elements.)

Because of transfer, sets $Y \in {}^*\mathfrak{F}(C)$ merely need to be in bijective correspondence with initial segments of $^*\mathbb{N}$, so once we see that $^\sigma\mathbb{N} \subset {}^*\mathbb{N}$, we will see that there must be "*-finite" subsets of any infinite set. Nelson's axiom of idealization gives us these facts and more.

The Finite Saturation Principle

Let $\phi(x, y, b)$ be an internal formula with only the variables x and y free and only internal constants $b = (b_1, \ldots, b_n)$. Let $B =^* C$ and D be standard constants. Then

$$(\forall Y \in^\sigma \mathfrak{F}(C)(\exists x \in D(\forall y \in Y \ \phi(x, y, b)))) \Leftrightarrow (\exists x \in D(\forall y \in^\sigma B \ \phi(x, y, b))))$$

abbreviated

$$(\forall^{st}_{finite} Y)(\exists x(\forall y \in Y \ \phi))) \Leftrightarrow (\exists x \ (\forall^{st} y \ \phi))) \qquad \text{(Nelson's Axiom I)}$$

Henceforth we will assume Axiom (I). Our first observation is that we can include $^\sigma\mathbb{N}$ in a *-finite proper subset of $^*\mathbb{N}$. Consider the formula that makes the obvious real analysis statement, 'if Y is a finite set of natural numbers, then there is an element of $\mathfrak{F}(\mathbb{N})$ that contains all its elements, yet does not contain all natural numbers,'

$$(\forall Y \in^\sigma \mathfrak{F}(\mathbb{N})(\exists X \in {}^* \mathfrak{F}(\mathbb{N})(\forall y \in Y \ \phi(X, y, ^*\mathbb{N}))))$$

where $\phi(X, y, {}^*\mathbb{N}) = (y \in X \wedge (\exists z \in {}^*\mathbb{N} \ \neg(z \in X))$. Axiom (I) tells us that there is a single internal set $X \in {}^* \mathfrak{F}(\mathbb{N}))$ so that every element of $^\sigma\mathbb{N} = \mathbb{N}$ belongs to X, yet there is an element of $^*\mathbb{N}$ not contained in X. More simply put, we see that there are infinite *-natural numbers, n, satisfying

$$\mathbb{N} \subset \{0, 1, 2, \ldots, n\} \subset {}^*\mathbb{N}$$

Naturally, $\delta = \frac{1}{n}$ is infinitesimal.

Exercise

*(1) Show that all the standard elements of a standard set are always contained in a *-finite set.*

(2) Use idealization as follows to show that there are infinitesimals. For any finite set of positive reals, $\{\varepsilon_1, \ldots, \varepsilon_m\}$, there is a positive Δ less than all the epsilons. Formalize this as the left side of (I) and conclude by (I) that there is a single positive δ which is less than every standard positive hyperreal number.

*(3) Show that $^\sigma\mathbb{N} = \mathbb{N}$ can not be internal by deriving a contradiction. Suppose $A = \mathbb{N}$ is internal. Then $\forall n \in^\sigma \mathbb{N} \ (n \in A)$. Apply axiom (T) to show that there are no infinite *-natural numbers contrary to Axiom (I).*

17

We will return to a more set-theoretic view of Axiom (I) when we discuss saturated extensions in a later section. We have strayed from our original modest proposal with all this set theory, but have not forgotten.

Exercise
Use the formal language to justify the intuitive proof of the Fundamental Theorem of Integral Calculus given at the beginning of the article. Now you can simply say things like
$$Sum[f(x)\ \delta x\ :\ a \leq x \leq b, step \delta x] = F[b'] - F[a]$$
for the last value of $x = b'$ of the form $x = a + n \cdot \delta x$, with $n \in {}^*\mathbb{N}$.

We are going to show how to use Nelson's Axiom (I) as a quantifier manipulation rule to explain the approximation needed for derivable functions. The negation of Axiom (I) may be abbreviated:
$$(\exists^{st}_{finite} E\ (\forall \delta\ (\exists \varepsilon \in E\ (\phi(\varepsilon, \delta))))) \Leftrightarrow (\forall \delta\ (\exists^{st} \varepsilon\ (\phi(\varepsilon, \delta))))$$
This will allow us to replace a dependent quantifier pair
$$\forall x\ \exists^{st} \varepsilon \ldots \rightarrow \exists^{st}_{finite} E\ \forall x\ \exists \varepsilon \in E \ldots$$
Taking the minimum of the finite set will give the replaced quantifiers $\exists^{st} \varepsilon\ \forall x$.

Quantifier rules are quite confusing- even the ones from conventional logic. For example, you may need to review the following one (where $\phi(x)$ is free of y and $\psi(y)$ is free of x):
$$(\forall x \phi(x)) \Rightarrow (\forall y \psi(y)) \Leftrightarrow (\forall y (\exists x (\phi(x) \Rightarrow \psi(y)))) \Leftrightarrow (\exists x (\forall y (\phi(x) \Rightarrow \psi(y)))) \quad (\forall \exists)$$

Exercise
Show that if ϕ is free of y and ψ is free of x,
$$((\exists x\ \phi) \Rightarrow (\forall y\ \psi)) \Leftrightarrow (\forall x \forall y\ (\phi \Rightarrow \psi)) \qquad (\exists \forall)$$

Recall the definition of derivable given above for a standard function $F[x]$ defined on an open interval (a, b). We define a function of two variables by
$$\varepsilon(x, \Delta x) = \frac{F[x + \Delta x] - F[x]}{\Delta x} - f(x)$$
so the definition may be expressed formally by saying that if x is a standard distance from a and b and if δx is smaller than any standard positive real, then the extension $\varepsilon(x, \delta x)$ is also smaller than any standard positive real:

$(\forall x \in (a, b)\ (\forall \delta x \in {}^*\mathbb{R}$
$((\exists m \in^{\sigma} \mathbb{N}\ (|x - a| > 1/m \wedge |x - b| > 1/m)) \wedge (\forall k \in^{\sigma} \mathbb{N}\ (|\delta x| < 1/k))$
$$\Rightarrow (\forall h \in^{\sigma} \mathbb{N}\ |\varepsilon(x, \delta x)| < 1/h))))$$

Just giving quantifier types, we have
$(\forall x \forall \delta x$
$$((\exists^{st} m \forall^{st} k (\phi \wedge \psi)) \Rightarrow \forall^{st} h\ |\varepsilon| < 1/h))$$

We apply the ordinary rules of logic (∀∃) and (∃∀) above to obtain the equivalent statement:

$$(\forall(x, \delta x)$$
$$\forall^{st} m \; \forall^{st} h \exists^{st} k((\phi \wedge \psi) \Rightarrow |\varepsilon| < 1/h))$$

Universal quantifiers may be interchanged, so we have the equivalent statement

$$\forall^{st} m \; \forall^{st} h \; \forall(x, \delta x) \; \exists^{st} k((\phi \wedge \psi) \Rightarrow |\varepsilon| < 1/h)$$

Now we apply the negation of Axiom (I) as described above, take the minimum k and see the equivalent formula

$$\forall^{st} m \; \forall^{st} h \; \exists^{st} k \; \forall(x, \delta x) \; ((\phi \wedge \psi) \Rightarrow |\varepsilon| < 1/h)$$

which says, 'For every standard natural number m and every standard tolerance $1/h$, there is a closeness, $1/k$, so that if $x \in [a + 1/m, b - 1/m]$ and $|\delta x| < 1/k$, then $\varepsilon(x, \delta x)| < 1/h$, in other words, if $F[x]$ is standard,

$$\lim_{\Delta x \to 0} \frac{F[x + \Delta x] - F[x]}{\Delta x} = f(x) \quad \text{uniformly on compact subintervals of } (a, b)$$

This is not such a simple kind of approximation in this form (because of all the quantifiers), but it is "well known." The logical equivalence means that a standard function is derivable in the straightforward infinitesimal sense if and only if its difference quotients converge 'locally uniformly' (as given by the final complicated formula.)

This reformulation by rules of logic and Nelson's Axiom (I) makes it clear to classically trained analysts why the derivability approximation makes the results of calculus clearer and simpler. Uniform limits of continuous functions are continuous and uniform approximations can be interchanged with the integration limit. A little more thought will convince those same analysts that all the classical rules of calculus work "locally uniformly," so indeed one could simplify the basic results by starting with this definition of derivative. It is clear now why this is a better definition of derivative, but the other point of our modest proposal is that the notion of derivable is just as simple as the traditional approach when it is formulated with infinitesimals. Infinitesimals give the best of both worlds.

Exercise
Use quantifier manipulation to show that if x_1 and x_2 are in a compact interval $[a, b]$ and $x_1 \approx x_2$ implies $f(x_1) \approx f(x_2)$, then $f(x)$ is UNIFORMLY continuous in the traditional sense on $[a, b]$:

$$\forall x_1, x_2 (a \leq x_1, x_2 \leq b) \; (x_1 \approx x_2 \Rightarrow f(x_1) \approx f(x_2))$$
$$\forall x_1, x_2 (a \leq x_1, x_2 \leq b)$$
$$((\forall^{st} n \in \mathbb{N}(|x_1 - x_2| < 1/n)) \Rightarrow (\forall^{st} m \in \mathbb{N}(|f(x_1) - f(x_2)| < 1/m))$$

Nelson's Axiom (T) applies only to standard formulas (this is clear, because real analysis has no infinitesimals), his Axiom (I) only to internal formulas, but his final axiom applies to external formulas. For example, the predicate that says 'there exists a function $f : (a, b) \to^* \mathbb{R}$ so that $F[x]$ is derivable on (a, b)' with free variable F is formalizable as an external predicate $\Phi(F)$ of infinitesimal analysis. We know (from our quantifier manipulations) that the standard functions that satisfy Φ are exactly the continuously differentiable functions. Since Nelson's full formal version of infinitesimal analysis does not allow set formation with external formulas, he introduces an additional axiom partly so that he can discuss sets like, 'The standard set of all functions F satisfying $\Phi(F)$.' In a superstructure this is the set:

$$^s\{F : (a, b) \to \mathbb{R} \mid \Phi(F)\} =^{*\sigma} \{F : (a, b) \to \mathbb{R} \mid \Phi(F)\}$$

and this equals the extended standard set $^*C^1[(a, b)]$ of all internal $*$-smooth functions. Notice that an internal function G could satisfy $\Phi(G)$, but it turns out that transfer does not apply to the predicate defining derivability (such functions are called 'S-derivable'), so we can have $G \in^* C^1[(a, b)] \& \neg \Phi(G)$.

Exercise
Show that the function $G[x] = \sin[\pi\, x/n]$, with infinite $n \in^* \mathbb{N} \setminus \mathbb{N}$ is an element of $^*C^1[(a, b)]$. (Hint: Consider the real set

$$S = \{g[x] = \sin[\pi\, x/m] \mid a \le x \le b\ \&\ m \in \mathbb{N}\}$$

and its standard extension. Show that transfer implies that $S \subset\, ^*C^1[(a, b)]$.)

Show that $G[x]$ is not S-derivable, that is, $\neg \Phi(G)$. (Intuitively, it wiggles infinitely fast, so can't be infinitesimally linear. Just compute.)

Following is a statement of Nelson's Axiom (S). Notice that it can be viewed as a quantifier interchange rule.

The Standardization Axiom
Let $\phi(x, y)$ be a standard, internal, or external formula of infinitesimal analysis with only x and y free. Let A and B be standard sets. Then

$$(\forall x \in^\sigma A\ (\exists y \in^\sigma B\ \phi(x, y))) \Rightarrow (\exists y \in^\sigma B^A\ (\forall x \in^\sigma A\ \phi(x, y(x))))$$

abbreviated

$$\forall^{st} x\ \exists^{st} y \phi(x, y) \Rightarrow \exists^{st} y(\cdot)\ \forall^{st} x\ \Phi(x, y(x)) \qquad \text{(Nelson's Axiom S)}$$

Even from the point of view of superstructures the additional axiom is important, because Nelson [N-S] uses it to give a list of reduction rules that show:

Theorem
Every sentence of formal infinitesimal analysis reduces to a logically equivalent formula of the form

$$\forall^{st} x \exists^{st} y\ \phi(x, y)$$

where ϕ is internal (and one or both of the external quantifiers may be missing.)

If the formula ϕ of the reduced form only contains standard constants, then Axiom (T) and its negation show that the original formula is equivalent to the formula

$$\forall x \exists y\ \psi(x, y)$$

in $\mathfrak{R}$ where ψ equals ϕ with the stars removed. In this way, Nelson's reduction algorithm mechanically translates any formal sentence of hyperreal analysis into a real sentence. The point of 'reducing' the infinitesimal form of derivable to its standard equivalent of 'locally uniformly differentiable' is to show that the 'reduced' formal property may be more complicated than the infinitesimal one. In fact that example is not very complicated. When we need to use Axiom (S) in the reduction, the real formulas raise the 'type' of variables from points to functions even at the first level. These get COMPLICATED. However, Nelson's reduction algorithm is a marvelous solution to Robinson's metamathematical problem of giving a procedure to translate infinitesimal proofs to non-infinitesimal proofs. (See [N-S] and [HK].)

9. Saturated Superstructure Extensions

Irrespective of whether or not it is possible to represent a particular external set by an external predicate, $E = \{x \in H_m \mid \Phi(x)\}$, it may not be convenient to represent it this way. The external set theory of $\mathfrak{H}$ gives us somewhat more freedom. Another article in this volume shows how this additional freedom helps us to extend the simple intuitive integral from the beginning of the article to discontinuous functions by sketching the Loeb integral construction.

The starting point for this discussion is to view Nelson's Axiom (I) set theoretically. If $\phi(x,y)$ is an internal predicate with x and y free, for each $y \in C$, a standard set, we may define internal sets by $P_y = \{x \in C \mid \phi(x,y)\}$. In this form Axiom (I) says that if for every standard finite subset $Y \subset B$, finite intersections are non-empty,

$$\cap \{P_y \mid y \in Y\} \neq \emptyset$$

then standardly indexed intersections are non-empty,

$$\cap \{P_y \mid y \in^\sigma B\} \neq \emptyset$$

The various "saturation principles" in superstructures say that families of internal sets that have the finite intersection property must have non-empty intersection. The difference between these saturation principles and Axiom (I) is the way the families of internal sets are indexed. The most basic extension of Axiom (I) says any 'standard sized' family of internal sets with the finite intersection property has non-empty intersection. This is the case of the next definition when $\kappa = \text{card}(\mathfrak{R})$.

κ-Saturation
Let κ be an infinite cardinal (of set theory, not *$\mathbb{N}$). The superstructure $\mathfrak{H}$ is κ-saturated if whenever $\text{card}(\Gamma) < \kappa$ and $\{Q_\gamma \subset H_m : \gamma \in \Gamma\}$ is a family of internal sets with the finite intersection property, then

$$\cap [Q_\gamma : \gamma \in \Gamma] \neq \emptyset$$

If $\kappa = \text{card}(\mathfrak{R})$, $\mathfrak{H}$ is said to be polysaturated.

References

[C] N. Cutland, Editor, *Nonstandard Analysis and its Applications*, London Math Soc. Student Texts, vol 10, Cambridge, 1988

[HK] C. W.Henson & H. J. Keisler *The Strength of Nonstandard Analysis*, J. Symbolic Logic, vol 51, 377-386, 1986

[K] H. J. Keisler *Foundations of Infinitesimal Calculus*, Prindle Weber & Schmidt, Boston, 1976

[N-S] E. Nelson *The syntax of nonstandard analysis*, Ann. Pure Appl Logic, vol. ??, pp. ?? 19??

[N-I] E. Nelson *Internal Set Theory*, BAMS, vol. 83, pp. 1165-1198 1977

[R-B] A. Robinson *Non-Standard Analysis*, North Holland, Amsterdam, 1966, 2nd Ed., 1974

[R-M] A. Robinson *Metamathematical problems* J. Symbolic Logic, vol. 38, pp. 500-515, 1973

[S] K. D. Stroyan *Mathematical Background for Calculus using Mathematica*, Academic Press, Boston, 1993

[S&L] K. D. Stroyan and W. A. J. Luxemburg *Introduction to the Theory of Infinitesimals*, Academic Press, New York, 1976

University of Iowa
Iowa City, IA 52242
USA

E-mail stroyan@math.uiowa.edu

Mark McKinzie and Curtis Tuckey

Nonstandard methods in the precalculus curriculum

What should undergraduates get out of their calculus courses?

Is it possible to have a calculus course that provides the intuitions for the concepts and applications, while providing the facility with discrete mathematics that the computer scientist needs? Can a calculus course build on the skills taught in entry-level discrete mathematics courses in a meaningful way—not with rote computations, but in the process of developing the basic material in calculus? Furthermore, can a calculus course give students the impetus to explore the connections between continuous and discrete mathematics, as well as the history of both subjects?

We believe that the answer is yes. Astonishingly, we believe moreover that much can be done toward accomplishing these ends in a precalculus or calculus course, even prior to the definition of the derivative. Even more astonishingly, we claim that a textbook achieving these aims already exists, and has been around for quite some time.

In the first paragraph of the preface to his textbook, *Introductio in Analysin Infinitorum* [1748], Leonhard Euler wrote,

> Often I have considered the fact that most of the difficulties which block the progress of students trying to learn analysis stem from this: that although they understand little of ordinary algebra, still they attempt this more subtle art. From this it follows not only that they remain on the fringes, but in addition they entertain strange ideas about the concept of the infinite, which they must try to use. Although analysis does not require an exhaustive knowledge of algebra, even of all the algebraic techniques so far discovered, still there are topics whose consideration prepares a student for a deeper understanding. However, in the ordinary treatise on the elements of algebra, these topics are either completely omitted or are treated carelessly. For this reason, I am certain that the material I have gathered in this book is quite sufficient to remedy that defect.[1]

In this textbook, without even the barest mention of continuity or the derivative, Euler was nonetheless able to derive the series expansions for the exponential and logarithmic functions, prove the Eulerian relation $e^{i\theta} = \cos\theta + i\sin\theta$, compute the series expansions for the sine and cosine, prove the factorization formula for the sine,

---

This paper was presented to the International Conference on Nonstandard Mathematics (CIMNS), Aveiro, Portugal, July 1994. A revised and expanded version of this paper will appear elsewhere, under the title, "A concrete foundation for calculus."

Acknowledgments. Special thanks to Richard Askey, for pointing out historical references; to Michael Benedikt, for help with the manuscript; and to Jerome Keisler, for technical support.

[1] Euler [1], p. v.

and deduce his celebrated formula, $1 + 1/4 + 1/9 + 1/16 + \cdots = \pi^2/6$. "All this follows from ordinary algebra," Euler claimed, and this in a textbook for beginners!

How did he do it? Part of his success is due to this: even though Euler did not use the notion of the derivative (and certainly not theorems like Taylor's Theorem, which depend on the derivative), his notion of "ordinary algebra" went a bit beyond what our contemporaries would include. In particular, Euler explicitly included the arithmetic of infinite and infinitesimal quantities, and evidently used a nascent notion of convergence to simplify calculations involving infinitely many infinitesimals.

In this paper, we take inspiration from Euler's amazing achievements, and sketch a way that Euler's techniques and results might fruitfully be incorporated into a modern calculus or even precalculus course. We keep the development elementary by leaving the results in nonstandard form. The tools we use, other than the axiomatic introduction of infinite and infinitesimal numbers, are essentially these: mathematical induction, the truncated geometric series, and the binomial theorem for whole exponents. This suggests a way of providing a preliminary to calculus that puts continuous and discrete mathematics on even ground.[2]

We sketch modern axioms for infinite and infinitesimal numbers that can be used to give proofs very similar to Euler's original arguments. We then describe a simple notion of convergence for infinite sums, which is defined without reference to limits, and which appears in one of Euler's early papers. We use this result to provide a criterion for determining whether two infinite sums differ infinitesimally. These definitions and results are all made in a simple axiomatic system. We then use this foundation to deduce the infinite polynomial expansion for the exponential function from the axioms for exponentiation, deduce the Eulerian relations and the infinite polynomial expansions for sine and cosine using basic trigonometry, prove the Binomial Theorem for fractional exponents using algebra, and then deduce the infinite polynomial expansion for the logarithm.[3]

Our derivations would fit well in the calculus courses advocated by Keisler [4] or Stroyan [11], which are based on infinitesimals, or in the course in concrete mathematics for computer science advocated by Graham, Knuth, and Patashnik [2], which emphasizes the interplay between discrete and continuous mathematics. Although we diverge from Euler's exposition when we find it expedient, our methods preserve the concrete and intuitive appeal of the original. In addition to providing students with skills that are useful in other fields, this approach permits the development of calculus within a concrete framework that is at once intuitive, historically faithful, and mathematically rigorous.

---

[2]This approach emphasizes those topics in discrete mathematics, including predicate logic and formal proof, that are becoming increasingly important in some branches of engineering. For example, see Graham, Knuth, and Patashnik [2], and Gries and Schneider [3]—recent books geared for computer- and information-science undergraduates.

[3]Because of limitations of space, we have not been able to provide full proofs here. See McKinzie and Tuckey [7].

The arithmetic of the infinite

A positive infinite number is a number that is bigger than 1, 2, 3, and so on; a positive infinitesimal is a number that is greater than zero, yet less than 1, 1/2, 1/3, and so on. We argue briefly for the plausibility of an axiomatic presentation of the real numbers that assumes the standard axioms, but then draws a distinction between numbers that are finite and numbers that are not. (This could be a good setting for students to get a taste of the axiomatic method, but please note that it is not *necessary* to cover this material any more formally than the basic part of a standard precalculus course. Our approach *requires* no more logic than any other approach.)

What do we mean by an axiomatic presentation? The usual foundational approach is to start with axioms for the natural numbers, then either assume that there is a set of objects and associated operations that satisfy those axioms, or else construct such a set and operations within set theory. Then the rational and real numbers are constructed set-theoretically from the natural numbers as classes modulo specific equivalence relations.[4] This is axiomatic, but the axioms are the axioms of set theory. On the other hand, in a course like calculus or advanced calculus, most textbooks simply give axioms (more or less explicitly) for the relevant number systems, and proceed from there by proving theorems. We use the latter approach. Axioms which appear to refer to sets actually only mention predicates. For example, if $\mathbf{X}$ is a unary predicate letter and φ is a formula, we write $(\forall^{\mathbf{X}} x)\,\varphi(x)$ to abbreviate $(\forall x)[\mathbf{X}(x) \to \varphi(x)]$, and write $(\exists^{\mathbf{X}} x)\,\varphi(x)$ to abbreviate $(\exists x)[\mathbf{X}(x) \wedge \varphi(x)]$. If $\mathbf{X}$ and $\mathbf{Y}$ are unary predicate symbols and f is a unary function symbol, we write $f : \mathbf{X} \to \mathbf{Y}$ to abbreviate $(\forall x)[\mathbf{X}(x) \to \mathbf{Y}(f(x))]$, and similarly for function symbols with additional argument places.

We'll give a brief sketch of the axioms for the natural numbers, then give an even briefer sketch of the axioms for the rational, real, and complex numbers. We feel justified in being very sketchy here because we really mean to simply take the axioms one would use to axiomatize a standard precalculus course, but then to adjoin extra axioms for the notion of *finiteness*.

For the natural numbers, we use the axioms of primitive recursive arithmetic, which are as follows. The language has a predicate letter, $\mathbf{N}$, for the natural numbers, has symbols for zero, successor ($\mathbf{S}$), plus, and times, and symbols for each primitive recursive function that we please. (There are only a few of these.) The simplest axioms state that zero is a natural number, that every natural number has a unique natural successor, that zero is not the successor of any natural number, and that any primitive recursive function (including successor, plus, and times) applied to natural arguments returns a natural result. More complicated axioms give the recursion equations defining the primitive recursive functions. For example, the falling-power function, $\mathrm{Fall}(x, n)$, takes natural arguments and return a natural number, $\mathrm{Fall} : \mathbf{N} \times \mathbf{N} \to \mathbf{N}$, and is defined by

$$(\forall^{\mathbf{N}} x)\bigl[\mathrm{Fall}(x,0) = 1\bigr] \quad \text{and} \quad (\forall^{\mathbf{N}} k)[\mathrm{Fall}(x, k+1) = \mathrm{Fall}(x,k) \cdot (x-k)].$$

---
[4] Cf. Landau [5].

We usually write $x^{\underline{k}}$ for $\text{Fall}(x,k)$. Similarly, for each function a defined on the natural numbers by a term of the language, we have a function symbol, $\text{Sum}_a(n)$, with axioms $\text{Sum}_a : \mathbf{N} \to \mathbf{N}$ and

$$\text{Sum}_a(0) = a_0 \quad \text{and} \quad (\forall^{\mathbf{N}} k)[\text{Sum}_a(k+1) = \text{Sum}_a(k) + a_{k+1}].$$

We usually write $\sum_{k=0}^{n} a_n$ for $\text{Sum}_a(n)$.

Finally, we have the collection of axioms comprising the principle of *complete induction*. This principle says that for each formula $\varphi(x)$ in this language, if $\varphi(0)$ holds, and if for all natural x, the formula $\varphi(\mathbf{S}x)$ holds whenever $\varphi(x)$ does, then $(\forall^{\mathbf{N}} x)\,\varphi(x)$ holds. It turns out that inductions on equations and inequalities will be sufficient for our purposes here.

The rational, real, and complex numbers are introduced axiomatically, with predicate symbols $\mathbf{Q}$, $\mathbf{R}$, and $\mathbf{C}$. The axioms for $\mathbf{Q}$ say, for example, that every natural number is rational, and that every rational number is a quotient of natural numbers. There are also axioms extending $+$ and $\cdot$ to the rationals, and the definition of the new function symbol, $\div$. For $\mathbf{R}$, we need to know that every positive real number is less than some natural number, and that for every positive real x and every natural n, there is a real y such that $y^n = x$. (Cf. the algebraic and ordinal axioms given in Keisler [4].) Furthermore, we need symbols for the basic trigonometric functions, and axioms saying the basic identities hold. (For example, $\cos : \mathbf{R} \to \mathbf{R}$ and $(\forall^{\mathbf{R}} x)(\forall^{\mathbf{R}} y)[\cos(x+y) = \cos(x)\cos(y) - \sin(x)\sin(y)]$.) These, as well as axioms for the exponential and logarithmic functions, will be introduced below as necessary. For $\mathbf{C}$, we need to know that every real number is a complex number, that there is complex number i such that $i^2 = -1$, and that every complex number can be written as $a + bi$, where a and b are real. Finally, we allow recursive definitions of functions that take real or complex values (with the recursion occurring on some natural number parameter—as with the function $\text{Fall}(x,n)$ defined above), and have induction axioms for formulas that take natural arguments.

Binomial Theorem (Natural Exponents). *For all real a and b and all natural number n,*

$$(a+b)^n = \sum_{k=0}^{n} \binom{n}{k} a^{n-k} b^k.$$

Proof. By complete induction. □

So far these are standard axioms. Next, we introduce a new predicate symbol, **Finite**. We say that a natural number n is *finite* if $\textbf{Finite}(n)$ holds and otherwise *infinite*—but rather than having to rely on an intuitive notion, we define this new symbol axiomatically. This means that we have to add the axioms that zero is finite, that the successor of any finite number is finite, that any number less than a finite number is finite, and that any primitive recursive function applied to finite arguments is finite. For some applications, we need axioms for the principle of *finite induction*. This principle says that for each formula $\psi(x)$ in the language obtained by adjoining the predicate **Finite**, if $\psi(0)$ holds, and if for all finite natural x, the formula $\psi(\mathbf{S}x)$

holds whenever $\psi(x)$ does, then $\psi(x)$ holds for all finite natural numbers. In this system, one cannot prove that *all* natural numbers are finite—$(\forall^{\mathbf{N}} x)[\mathbf{Finite}(x)]$—because this would require complete induction on the formula "$\mathbf{Finite}(x)$"—an axiom that we have specifically excluded. We could adjoin this as a new axiom, but instead we adjoin its negation, $(\exists^{\mathbf{N}} N)[\text{not } \mathbf{Finite}(N)]$.[5]

Definition. *A real number is* finite *if its absolute value is less than some natural n such that* $\mathbf{Finite}(n)$ *holds, and* infinite *otherwise. A real number is* infinitesimal *if it is either zero or its reciprocal is infinite. We write "$r \simeq s$" to mean that r and s differ by an infinitesimal.*

This is very sketchy, but we hope it gives a hint of what we mean by the axiomatic approach. We could write down all the axioms ahead of time, but we prefer to jump into our story. The point of this section is not what the specific axioms are, but rather that they are the same as the standard axioms, if one were to axiomatize a number system suitable for precalculus, but with the axiomatic addition of the new notion of *finite*, and the derived notions of *infinite* and *infinitesimal*.

The summation comparison theorem

Euler's original derivation of the series for the exponential function seems to rely on the fact that one can neglect infinitely many infinitesimals in a summation. There are easy examples to the contrary, so to be rigorous, one must have a criterion for separating the valid from the invalid cases. We say that an infinite sequence $a_0, a_1, a_2, \ldots$ satisfies *Euler's criterion* if whenever the sum is carried out to an infinite number of terms, $a_1 + a_2 + \cdots + a_H$, the value is finite, and if whenever the sum is carried out still farther, $a_1 + a_2 + \cdots + a_H + \cdots + a_K$, only an infinitesimal amount more is added or subtracted. The second part is equivalent to saying that $a_H + \cdots + a_K$ is infinitesimal whenever H and K are infinite. This criterion, which can be found explicitly in Euler's work, is an early version of "Cauchy's" criterion for convergence of series.[6]

Euler stated explicitly at the beginning of the *Introductio* that by a "function of a variable quantity" he meant a function given by an "analytic expression" (or formula) and gave examples to show what he meant. We say that a formula is *intrinsic* if it is given by an analytic or logical expression that does not mention the extra predicate **Finite**, or terms like "infinite" or "infinitesimal," the meanings of which derive from the meaning of **Finite**. We say that a function or sequence is *intrinsic* if it is definable by an intrinsic formula. All the functions Euler discussed are intrinsic.

Definition. *A sequence* $s_1, s_2, s_3, \ldots$ *is S-convergent* iff *it is intrinsic, all the terms are finite, and for every infinite J and K, $s_J \simeq s_K$.*

Example. If $a_k = 1/k^2$, then the sequence of partial sums $a_1, a_1 + a_2, a_1 + a_2 + a_3$, ... is S-convergent. Similarly, if x is finite and $a_k = x^k/k!$, then the sequence of partial sums is S-convergent.

---

[5] This approach, in a restricted context, is outlined in Nelson [9].
[6] See Laugwitz [6] p. 265.

Proof. We first show that all the partial sums are finite. For all N, $\sum_{k=1}^{N} \frac{1}{k^2} < 1 + \sum_{k=2}^{N} \frac{1}{k^2 - 1/4} = 1 + \sum_{k=2}^{N} \left(\frac{1}{k-1/2} - \frac{1}{k+1/2} \right) = 1 + \frac{2}{3} - \frac{1}{N+1/2} < \frac{5}{3}$, so the partial sums are all less than $5/3$. Similarly, if $H > N > 1$, then $\sum_{k=N}^{H} \frac{1}{k^2} < \sum_{k=N}^{H} \frac{1}{k^2 - 1/4} = \frac{1}{N-1/2} - \frac{1}{H+1/2}$, so if both N and H are infinite, this sum is infinitesimal. This shows that the sequence of partial sums of $1/k^2$ is S-convergent, as required. The second result is left to the reader. □

Thus an infinite sum $a_1 + \cdots + a_H$ satisfies Euler's criterion if and only if its sequence of partial sums is S-convergent. It turns out that any sequence that is convergent in the standard sense is also S-convergent. Furthermore, analogs of the usual results about convergence of series can be easily proved in this context. The sequence of partial sums of a sequence is S-convergent if the sequence of partial sums of the absolute values is S-convergent. The comparison test holds: if a and b are sequences of positive terms, and $a_n \leq b_n$ for all n, and the sequence of partial sums of b is S-convergent, then the sequence of partial sums of a is S-convergent as well. (The comparison test will be used in later proofs without explicit comment.) Analogs of the other standard convergence tests are also available.

The next three results are elementary but important consequences of complete induction.

Complete Induction (Least Counterexample Form). *Let $\varphi(n)$ be an intrinsic formula. Either $\varphi(n)$ holds for all natural n, or else there is a natural m such that $\varphi(m)$ fails but such that $\varphi(n)$ holds for all natural numbers less than m.*

Overspill Principle. *Let $\varphi(n)$ be an intrinsic formula. Then $\varphi(j)$ holds for all finite j greater than some finite m iff $\varphi(J)$ holds for all infinite J less than some infinite M.*

Proof. For the first direction, assume that m is finite and that $\varphi(j)$ holds for all finite j greater than m. If $\varphi(j)$ holds *for all* j greater than m then we are done. Otherwise, there is a least counterexample; i.e., there is a natural number M greater than m such that $\varphi(M)$ fails but $\varphi(j)$ holds for all j greater than m and less than M. This M cannot be finite. The other direction is left as an exercise. □

Sequential Lemma. *If $a_1, a_2, a_3, \ldots$ is intrinsic and $a_k \simeq 0$ for all finite k, then there is an infinite N such that for all J smaller than N, $a_J \simeq 0$.*

Proof. Since the relation, $\simeq$, is defined using the predicate **Finite**, one cannot apply the Overspill Principle directly. Instead, we note that if $a_k \simeq 0$ for all finite k, then it is also true that $|a_k| < 1/k$ for each finite k. By the Overspill Principle applied to this inequality, there is an infinite N such that for all J smaller than N, $|a_J| < 1/J$. Since $1/J$ is infinitesimal for infinite values of J, this implies that $a_J \simeq 0$ for all infinite J smaller than N, and hence for *all* J smaller than N. □

This next theorem will be our principal tool in deriving Euler's results about series. It follows easily from the Sequential Lemma.

Summation Comparison Theorem. *If the sequences of partial sums $a_1, a_1 + a_2, a_1 + a_2 + a_3, \ldots$ and $b_1, b_1 + b_2, b_1 + b_2 + b_3, \ldots$ are S-convergent, and if for each finite k, $a_k \simeq b_k$, then for all infinite J, $a_1 + a_2 + \cdots + a_J \simeq b_1 + b_2 + \cdots + b_J$.*

Proof. If $a_k \simeq b_k$ for all finite k, then $a_1 + \cdots + a_k \simeq b_1 + \cdots + b_k$ for all finite k as well. By the Sequential Lemma, there is an infinite N such that for all K less than N, $a_1 + \cdots + a_K \simeq b_1 + \cdots + b_K$. Now let J be greater than K. If the sequences of partial sums are S-convergent, then by definition, $a_{K+1} + \cdots + a_J$ and $b_{K+1} + \cdots + b_J$ are both infinitesimal, and hence *for all J, $a_1 + \cdots + a_J \simeq b_1 + \cdots + b_J$.* □

The exponential function

As Euler did in his discussion of exponentials, we start with the algebraic axioms that for $a > 0$ and m, n positive integers, a^0 is 1, $a^{m/n}$ is the mth power of the positive n^{th} root of a, and $a^{-m/n}$ is $1/a^{m/n}$. From this definition we can prove, for example, that if $a > 1$ and $\frac{m}{n} > 0$ then $a^{m/n} > 1$.

In the following, a and b will be finite and greater than 1; m, n, p and q will be integers.

Proposition. *Assume that $a, b > 1$. If $\frac{m}{n}$ is infinitesimal then $a^{m/n} \simeq 1$. If $\frac{m}{n}$ and $\frac{p}{q}$ are finite and $a \simeq b$ and $\frac{m}{n} \simeq \frac{p}{q}$, then $a^{m/n} \simeq b^{p/q}$.*

This proposition shows that the definition of the exponential function a^x could be extended from the rationals to the reals, using equivalence classes. For our purposes, it will be sufficient to assume the following axiom for real exponentiation.

Axiom. *If $a > 1$ and $r > s$, then $a^r > a^s$. If r is finite and $r \simeq \frac{m}{n}$, then $a^r \simeq a^{m/n}$.*

Let N be an infinite integer, so that $1/N$ is infinitesimal. Then as above, $a^{1/N}$ will be greater than 1 by an infinitesimal amount, and hence we may write $a^{1/N} = 1 + \lambda \frac{1}{N}$, where $\lambda > 0$ and $\lambda/N \simeq 0$. Let x be positive and finite, and let J be $\lfloor Nx \rfloor$, the greatest integer less than Nx, so that the difference between x and J/N is less than $1/N$, an infinitesimal. Then we may compute

$$
\begin{align}
(1) \quad a^x &\simeq a^{J/N} \\
(2) \quad &= (a^{1/N})^J \\
(3) \quad &= \left(1 + \lambda \frac{1}{N}\right)^J \\
(4) \quad &= 1 + J\lambda \frac{1}{N} + \frac{J^2}{2!}\left(\lambda \frac{1}{N}\right)^2 + \frac{J^3}{3!}\left(\lambda \frac{1}{N}\right)^3 + \cdots + \left(\lambda \frac{1}{N}\right)^J \\
(5) \quad &= 1 + \frac{J}{N}\lambda + \frac{J^2 \lambda^2}{J^2\, 2!}\left(\frac{J}{N}\right)^2 + \frac{J^3 \lambda^3}{J^3\, 3!}\left(\frac{J}{N}\right)^3 + \cdots + \frac{J^J \lambda^J}{J^J\, J!}\left(\frac{J}{N}\right)^J \\
(6) \quad &\simeq 1 + \lambda x + \frac{\lambda^2}{2!}x^2 + \frac{\lambda^3}{3!}x^3 + \cdots + \frac{\lambda^J}{J!}x^J,
\end{align}
$$

using the falling-power notation, $n^{\underline{k}} = n(n-1)(n-2)\cdots(n-k+1)$. The relation in (1) follows from the axiom, and equations (2) through (5) follow from ordinary algebra. The step from (5) to (6) is justified precisely by the Summation Comparison Theorem. To see this, first note that by the example and the comparison test, the two sums have S-convergent partial sums. Second, note that $J^{\underline{k}}/J^k \simeq 1$ and $(J/N)^k \simeq x^k$ for all finite k, and hence

$$\frac{J^{\underline{k}}}{J^k}\left(\frac{J}{N}\right)^k \frac{\lambda^k}{k!} \simeq \frac{\lambda^k}{k!}$$

for all finite k as well. *Voilá.*

The natural exponential function

Recall that for a fixed infinite N, λ was defined by the equation $a^{1/N} = 1 + \lambda/N$. As a function of λ, then, we have $a = (1 + \lambda/N)^N$, and we may ask which value of a corresponds to λ being set equal to 1. We first note that the corresponding value of a depends on the choice of N only to within an infinitesimal, so long as N is infinite.

Proposition. *For all infinite M and N,* $\left(1 + \frac{1}{M}\right)^M \simeq \left(1 + \frac{1}{N}\right)^N.$

Proof. By the Binomial Theorem and the Summation Comparison Theorem. □

We could define e by fixing some infinite N and taking λ to be 1; namely, $e = (1 + \frac{1}{N})^N$. For our purposes here, however, all we need is the following axiom for e.

Axiom. *For all infinite M,* $e \simeq \left(1 + \frac{1}{M}\right)^M.$

With this axiom, we can prove the following.

Proposition. *For all finite x and infinite N,*

(7) $$e^x \simeq 1 + x + \frac{x^2}{2!} + \frac{x^3}{3!} + \cdots + \frac{x^N}{N!},$$

and

(8) $$e^x \simeq \left(1 + \frac{x}{N}\right)^N.$$

Proof. By the Binomial Theorem and the Summation Comparison Theorem. □

Ordinary algebra reveals that the inverse of the function given by $(1+x/N)^N$ is given by $N[y^{1/N} - 1]$. Euler applied the Binomial Theorem for Fractional Exponents to this latter quantity, with $1 + y$ substituted for y, to obtain the series for the natural logarithm. We will derive the series for the logarithm shortly, but first we will derive the Eulerian relations and the series for sine and cosine.

The Eulerian relations and the series for sine and cosine

We prove the Eulerian relations in the form $\cos x \simeq [e^{ix} + e^{-ix}]/2$ and $\sin x \simeq [e^{ix} - e^{-ix}]/2i$, using the representation $e^x \simeq (1 + x/N)^N$ derived in the previous section.

Theorem. *For all finite x,* $\quad \cos x \simeq \dfrac{e^{ix} + e^{-ix}}{2} \quad$ *and* $\quad \sin x \simeq \dfrac{e^{ix} - e^{-ix}}{2i}.$

Proof. Using the familiar formulas for the sine and cosine of a sum of angles, one can show that for all n and θ, $\cos n\theta = \frac{1}{2}[(\cos\theta + i\sin\theta)^n + (\cos\theta - i\sin\theta)^n]$ and $\sin n\theta = \frac{1}{2i}[(\cos\theta + i\sin\theta)^n - (\cos\theta - i\sin\theta)^n]$. Letting N be infinite, and substituting N for n and x/N for θ yields the equations,

(9) $$\cos x = \frac{1}{2}\left[\left(\cos\frac{x}{N} + i\sin\frac{x}{N}\right)^N + \left(\cos\frac{x}{N} - i\sin\frac{x}{N}\right)^N\right]$$

(10) $$\sin x = \frac{1}{2i}\left[\left(\cos\frac{x}{N} + i\sin\frac{x}{N}\right)^N - \left(\cos\frac{x}{N} - i\sin\frac{x}{N}\right)^N\right].$$

If x is finite and N infinite then $\cos\frac{x}{N} \simeq 1$ and $\sin\frac{x}{N} \simeq \frac{x}{N}$. One might guess that making these substitutions in (9) and (10) should yield the formulas,

(11) $$\cos x \simeq \frac{1}{2}\left[\left(1 + \frac{ix}{N}\right)^N + \left(1 - \frac{ix}{N}\right)^N\right]$$

(12) $$\sin x \simeq \frac{1}{2i}\left[\left(1 + \frac{ix}{N}\right)^N - \left(1 - \frac{ix}{N}\right)^N\right].$$

This is correct, but to verify it we must show that the error incurred by taking these substitutions to the Nth power are negligible in the final answer. For this, we require elementary polynomial bounds on $\sin\theta$ and $\cos\theta$ for small θ; in particular, we use the fact that for θ between 0 and π, $\theta - \frac{1}{4}\theta^3 < \sin\theta < \theta$ and $1 - \frac{1}{2}\theta^2 < \cos\theta < 1$. These inequalities can be derived from the inequality $\sin\theta < \theta < \tan\theta$ (for $0 < \theta < \pi/2$) using elementary trigonometric identities.[7] $\quad\square$

The familiar series for the sine and cosine can now be obtained by multiplying out the quantities in (11) and (12).

Theorem. *For all finite x and infinite odd H,*

(13) $$\cos x \simeq 1 - \frac{x^2}{2!} + \frac{x^4}{4!} - \cdots \pm \frac{x^{H+1}}{(H+1)!}$$

(14) $$\sin x \simeq x - \frac{x^3}{3!} + \frac{x^5}{5!} - \cdots \pm \frac{x^H}{H!}$$

Proof. By the Binomial Theorem and the Summation Comparison Theorem. $\quad\square$

---

[7] See McKinzie and Tuckey [7].

The binomial theorem

So far we have only required the Binomial Theorem for whole exponents, a result which is easily verified using complete induction. For fractional exponents, the proof requires a bit more algebra.

Binomial Theorem (Fractional Exponents). *If $|x| < 1$, m and n are finite, and H is infinite, then*
$$(1+x)^{m/n} \simeq 1 + \tfrac{m}{n}x + \frac{(\tfrac{m}{n})^{\underline{2}}}{2!}x^2 + \cdots + \frac{(\tfrac{m}{n})^{\underline{H}}}{H!}x^H.$$

Proof. Define the function bin by $\mathrm{bin}(r) = 1 + rx + \tfrac{r^{\underline{2}}}{2!}x^2 + \cdots + \tfrac{r^{\underline{H}}}{H!}x^H$. We must show that $(1+x)^{m/n} \simeq \mathrm{bin}(\tfrac{m}{n})$. The key to the proof is the fact that if r and s are finite, then $\mathrm{bin}(r)\,\mathrm{bin}(s) \simeq \mathrm{bin}(r+s)$. This follows from the formula $\sum_{i=0}^{k} \frac{r^{\underline{i}}\, s^{\underline{(k-i)}}}{i!\,(k-i)!} = \frac{(r+s)^{\underline{k}}}{k!}$ (Vandermonde's Convolution, which may be verified by complete induction) and the Summation Comparison Theorem. If m and n are finite, then
$$(1+x)^m = \mathrm{bin}(m) = \mathrm{bin}(\underbrace{\tfrac{m}{n} + \cdots + \tfrac{m}{n}}_{n}) \simeq \left(\mathrm{bin}(\tfrac{m}{n})\right)^n,$$
and hence $(1+x)^{m/n} \simeq \mathrm{bin}(\tfrac{m}{n})$, as required. □

This is easily extended to negative rational exponents by a similar argument, and, by the Sequential Lemma, to the case where m and n are infinite, so long as m/n is finite. The axiom for the exponential function and the Summation Comparison Theorem could then be used to obtain the result for irrational exponents, if desired.

The logarithm

The polynomial expansion of the logarithm takes a slightly more delicate approach than the others, and perhaps its derivation would be easier after more analysis is developed. (Euler appears to use more analysis than we do.) Our approach is as follows. For each natural n, let exp be the function defined by $\exp(n,x) = (1+x/n)^n$. The inverse of $\exp(n,\cdot)$ is given by $\log(n,y) = n(y^{1/n} - 1)$, defined for all positive y. We first note that for finite, positive, noninfinitesimal y, the value of $\log(H,y)$ is independent, up to an infinitesimal, of the value of H, so long as H is infinite.

Proposition. *For all finite, positive, noninfinitesimal y, and all infinite H and K,*
$$\log(H,y) \simeq \log(K,y).$$

Proof. Let H and K be infinite. It is fairly easy to verify (i) that for all finite x, the values $\exp(H,x)$ and $\exp(K,x)$ are finite, positive, and noninfinitesimal, and (ii) that whenever r and s are finite and $\exp(H,r) \simeq \exp(K,s)$, we may conclude that $r \simeq s$ as well. Now let y be finite, positive, and noninfinitesimal. By (i), both $\log(H,y)$ and $\log(K,y)$ are finite, and hence from the equation $\exp(H,\log(H,y)) = y = \exp(K,\log(K,y))$ and (ii), we conclude that $\log(H,y) \simeq \log(K,y)$. □

This result makes it reasonable to add the following axiom for a one-placed log function.

Axiom. *For all finite, positive, noninfinitesimal y, and all infinite H, $\log(y) \simeq H\left[y^{1/H} - 1\right]$.*

"For convenience," (as Euler would often say) we consider the related function, $\log(1+y)$, and restrict the values of y to those that give rise to an S-convergent series.

Proposition. *For all y with $|y| < 1$ but $y \not\simeq -1$, and all infinite H,*

$$\log(1+y) \simeq y - \frac{1}{2}y^2 + \frac{1}{3}y^3 - \cdots \pm \frac{1}{H}y^H.$$

Proof. By the Binomial Theorem, the Sequential Lemma, and the Summation Comparison Theorem. □

In the *Introductio,* Euler went on to exhibit series expansions for other transcendental functions, including the tangent, cotangent, and arctangent, and then turned to the problem of factoring polynomials and series. Such factorizations naturally give rise to infinite products.

We don't have space to go into this here, but two of Euler's major results in the *Introductio* were the factorization of the sine and the summation of the reciprocals of the squares. These results are within reach of the elementary methods outlined here, but the arguments are slightly harder than those we have encountered so far, in that they involve a few tricky trigonometric substitutions, and a bit from the theory of equations. Perhaps these results could be left to an "advanced precalculus" class. For details see McKinzie and Tuckey [7], [8].

Precalculus reform

Euler's *Introductio* was expressly intended to be a precalculus textbook, its goal being to see how far one could get in the foundations of analysis using only elementary methods. The point was not to give short and slick derivations from an extensive body of knowledge, but rather to educate beginners. Euler said,

> Although all of these nowadays are accomplished by means of differential calculus, nevertheless, I have here presented them using only ordinary algebra, in order that the transition from finite analysis to analysis of the infinite might be rendered easier. ... At the same time I readily admit that these matters can be much more easily worked out by differential calculus.[8]

In the standard treatments, discrete mathematics is held largely disjoint from calculus, and interesting and useful series are studied only after Taylor's Theorem is proved—usually at the end of the lessons on convergence of sequences and series, well after the derivative is thoroughly discussed. In our approach—Euler's approach—beginners get their hands on concrete examples of sequences and series even before the

---

[8] Euler [1], pp. ix–x.

derivative is defined. Our presentation is simplified by leaving the results in nonstandard form.[9] This approach also gives students practice with important topics from discrete mathematics—mathematical induction, recursion, geometric sums, axiomatics, and basic combinatorics—in the course of proving elementary analogs of real theorems of analysis. Finally, this approach points out the long and fruitful relationship between notions of the discrete and the continuous, as they have jointly contributed to the successful understanding and application of mathematics in science, engineering, and computing.

References

[1] L. Euler. *Introductio in Analysin Infinitorum*, Tomus primus, Lausanne, 1748. Reprinted in *Opera Omnia*, series 1, volume 8. Translated from the Latin by J. D. Blanton, *Introduction to Analysis of the Infinite*, Springer–Verlag, 1988.
[2] R. L. Graham, D. E. Knuth, and O. Patashnik. *Concrete Mathematics: A Foundation for Computer Science*, 2d ed. Addison–Wesley, 1989.
[3] D. Gries and F. B. Schneider, *A logical approach to discrete math*, Springer–Verlag, 1993.
[4] H. J. Keisler. *Elementary Calculus: An Infinitesimal Approach*, PWS Publishers, 2d edition, 1986.
[5] E. Landau. *Foundations of analysis; the arithmetic of whole, rational, irrational and complex numbers*, 2d ed. Chelsea, 1960.
[6] D. Laugwitz. Hidden lemmas in the early history of infinite series, *Æquationes Mathematicæ*, 34:264–276, 1987.
[7] M. McKinzie and C. Tuckey. A concrete foundation for calculus. Technical Memorandum BL0112650/931001/20TM, AT&T Bell Laboratories, 1993.
[8] M. McKinzie and C. Tuckey. Hidden lemmas in Euler's summation of the reciprocals of the squares. *Archive for the History of Exact Sciences*, to appear.
[9] E. Nelson. *Predicative Arithmetic*, Princeton University Press, 1986.
[10] *Radically elementary probability theory*, Princeton University Press, 1987.
[11] K. D. Stroyan. *Calculus using Mathematica*, Academic Press, 1993.

MARK MCKINZIE
Department of Mathematics
University of Wisconsin
E-mail mckinzie@math.wisc.edu

CURTIS TUCKEY
Software and Systems Research Center
AT&T Bell Laboratories
E-mail tuckey@research.att.com

---

[9] Nelson makes a similar point in [10] about nonstandard analogs of theorems in probability. Our nonstandard analogs are equivalent to the standard versions, and moreover this equivalence could easily be proved in a calculus course based on Keisler's book, [4], if one wished.

Vítor Neves
Difference quotients and smoothness
(a conjecture of Keith Stroyan)

1. Introduction

Part of the main theorem in this article (theorem 2.1) states that a standard curve f in a locally convex complete HM space is of class C^k iff its interpolating polynomials at any set of $k+1$ infinitely close nearstandard points are infinitely near for compact convergence, their common standard parts being k-th Taylor polynomials of f; this was a conjecture of Keith Stroyan which we proved by means of the theory of Difference Quotients as given in [1].

Actually we also show that a standard curve in a locally convex HM space is of class C^k iff its p-th ($0 \leq p \leq k$) difference quotients at nearstandard points are S-continuous, even when only equally spaced interpolating points are considered. This is a consequence of Lemma 4.1.2 proved in section 5.

The theory of differentiation underlying the calculations below is presented in [4] and compared with standard theories in [2]; an account of theorem 2.1 for real valued functions is given in [3].

Notation

What we need from the Theory of Infinitesimals and non-standard extensions $^*(.)$ can be found in [5] or [6]; nevertheless we describe some notation for the reader's convenience.

The set of standard elements of an extension $^*X$ is denoted $^\sigma X$. Given a uniform topological space X and elements $x, y \in {}^*X$ we write $x \approx y$ iff x is *infinitely close to* y.

For a real locally convex space E, if $x, y \in {}^*E$, $x \approx y$ iff $x - y$ is infinitely close to zero; $\mu(0)$ denotes the *monad* of zero, i.e., the set of elements of $^*E$ infinitely close to zero; for each $a \in E$, $\mu(a) := a + \mu(0)$; the set of *finite* elements of $^*E$ is denoted $fin(^*E)$. When $E = {}^*\mathbb{R}$ we denote the monad of 0 by o and the set of finite elements by O.

A locally convex space is a *complete* HM space if any finite element of $^*E$ is infinitely close to a standard element.

For a non-empty open interval I of $\mathbb{R}$ and each $k \in \mathbb{N}$, $I^{<k>} = \{(t_0, ..., t_k) \in I^{k+1} : t_i \neq t_j, (0 \leq i \neq j \leq k)\}$; τ, τ_k, σ or σ_k denote elements of $I^{<k>}$ or $^*I^{<k>}$ and $a_{(k)} := (a, ..., a) \in I^{k+1}$.

The author was partially supported by scholarship 10/c/92/PO of JNICT-INVOTAN.

Given a function $f : I \to E$, the interpolating polynomial of f with degree at most k at the coordinates of $\tau \in I^{<k>}$ evaluated at t is denoted $p(\tau, t)$; $T_a^k f(t)$ denotes the Taylor polynomial of degree k of f at a evaluated at t.

Let $f : I \to E$ be a curve in the locally convex complete HM space E.

Definition 1.1. *The k-th difference quotient of f at τ is the element of E, $\delta^k f(\tau)$ ($k \in \mathbb{N}$), given recursively by the following identities*

$$\delta^0 f(t_0) = f(t_0)$$

$$\delta^{k+1} f(\tau_{k+1}) = \frac{k+1}{t_0 - t_{k+1}} \left[\delta^k f(t_0, ..., t_k) - \delta^k f(t_1, ..., t_{k+1}) \right]$$

The equally spaced k-th difference quotient of f (with step $s > 0$) is the element $\delta^k_{eq} f(t; s)$ of E given by

$$\delta^k_{eq} f(t; s) = \delta^k f(t, t+s, ..., t+ks)$$

Definition 1.2. *Let X be a uniform topological space. A function $g : A \subseteq {}^*X \to {}^*E$ is S-continuous on A if $g(A) \subseteq \text{fin}(^*E)$ and $x \approx y \Rightarrow g(x) \approx g(y)$, for all $x, y \in A$.*

2. The main Theorem

We shall prove the following

Theorem 2.1. *When E is a complete HM space the following conditions are equivalent for all functions $f : I \to E$.*
  i. f is of class C^k,
  ii. For all $a \in {}^\circ I$, $t \in {}^*I$, $s \in {}^*\mathbb{R}$, $a \neq t \approx a$ & $0 \neq s \approx 0 \Rightarrow \delta^k_{eq} f(t; s) \approx f^{(k)}(a)$
  iii. For all $a \in {}^\circ I$, $\tau \in \mu(a_{(k)}) \cap {}^*I^{<k>}$, $\delta^k f(\tau) \approx f^{(k)}(a)$
  iv. For all $a \in {}^\circ I$, $\delta^k f$ is S-continuous on $\mu(a_{(k)}) \cap {}^*I^{<k>}$
  v. For all $a \in {}^\circ I$, $\tau, \sigma \in {}^*I^{<k>}$, $t \in O$, $\tau \approx \sigma \approx a_{(k)} \Rightarrow p(\tau, t) \approx p(\sigma, t)$
  vi. For all $a \in {}^\circ I$, $\tau \in \mu(a_{(k)})^* \cap I^{<k>}$, $t \in O$, $p(\tau, t) \approx T_a^k f(t)$, i.e., $p(\tau, .)$ is infinitely close to $T_a^k f$ for the topology of compact convergence in $C(\mathbb{R}, E)$.

The cases $k = 0$ and $k = 1$ are treated in section 3, the remaining cases are studied in section 4. Section 5 contains proofs of results presented in section 4.

3. Proof of Theorem 2.1. in the cases $k = 0$ and $k = 1$

3.1. Case $k = 0$

It is almost immediate that when $k = 0$ all conditions simply state the continuity of f, so the theorem holds when $k = 0$.

3.2. Case $k = 1$

Proposition 3.2.1. *When $k = 1$ conditions i, ii, iii and iv are equivalent.*

Proof. Conditions ii, iii and iv take the following form

ii. $\frac{f(t+s)-f(t)}{s} \approx f'(a)$, when $0 \neq s \approx 0$ and $a \neq t \approx a \in {}^\sigma\! I$

iii. $\frac{f(t_0)-f(t_1)}{t_0-t_1} \approx f'(a)$, when $t_0 \neq t_1$ and $t_0 \approx t_1 \approx a$.

iv. The quotients $\frac{f(t_0)-f(t_1)}{t_0-t_1}$ are all finite and infinitely close to each other, when $t_0 \neq t_1$ and $t_0 \approx t_1 \approx a$.

All of these are reformulations of uniform differentiability of f at a, therefore they are equivalent to i. The assumption that E is an HM space is used to deduce from iv that the standard parts of the ratios exist and are the same when $t_0 \neq t_1$ and $t_0 \approx t_1 \approx a$, so that f' exists at all $a \in I$ and f is uniformly differentiable. □

Proposition 3.2.2. *When $k = 1$ conditions iii, v and vi are equivalent.*

Proof. We show that the chain of implications iii ⇒ vi ⇒ v ⇒ iii holds.

(iii ⇒ vi) Assume iii holds. By 3.2.1, iii is equivalent to i, so f is continuous therefore $f(t_0) \approx f(a)$ if $t_0 \approx a$; moreover by assumption, $\frac{f(t_0)-f(t_1)}{t_0-t_1} \approx f'(a)$ and $t - t_0 \approx t - a \in O$, when $t_0 \neq t_1$ and $t_0 \approx t_1 \approx a \in \mathcal{R}$. Thus, for all $t \in O$,

$$p(\tau, t) = f(t_0) + (t - t_0)\delta^1 f(t_0, t_1) \approx f(a) + (t - a)f'(a) = T_a^1 f(t)$$

as required.

(vi ⇒ v) This is a consequence of symmetry and transitivity of $\approx$.

(v ⇒ iii) By hypothesis, if $t_0 \approx a \approx s_0$ and $t_0 \neq a \neq s_0$ one has, for all finite t

$$f(a) + (t-a)\frac{f(t_0) - f(a)}{t_0 - a} \approx f(a) + (t-a)\frac{f(s_0) - f(a)}{s_0 - a}$$

Taking t equal to $a + 1$ say, one obtains

$$\frac{f(t_0) - f(a)}{t_0 - a} \approx \frac{f(s_0) - f(a)}{s_0 - a}$$

if $t_0 \approx a \approx s_0$ and $t_0 \neq a \neq s_0$; this is equivalent to the existence of $f'(a)$ and the following holds

$$\delta^1 f(t, a) = \frac{f(t) - f(a)}{t - a} \approx f'(a) \quad (a \neq t \approx a)$$

in particular the left side is finite when $a \neq t \approx a$. Again by hypothesis, one has, for all $t \in O$ and all $t_0, t_1 \approx a$ such that $t_0 \neq t_1$,

$$f(t_0) + (t - t_0)\delta^1 f(t_0, t_1) \approx f(t_0) + (t - t_0)\delta^1 f(t_0, a)$$

Taking t as $t + 1$ say, one obtains

$$\delta^1 f(t_0, t_1) \approx \delta^1 f(t_0, a) \approx f'(a)$$

as required. □

It follows that theorem 2.1 holds when $k \in \{0, 1\}$.

4. Proof of Theorem 2.1. in the case k≥2

Again we study separately two chains of implications: first i⇒ii⇒iii⇒iv⇒i and then i(+iii)⇒vi⇒v⇒iv; implications ii⇒iii and iv⇒i are the hardest to prove.

Proposition 4.1. *Conditions i through iv are equivalent when $k \geq 2$.*

Proof. (i⇒ii) We show that, under hypothesis i,
$$\delta^k f(t;s) \approx f^{(k)}(t) \qquad (a \neq t \approx a \in {}^\sigma I \, \& \, 0 \neq s \in o)$$

First one can easily prove by induction that

(1) $$\delta^k_{eq} f(t;s) = s^{-k} \sum_{i=0}^{k} (-1)^{k-i} \binom{k}{i} f(t+is)$$

By [4;3.3] one has

Lemma 4.1.1. *Under the hypothesis of theorem 2.1, f is of class C^k iff there exist standard functions $D^i f : I \to E$ such that, when $t \in {}^\sigma I + o$ and $s \approx t$*
$$f(s) = f(t) + \sum_{i=1}^{k} \frac{1}{i!}(s-t)^i D^i f(t) + |s-t|^k \eta$$
with $\eta \approx 0$ in ${}^*E$. The functions $D^i f$ are the derivatives $f^{(i)}$ of f.

Expand the terms $f(t+is)$ in (1) according to this lemma and arrange the summands in order to apply the following
$$\sum_{i=1}^{k} (-1)^i \binom{k}{i} i^p = \begin{cases} -1 & p = 0 \\ 0 & 1 \leq p < k \\ (-1)^k k! & p = k \end{cases}$$
–which again is provable by (double) induction – and obtain
$$\delta^k f(t;s) = f^{(k)}(t) + \sum_{i=1}^{k} (-1)^i \binom{k}{i} i^k \eta_i$$
with all the η_i infinitesimal in ${}^*E$, so that
$$\delta^k f(t;s) \approx f^{(k)}(t)$$
as required.

(ii⇒iii) This is a simple consequence of the following lemma proved in section 5.

Lemma 4.1.2. *Let E be a locally convex space, J an interval in ${}^*\mathbb{R}$, $g : J \to {}^*E$ an internal function, $a \in {}^\sigma \mathbb{R}$ such that $\mu(a) \subseteq J$. If g is S-continuous on $\mu(a)$ and $k \in {}^\sigma \mathbb{N}$ the following conditions are equivalent*

(2) $$\delta^k_{eq} g(t;s) \approx \delta^k_{eq} g(t';s') \qquad (t, t' \approx a; \, s, s' \approx 0)$$

(3) $$\delta^k g(\tau) \approx \delta^k g(\sigma) \qquad (\tau, \sigma \in \mu(a_{(k)}) \cap J^{<k>})$$

(iii ⇒iv) This is immediate from transitivity and symmetry of $\approx$.

(iv ⇒i) This is already proven when $k \in \{0,1\}$ in section 3. Assume the implication holds for $k \geq 1$. We want to show that condition

(4) $\qquad \delta^{k+1} f$ is $S-$ continuous in $\mu(a_{(k+1)}) \cap {}^*I^{<k+1>}$, for all $a \in {}^\sigma I$

implies f is of class C^{k+1}. Actually we show that

(5) $\qquad\qquad\qquad (4) \Rightarrow f'$ is of class C^k

applying the induction hypothesis to f'.

The following is proved in [1;1.3.13]

Proposition 4.1.3. *If T and S are distinct subsets of $\mathbb{R}$ with $k+1$ elements, there exist orderings of T, $t_0, t_1, ..., t_k$ and of S, $s_0, s_1..., s_k$, such that $t_i \neq s_i$ if $i \leq j$ and*

$$\delta^k f(\tau) - \delta^k f(\sigma) = \frac{1}{k+1} \sum_{i=0}^{k} (t_i - s_i) \delta^{k+1} f(t_0, ..., t_i, s_i, ..., s_k)$$

Therefore, under hypothesis (4), for all $a \in {}^\sigma I$

$$\delta^k f(\tau) - \delta^k f(\sigma) \approx 0 \qquad if\ \tau, \sigma \approx a_{(k)}$$

so that the standard function $\delta^k f$ is S-continuous in $\mu(a_{(k)}) \cap {}^*I^{<k>}$, for all $a \in {}^\sigma I$; by induction hypothesis f is of class C^k, hence of class C^1 and, in particular f' exists. But the following holds by [1;1.3.16].

Proposition 4.1.4. *If f is differentiable at t_i*

$$\delta^{k+1} f(t_0, ..., t_i, ..., t_k, t_i) := \lim_{t \to 0} \delta^{k+1} f(t_0, ..., t_i, ..., t_k, t_i + t)$$

exists and

(6) $\qquad \delta^k f'(t_0, ..., t_k) = \dfrac{1}{k+1} \sum_{i=0}^{k} \delta^{k+1} f(t_0, ..., t_k, t_i)$

We are ready to show that condition (5) holds. We saw that condition (4) implies f' exists, so that we may use this last proposition: assume (4) holds and let τ and σ be given in $\mu(a_{(k)}) \cap {}^*I^{<k>}$; by proposition 4.1.4 above, if t and s are sufficiently small

$$\delta^{k+1} f(t_0, ..., t_k, t_i) \approx \delta^{k+1} f(t_0, ..., t_k, t_i + t)$$

and

$$\delta^{k+1} f(s_0, ..., s_k, s_j) \approx \delta^{k+1} f(s_0, ..., s_k, s_j + s)$$

Therefore, using t_i and s_j conveniently chosen in o and S-continuity of $\delta^{k+1} f$, one obtains the S-continuity of $\delta^k f'$ from (6): all addends on the right sides for τ and σ can be made infinitely close to each other in pairs, if $\tau \approx \sigma$.

This ends the proof of proposition 4.1. $\qquad\square$

Proposition 4.2. *When $k \geq 2$ the chain of implications $i \Rightarrow vi \Rightarrow v \Rightarrow iv$ holds.*

Proof. It is immediate that vi $\Rightarrow$ v; from [1;1.3.7] we have

(7) $$p(\tau, t) = f(t_0) + \sum_{i=0}^{k} \frac{1}{i!} \left[\prod_{j=0}^{i-1}(t - t_j)\right] \delta^i f(\tau_i)$$

Condition v implies that the coefficients of degree k in $p(\tau, t)$ and $p(\sigma, t)$ are infinitely close when $\tau \approx \sigma \approx a_{(k)}$; but these are respectively $\delta^k f(\tau)$ and $\delta^k f(\sigma)$, hence v $\Rightarrow$ iv.

Only i $\Rightarrow$ vi remains to be shown. Assume i holds, i.e., f is of class C^k. We have that f is of class C^i, for all $i \le k$. We saw that i $\Leftrightarrow$ iii, thus $\delta^i f(\tau_i) \approx f^{(i)}(a)$ $(0 \le i \le k)$ when $\tau \approx a_{(k)}$. It follows from condition (7) that, when $\tau \approx a_{(k)}$, the coefficients of similar terms in $p(\tau,.)$ and $T_a^k f(.)$ are infinitely near, i.e., $p(\tau, t) \approx T_a^k f(t)$ for all $t \in O$. We proved that i $\Rightarrow$ vi and finished the proof of 4.2. □

This ends the proof of theorem 2.1 when $k \ge 2$ modulo the proof of lemma 4.1.2, which we proceed to present.

5. A proof of lemma 4.1.2.

It is clear that condition (3) implies condition (2). To show that (2)$\Rightarrow$(3) consider the following proposition proved in [1;1.3.2 & 1.3.3]

Proposition 5.1.

$$\delta^k f(\tau) = k! \sum_{i=0}^{k} \beta_i^k(\tau) f(t_i) \quad \text{with} \quad \beta_i^k(\tau) = \left[\prod_{\substack{j=0 \\ j \ne i}}^{k}(t_i - t_j)\right]^{-1} \quad (i = 0, 1, ..., k)$$

In particular $\delta^k f$ is symmetrical.

Take $\tau \in \mu(a_{(k)}) \cap {}^*J^{<k>}$. As the functions β_i^k above are *-continuous in ${}^*J^{<k>}$, if $\sigma \in {}^*J^{<k>}$ and σ is sufficiently near $\tau \in \mu(a_{(k)}) \cap {}^*J^{<k>}$ one has $\delta^k g(\tau) \approx \delta^k g(\sigma)$. Now take $\sigma \in \mu(a_{(k)}) \cap {}^*J^{<k>} \cap {}^*\mathbb{Q}^{k+1}$ satisfying the following conditions

(8) $$\tau \approx \sigma = \left(\frac{p_i}{q_i}; 0 \le i \le k\right) \quad \text{with} \quad p_i, q_i \in {}^*\mathbb{Z}$$

(9) $$\delta^k g(\tau) \approx \delta^k g(\sigma)$$

By Proposition 5.1, we may assume that $\sigma = (s_0, ..., s_k)$ with $s_0 < ... < s_k$. One of the q_i is infinite (otherwise all coordinates of σ would be a) therefore the least common multiple of the q_i, q, is infinite, that is $1/q \approx 0$ and

$$s_i = s_0 + \frac{r_i}{q} \quad \& \quad 0 \le r_0 < r_1 < ... < r_k = q\left(\frac{p_k}{q_k} - \frac{p_0}{q_0}\right) = n \in {}^*\mathbb{N}$$

It follows that
$$a \approx s_0 < s_0 + \frac{i}{q} < s_k \approx a \qquad (0 \le i \le n)$$
$$\sigma^i = \left(s_0 + \frac{i}{q}, ..., s_0 + \frac{i+k}{q}\right) \approx a_{(k)} \qquad (0 \le i \le n-k)$$

But the following is proved in [1;1.3.10]

Proposition 5.2. *If $t_0 < t_1 < ... < t_n$ and $0 = i_0 < i_1 < ... < i_k = n$ $(k \le n)$, there exist nonnegative real numbers β_r $(r = 0, ..., n-k)$ such that $\sum_{r=1}^{n-k} \beta_r = 1$ and for any f*
$$\delta^k f(t_{i_0}, t_{i_1}, ..., t_{i_k}) = \sum_{r=0}^{n-k} \beta_r \delta^k f(t_r, t_{r+1}, ..., t_{r+k})$$

We may conclude that $\delta^k g(\sigma)$ is an (internal) hyperconvex combination of the $g(\sigma^i)$ and consequently
$$\delta^k g(\sigma) \approx \delta^k g(\sigma^i) \qquad (0 \le i \le n-k)$$

By condition (9)
$$\delta^k g(\tau) \approx \delta^k g(\sigma^0) = \delta^k_{eq} g(s_0; \frac{1}{q})$$

Therefore the (arbitrary) difference quotient $\delta^k g(\tau)$ is infinitely close to one of the $\delta^k_{eq} g(t;s)$ which are infinitely near each other by hypothesis. It follows that the $\delta^k g(\tau)$ are infinitely near each other, as we wanted to show. This ends the proof of lemma 4.1.2, ii $\Rightarrow$iii holds and theorem 2.1 is proven. $\square$

References

[1] Frolicher, A. and Kriegl, A.: *Linear spaces and differentiation theory*, John Wiley & Sons, 1988.
[2] Neves, V.: *Infinitesimal calculus in HM spaces*, Bull. Soc.Math. Belgique, Vol. 40 (1988), 177-198.
[3] Neves, V. and Stroyan, K. D.: *A discrete condition for higher order smoothness*, 1994, (submitted)
[4] Stroyan, K. D.: *Infinitesimal calculus in locally convex spaces I, Fundamentals*, Trans.Amer. Math. Soc., 240 (1978), 363-383.
[5] Stroyan, K. D.: *The infinitesimal rule of three*, (these proceedings)
[6] Stroyan, K. D. and Luxemburg, W. A. J.: *Introduction to the theory of infinitesimals*, Acad. Press, Mathematics 72, 1976.

Dep. Matemática
Univ. da Beira Interior
R. Marquês d'Ávila e Bolama
6200 Covilhã Portugal

E-mail v_neves@ubivms.ubi.pt

HERMANN RENDER
Continuous maps with special properties

0. Introduction

Let $^*X$ be a nonstandard model of the topological space (X, τ). Fundamental in nonstandard topology is the notion of the *monad of a point* $x \in X$ which is by definition the set $m(x) := \cap_{U \in \tau, x \in U} {}^*U$. Instead of $y \in m(x)$ we use as well the more intuitive notion $y \approx x$. The set $ns^*X := \cup_{x \in X} m(x)$ is called the *set of all nearstandard points*. If A is a subset of the topological space (X, τ) the *monad of* A is the set $m_\tau(A) := \cap_{U \in \tau, A \subset U} {}^*U$. If K is a compact subset of X then the relation $m_\tau(K) = \cup_{x \in K} m(x)$ holds. If $\mathcal{F}$ is a filter on X the *monad of the filter* $\mathcal{F}$ is the set $m(\mathcal{F}) := \cap_{F \in \mathcal{F}} {}^*F$. For a general treatment of monads we refer to [4]. Let k be the system of all compact subsets of the topological space X. Then the set $cpt^*X := \cup_{K \in k} {}^*K$ is called the set of all *compact points*. It is well known that the equality $ns^*X = cpt^*X$ means that X is *locally compact*, i.e., that every point possesses a compact neighborhood.

In the first section we give a standard formulation of the property $^*f(ns^*X) = ns^*Y$ which leads to the concept of so-called biquotient maps. Some properties of biquotient maps are proved by nonstandard methods. The second section is devoted to perfect maps and contains some nonstandard proofs of known results. Section 3 deals with sufficient conditions for the validity of the inclusion $ns^*Y \subset {}^*f(\cup_{n=1}^{\infty} {}^*A_n)$ for a sequence of subsets A_n with $ns^*X \subset \cup_{n=1}^{\infty} {}^*A_n$. In section 4 we introduce the notion of a nearly biquotient map and prove some elementary results. Finally let us introduce some terminology: A topological space is a *k-space* if a subset A is closed whenever $A \cap K$ is a closed subset of K for all compact subspaces K. A space X is a T_1-*space* if each point is closed and it is *completely regular* if for each closed set A and $x \in X \setminus A$ there exists a continuous real-valued function f such that $f(A) = \{0\}$ and $f(x) = 1$. For further unexplained topological definitions we refer to [21].

1. Biquotient maps

Let X and Y be topological spaces. Recall that a continuous surjective map $f: X \to Y$ is *open* if for each $x \in X$ and every neighborhood U of x the image $f(U)$ is a neighborhood of $f(x)$. It is easy to see that f is open iff $^*f(m(x)) = m(f(x))$ for all $x \in X$. In particular we obtain the (weaker) condition $ns^*Y = {}^*f(ns^*X)$. In fact, we show that this condition corresponds to the notion of a surjective biquotient map which was introduced by O. Hájek, and independently by E. Michael, see [5]. We use a slight modification of the definition which is equivalent for Hausdorff spaces, cf. Theorem 1.4.

1.1 Definition
Let X, Y be topological spaces and $f: X \to Y$ be continuous. Then f is a biquotient map if for every $y \in Y$ and for every open covering $\mathcal{U}$ of X there exists finitely many $U_{i_1}, \ldots, U_{i_n} \in \mathcal{U}$ and a neighborhood V of y such that $V \subset f(U_{i_1}) \cup \cdots \cup f(U_{i_n})$.

1.2 Theorem
Let X, Y be topological spaces. Then $f: X \to Y$ is a biquotient map if and only if $ns^*Y = {}^*f(ns^*X)$ holds.

Proof Suppose that there exists $y \in ns^*Y$ with $y \notin {}^*f(m(x))$ for all $x \in X$. By saturation there exists an open neighborhood U_x of x with $y \notin {}^*f({}^*U_x)$. Then $(U_x)_{x \in X}$ is an open covering of X. Choose $y_0 \in Y$ with $y \approx y_0$. Since f is a biquotient map there exists an open neighborhood V of y_0 and $U_{x_1}, \ldots, U_{x_n} \in (U_x)_{x \in X}$ with $V \subset f(U_{x_1}) \cup \cdots \cup f(U_{x_n})$. For $y \in {}^*V$ we have $y \in {}^*f({}^*U_{x_i})$ for some x_i, a contradiction. For the converse let $y_0 \in Y$ and $(U_i)_{i \in I}$ be an open covering of X. For $y \in m(y_0)$ there exists by assumption $x \in ns^*X$ with ${}^*f(x) = y$. It follows that $m(y_0) \subset {}^*f(\cup_{i \in I} {}^*U_i) = \cup_{i \in I} {}^*f({}^*U_i)$. By saturation there exists an open neighborhood V of y_0 and $U_{i_1}, \ldots, U_{i_n} \in (U_i)_{i \in I}$ with ${}^*V \subset \cup_{k=1}^n {}^*f({}^*U_{i_k})$. An application of the transfer principle completes the proof. □

A continuous map $f: X \to Y$ is *closed* if $f(A)$ is closed for each closed subset A of X. In [20] it is shown that f is closed iff ${}^*f^{-1}(m(y)) \subset m_\tau(f^{-1}\{y\})$ for all $y \in Y$. A biquotient map is usually neither open nor closed as the following simple example shows: define a continuous function $f: \mathbb{R} \to [0, \infty)$ on $(-\infty, 0]$ by $f(x) = -x$ and on $[0, \infty)$ suitably. Then f is a biquotient map; if f on $[0, \infty)$ is defined such that $f((0, \infty)) = (0, 1]$ and $f([0, \infty)) = [0, 1)$ respectively then f is not open and closed respectively. Moreover a closed map is in general not a biquotient map as Example 8.1 in [5] shows even if the preimage space is a locally compact separable metric space.

1.3 Proposition
Let $f: X \to Y$ be a biquotient map and Y Hausdorff. Then f is a quotient map.

Proof Let τ_Y be the topology on Y and $V \subset Y$ be open with respect to the quotient topology. It suffices to show that $m_{\tau_Y}(y) \subset {}^*V$ for all $y \in V$. Note that $U := f^{-1}(V)$ is an open subset of X. For $z \approx_{\tau_Y} y \in V$ there exists a $x \in ns^*X$ with $z = {}^*f(x)$. For $x_0 \in X$ with $x \approx x_0$ it follows that $z = {}^*f(x) \approx_{\tau_Y} f(x_0)$. The Hausdorff property yields $f(x_0) = y$. Hence $x_0 \in f^{-1}(V)$ and $x \in {}^*U$ by openness. Thus $z = {}^*f(x) \in {}^*V$. □

1.4 Theorem
Let Y be a Hausdorff space and let $f: X \to Y$ be continuous and surjective. Then the following statements are equivalent:

  a) f is a biquotient map.
  b) For every $y \in Y$ and every open covering $\mathcal{U}$ of $f^{-1}(\{y\})$ there exists $U_1, \ldots, U_n$ in $\mathcal{U}$ and an open neighborhood V of y with $V \subset f(U_1 \cup \cdots \cup U_n)$.
  c) $m(y) \subset {}^*f(\cup_{x \in f^{-1}(y)} m(x))$ for every $y \in Y$.

We omit the proof and refer to [5] where standard proofs of the next three results are given as well.

1.5 Theorem
Let $f_i\colon X_i \to Y_i$ be biquotient maps for every $i \in I$ and $X = \prod_{i\in I} X_i$ and $Y := \prod_{i\in I} Y_i$. Then $f\colon X \to Y$ defined by $f((x_i)_{i\in I}) := (f_i(x_i))_{i\in I}$ is a biquotient map.

Proof Note that $ns^*X = \{x \in {}^*X : x_i \in ns^*X_i \text{ for every } i \in I\}$. Let $y \in ns^*Y$. Then $y_i \in ns^*Y_i$ for every $i \in I$ and therefore there exists $x_i \in ns^*X_i$ with $y_i = {}^*f_i(x_i)$ for $i \in I$. Hence, for every finite subset E of I, there exists $z \in {}^*X$ with $z_i = x_i$ (hence ${}^*f_i(z_i) = y_i$) for all $i \in E$. A saturation argument yields the existence of an element $z \in {}^*X$ with ${}^*f_i(z_i) = y_i$ for all $i \in I$. $\square$

1.6 Proposition
Let $f\colon X \to Y$ be a biquotient map. If X locally compact then as well Y.

Proof $ns^*Y \subset {}^*f(ns^*X) = {}^*f(cpt^*X) \subset cpt^*Y$. $\square$

1.7 Proposition
Let $f\colon X \to Y$ and $g\colon Y \to Z$ be surjective biquotient maps. Then $f \circ g$ is a biquotient map.

Proof $ns^*Z \subset {}^*g(ns^*Y) \subset {}^*g({}^*f(ns^*X))$. $\square$

A continuous map $f\colon X \to Y$ is called *compact-covering* if for each compact subset L of Y there exists a compact subset K such that $L \subset f(K)$. The proof of the following proposition is a simple consequence of the saturation principle:

1.8 Proposition
$f\colon X \to Y$ is compact-covering iff ${}^*f(cpt^*X) = cpt^*Y$.

A map $f\colon X \to Y$ is called *semi-proper* if for each $y_0 \in Y$ there exists for each $y_0 \in Y$ a neighborhood V of y_0 and a compact subset $K \subset X$ such that for all $y \in V \cap f(X)$ the set $f^{-1}(y) \cap K$ is non-empty.

1.9 Theorem.
Let $f\colon X \to Y$ be continuous. Then the following statements are equivalent:
 a) f is semi-proper.
 b) ${}^*f^{-1}(y) \cap cpt^*X \neq \emptyset$ for all $y \in ns^*Y \cap {}^*f(^*X)$.
 c) ${}^*f(cpt^*X) = ns^*Y \cap {}^*f(X)$.
 d) For each $y \in Y$ there exists a compact subset K of X and a neighborhood V of x such that $V \cap f(X) \subset f(K)$.

Proof For a) $\Rightarrow$ b) let $y \in {}^*f(^*X)$ and $y \approx y_0 \in Y$. Let V be a neighborhood of y_0 and $K \subset X$ compact with $f^{-1}(z) \cap K \neq \emptyset$ for all $z \in V$. Then $y \in {}^*V$ and ${}^*f^{-1}(y) \cap {}^*K \neq \emptyset$.

For b) $\Rightarrow$ c) let $y \in ns^*Y \cap {}^*f(^*X)$. Then there exists $y_0 \in Y$ with $y \approx y_0$. Since $f^{-1}(y) \cap cpt^*X \neq \emptyset$ there exists $x \in cpt^*X$ with $f(x) = y$. Hence $y \in f(cpt^*X)$.

For c) $\Rightarrow$ d) let k be the system of all compact subsets of x and τ_{y_0} be the system of all neighborhoods of y_0. Choose $K \in {}^*k$ with $cpt^*X \subset K$ and choose $V \in {}^*\tau_{y_0}$ such that $V \subset m(y_0) \subset ns^*Y$. Then the following statement is true: $(\exists V \in {}^*\tau_{y_0})(\exists K \in {}^*k)({}^*f(^*K) \subset V \cap {}^*f(^*X))$. Via Transfer we infer d).

The implication d) $\Rightarrow$ a) is trivial. $\square$

1.10 Corollary
Let $f\colon X \to Y$ be continuous and surjective. Then f is semi-proper if and only if Y locally compact and f is compact-covering.

Proof By Theorem 1.9 we have $ns^*Y = f(cpt^*X) \subset cpt^*Y$. Hence Y is locally compact and clearly f is compact-covering. The converse is clear since $ns^*X = cpt^*X = f(cpt^*X)$. □

1.11 Lemma
Let $f\colon X \to Y$ be semi-proper and Y Hausdorff. Then $f(X)$ is closed in Y.

Proof Let $y = {}^*f(x) \in {}^*f(^*X)$ and let $y \approx y_0$. By 1.9 c) there exists $x \in cpt^*X$ with $y = {}^*f(x)$. Choose $x_0 \in X$ with $x \approx x_0$. By continuity we obtain $y = {}^*f(x) \approx f(x_0)$. Now the Hausdorff property implies that $y_0 = f(x_0) \in f(X)$. □

2. Perfect maps

Let X, Y be topological spaces. A continuous map $f\colon X \to Y$ is called *perfect* if it is a closed map and every fiber $f^{-1}(y)$ is compact. The following result is well known and we include the proof only for the convenience of the reader. The equivalence of a) and d) (valid for k-spaces) was proved in [11].

2.1 Theorem
Let $f\colon X \to Y$ be continuous. If Y is a Hausdorff space then a) and b) are equivalent, if Y is in addition a k-space then all statements are equivalent:
 a) f is perfect.
 b) ${}^*f^{-1}(ns^*Y) \subset ns^*X$.
 c) ${}^*f^{-1}(cpt^*Y) \subset ns^*X$.
 d) $f^{-1}(K)$ is compact for each compact subset K of Y.

Proof. a) $\Rightarrow$ b). Let $y \approx {}^*y_0$. Then $K := f^{-1}(y_0)$ is compact. Let $x \in {}^*f^{-1}(y)$. Suppose that $x \notin m_\tau(K) = \bigcup_{x \in K} m(x) \subset ns^*X$. Then there exists an open set U with $x \notin {}^*U$ and $K \subset U$. Hence $x \in {}^*(X \setminus U)$ and $f(X \setminus U)$ is closed by a). By continuity ${}^*f(x) = y \approx {}^*y_0$ and therefore $y_0 \in f(X \setminus U)$. Hence there exists $x_0 \in X \setminus U$ with $y_0 = f(x_0)$. This contradicts to the fact that $x_0 \in f^{-1}(y_0) = K$.

For b) $\Rightarrow$ a) let $y \in Y$. Then ${}^*f^{-1}(y) \subset ns^*X$. Since $f^{-1}(y)$ is closed we infer the compactness of $f^{-1}(y)$. Now let $A \subset X$ be closed. Let $y \in \overline{f(A)}$. Then there exists $x \in {}^*A$ with ${}^*f(x) \approx {}^*y$. By b) we have $x \in {}^*f^{-1}(ns^*Y) \subset ns^*X$. Hence there exists $x_0 \in X$ with $x \approx {}^*x_0$; since A is closed x_0 is actually in A. It follows that ${}^*f(x) \approx {}^*f(^*x_0)$ and ${}^*f(x) \approx {}^*y$, hence $y = f(x_0) \in f(A)$ by the Hausdorff property.

The implication b) $\Rightarrow$ c) is trivial. Now let Y be a k-space. We show c) $\Rightarrow$ a). Let $A \subset X$ be closed. It suffices to show that $f(A) \cap K$ is closed in K for every compact subset $K \subset Y$. Let ${}^*f(x) \approx {}^*y_0$ with $x \in {}^*A$ and ${}^*f(x) \in {}^*K \subset cpt^*Y$. Then $x \in {}^*f^{-1}(cpt^*Y) \subset ns^*X$ and hence there exists $x_0 \in X$ with $x \approx {}^*x_0$. Since A is closed we have $x_0 \in A$. Then ${}^*f(x) \approx {}^*f(^*x_0)$ and as before $y_0 = f(x_0) \in f(A) \cap K$. The equivalence of c) and d) is straightforward. □

The next two results can be found in [6].

2.2 Corollary
Let $f\colon X \to Y$ be perfect. If $g\colon X \to Z$ is continuous and Z is Hausdorff then $(f,g) \colon X \to Y \times Z$ defined by $(f,g)(x) := (f(x), g(x))$ is closed.

Proof. Let $A \subset X$ be closed and $({}^*f(x), {}^*g(x)) \approx {}^*(y_1, y_2)$ with $x \in {}^*A$. Then ${}^*f(x) \in ns^*Y$. Hence we can find $x_0 \in X$ with $x \approx {}^*x_0$ and $x_0 \in A$ since A is closed. By continuity of g we have ${}^*g(x) \approx {}^*g({}^*x_0)$ and ${}^*g(x) \approx {}^*y_2$. The Hausdorff property yields $g(x_0) = y_2$. Hence $(y_1, y_2) = (f(x_0), g(x_0))$. □

2.3 Corollary
Let $f\colon X \to Y$ be perfect. If Y is Hausdorff and X is embedded in a space Z then X is homeomorphic to a closed subspace of $Y \times Z$.

Proof Let $i\colon X \to Z$ be an embedding. Then $(f,i)\colon X \to Y \times Z$ is obviously continuous and injective. Corollary 2.2 shows that (f,i) is a closed mapping. Hence the image is a closed subset and an embedding. □

2.4 Theorem
Let Y be a Hausdorff space and P a topological property which is closed-hereditary. Then the following statements are equivalent:

a) For all compact Hausdorff spaces Z holds: if Y has property P then the product space $Z \times Y$ has property P.

b) For all completely regular Hausdorff spaces X and all perfect maps $f\colon X \to Y$ holds: if Y has property P then X has property P.

Proof. For a) $\Rightarrow$ b) let $f\colon X \to Y$ be perfect. Let Z be a compact Hausdorff space and $i\colon X \to Z$ be an embedding. By Corollary 2.2 we know that X is a closed subset of $Z \times Y$. Hence X has property P.

For the converse let Z be compact Hausdorff. We consider the projection $p\colon Z \times Y \to Y$. Then ${}^*p^{-1}(ns^*Y) = {}^*Z \times ns^*Y = ns^*(Z \times Y)$. Hence p is perfect. By b) it follows that if Y has property P then also $Z \times Y$. □

3. Sufficient conditions

Recall that a topological space X is *first countable* if each point $x \in X$ possesses a countable neighborhood base. A space is called *hemicompact* if there exists a sequence of compact sets K_n such that for each compact subset K there exists $n \in \mathbb{N}$ with $K \subset K_n$.

3.1 Theorem
Let X be a topological space and Y be a T_1-space and $f\colon X \to Y$ be a quotient mapping. Suppose that $(A_n)_{n \in \mathbb{N}}$ are subsets of X with $ns^*X \subset \bigcup_{n=1}^{\infty} {}^*A_n$. If y possesses a countable neighborhood base then $m(y) \subset {}^*f(\bigcup_{n=1}^{\infty} {}^*A_n)$.

Proof We can assume that y is not isolated and that $A_n \subset A_{n+1}$ for all $n \in \mathbb{N}$. Let $(V_n)_{n \in \mathbb{N}}$ be a neighborhood base of y with $V_{n+1} \subset V_n$. By the transfer principle it suffices to show that $V_n \subset f(A_n)$ for some $n \in \mathbb{N}$. Suppose the contrary: then for every neighborhood V_n there exists $y_n \in V_n \setminus \{y\}$ such that $y_n \notin f(A_n)$, i.e., that ${}^*A_n \cap {}^*f^{-1}(\{{}^*y_n\}) = \emptyset$ for all $n \in {}^*\mathbb{N}$ by transfer. It suffices to show that $M := f^{-1}\{y_n : n \in \mathbb{N}\}$ is a closed set: Then $X \setminus M = f^{-1}(Y \setminus \{y_n : n \in \mathbb{N}\})$ is open. Since f is quotient $Y \setminus \{y_n : n \in \mathbb{N}\}$ is an open neighborhood of y. Hence there exists $n_0 \in \mathbb{N}$ such that $y \in V_{n_0} \subset Y \setminus \{y_n : n \in \mathbb{N}\}$, a contradiction to $y_{n_0} \in V_{n_0}$. Now let us show that M is closed: Let $x_0 \in X$ and $x \in {}^*M$ with $x \approx x_0$. Then $x \in {}^*f^{-1}(\{{}^*y_N\})$ for some $N \in {}^*\mathbb{N}$. If N is infinitely large then ${}^*A_N \cap {}^*f^{-1}(\{{}^*y_N\}) = \emptyset$ and ${}^*A_N \subset {}^*A_n$ for all $n \in \mathbb{N}$ show that $x \notin ns^*X$, a contradiction. Hence $N \in {}^\sigma\mathbb{N}$ and ${}^*f(x) \approx {}^*(f(x_0))$ by continuity. As Y is a T_1-space we infer ${}^*f(x_0) = {}^*y_N$, i.e., that $x_0 \in M$. □

3.2 Theorem
Let X be a k-space and Y a first countable T_1-space and $f: X \to Y$ be quotient. If $(A_n)_{n \in \mathbb{N}}$ are subsets of X with $cpt^*X \subset \cup_{n=1}^\infty {}^*A_n$ then $m(y) \subset {}^*f(\cup_{n=1}^\infty {}^*A_n)$.

Proof In view of the last proof it suffices to show that M (in the last proof) is closed. Since X is a k-space it suffices to show that $M \cap K$ is closed for each compact set K. This can be done similarly as in the last part of the proof of 3.1. □

In general the continuous image of a hemicompact space is not hemicompact (consider $\mathbb{Q}$ with the discrete and the usual topology and f as the identity).

3.3 Corollary
Let X be a hemicompact k-space and Y be a first countable T_1-space. Then a quotient map $f: X \to Y$ satisfies $ns^*Y \subset {}^*f(cpt^*X)$. In particular, Y is a hemicompact locally compact space.

Proof It is easy to see that X is hemicompact iff there exists compact sets K_n such that $cpt^*X = \cup_{n=1}^\infty {}^*K_n$. Theorem 3.2 shows that $ns^*Y = \cup_{n=1}^\infty {}^*f(K_n) \subset cpt^*Y$. □

The next result can be found in [5, Proposition 3.3].

3.4 Corollary
Let X be a Lindelöf space and Y be a first countable T_1-space. Then a quotient map $f: X \to Y$ is biquotient.

Proof Let $(U_i)_{i \in I}$ be an open covering and let $(U_n)_{n \in \mathbb{N}}$ be a countable subcovering. By Theorem 3.2 we have $m(y) \subset \cup_{n=1}^\infty {}^*f({}^*U_n)$. By saturation there exists $n \in \mathbb{N}$ and a neighborhood V of y such that ${}^*V \subset {}^*f({}^*U_1 \cup \cdots \cup {}^*U_n)$. Now Transfer completes the proof. □

A review of the proof of Theorem 3.1 shows that the assumption of first countability can be weakened (without changing the proof): suppose that for each $z \approx y$ there exists a sequence of sets B_n such that $z \in \cap_{n=1}^\infty {}^*B_n \subset m(y)$. Suppose that B_n is not contained in $f(A_n)$ and construct y_n as above. Then M is closed and $\cap_{n=1}^\infty {}^*B_n \subset {}^*X \setminus {}^*M$. By saturation there exists $n_0 \in \mathbb{N}$ with ${}^*B_{n_0} \subset {}^*X \setminus {}^*M$, a contradiction. Let us call a space *bisequential* if for each $z \approx y$ there exists a sequence of sets B_n such that $z \in \cap_{n=1}^\infty {}^*B_n \subset m(y)$. A standard description runs as follows (see [16]): for every ultrafilter $\mathcal{F}$ converging to the point y there exists a sequence of decreasing sets $B_n \in \mathcal{F}$ converging to the point y. This leads to the following

3.5 Remark. The results 3.1, 3.2, 3.3 and 3.4 are valid for bisequential T_1-spaces (instead of first countable T_1-spaces).

4. Nearly biquotient maps

Recall that a continuous map $f\colon X \to Y$ is *nearly open* if for every open neighborhood U of $x \in X$ the set $\overline{f(U)}$ is a neighborhood of $f(x)$.

4.1 Definition
Let X, Y be topological spaces. A continuous map $f\colon X \to Y$ is *nearly biquotient* if for every $y \in Y$ and for every open covering $\mathcal{U}$ there exists $U_1, \ldots, U_n \in \mathcal{U}$ such that $\overline{f(U_1)} \cup \cdots \cup \overline{f(U_n)}$ is a neighborhood of y.

Clearly every nearly open map is nearly biquotient.

In the next theorem let $\mathcal{B}$ be a system of (arbitrary) subsets of a topological space with the following property: if $x \in X$ and U is open then there exists a *neighborhood* $B \in \mathcal{B}$ with $x \in B \subset U$. We call $\mathcal{B}$ a *generalized base*.

4.2 Proposition
Let $f\colon X \to Y$ be a nearly biquotient map. If $f(Z)$ is closed for all Z in a generalized base $\mathcal{B}$ then f is a biquotient map, i.e., that $ns^*Y \subset {}^*f(ns^*X)$.

Proof Let $y \in Y$ and $(U_x)_{x \in X}$ be an arbitrary open covering with $x \in U_x$. For $x \in U_x$ there exists an open set V_x and a set $B_x \in \mathcal{B}$ with $x \in V_x \subset B_x \subset U_x$. Since f is nearly biquotient there exists an open neighborhood V of y with $V \subset \overline{f(V_{x_1})} \cup \cdots \cup \overline{f(V_{x_n})} \subset f(B_{x_1} \cup \cdots \cup B_{x_n})$ for some $x_1, \ldots, x_n$. The proof is complete. □

4.3 Corollary
Let X be a regular space. Then a closed, nearly biquotient map is biquotient.

4.4 Corollary
Let X be a locally compact regular space and Y be Hausdorff. Then a nearly biquotient map is biquotient.

Proof Apply Proposition 4.2 to the system of all compact subsets. □

Let X be a topological space. Then $psns^*X := \{x \in {}^*X : {}^*f(x) \in ns^*\mathbb{R}$ for all $f\colon X \to \mathbb{R}$ continuous$\}$ is called the set of all *pseudo-nearstandard points*. For a detailed discussion of this set and its relationship to realcompactness we refer to [12, 15, 16, 18, 19]. Let $\upsilon(X)$ be the real-compactification of the completely regular Hausdorff space X. It is well-known that each continuous function $f\colon X \to Y$ possesses a continuous extension $f^\upsilon\colon \upsilon(X) \to \upsilon(Y)$.

4.5 Corollary
Let X, Y be Tychonoff spaces and let $f\colon X \to Y$ be continuous and surjective. If $\upsilon(X)$ is locally compact then ${}^*f(psns^*X) = psns^*Y$ implies that $f^\upsilon\colon \upsilon(X) \to \upsilon(Y)$ is a biquotient map.

Proof The equation $^*f(psns^*X) = psns^*Y$ is equivalent to the statement that f^υ a biquotient map with respect to (X,Y) (for definition see [16]). It follows that f^υ is nearly biquotient. Corollary 4.4 completes the proof. □

References

[1] H. Grauert, *Set theoretic complex equivalence relations.* Math. Annalen **265** (1983) 137-148.
[2] R.A. Herrmann, *A nonstandard characterization for perfect maps.* Zeitsch. f. math. Logik u. Grundlagen der Math. **23** (1977), 223-236.
[3] T. Isiwata, *Mappings and spaces.* Pacific J. Math. **20** (1967), 455-480.
[4] W.A.J. Luxemburg, *A general theory of monads,* in "Applications of Model theory to Algebra, Analysis and Probability" Holt, Rinehart and Winston, New York 1969, pp. 18-86.
[5] E.A. Michael, *Bi-quotient maps and cartesian products of quotient maps.* Ann. Inst. Fourier, Grenoble, **18** (1968) 287-302.
[6] − , *A Theorem on perfect maps.* Proc. Amer. Math. Soc. **28** (1971) 633-634.
[7] − , *A quintuple quotient quest.* Gen. Top. Appl. **2** (1972) 91-138.
[8] K. Nagami, *Ranges which enable open maps to be compact covering.* General Top. and its Appl. **3** (1973) 355-367.
[9] D. Noll, *Open mapping theorems in topological spaces.* Czech. Math. J. **35** (110) (1985) 373-384.
[10] D. Noll, *On the theory of B- and B_r-spaces in general topology.* Czech. Math. J. **39** (114) (1989) 589-594.
[11] R.S. Palais, *When proper maps are closed.* Proc. Amer. Math. Soc. **24** (1970) 835-836.
[12] H. Render , *Nonstandard topology on function spaces with applications to hyperspaces.* Trans. Amer. Math. Soc., **336** (1993), 101-119.
[13] − , *On the continuity of internal functions.* Publications de l'Institut Math. **53** (67) (1993) 139-143.
[14] − , *Countably determined sets and a conjecture of C.W Henson.* to appear in Math. Annalen.
[15] − , *Covering functions and the inverse standard part map.* to appear in Math. Japonica.
[16] − , *A nonstandard approach to realcompactness with applications to the equation* $\upsilon(X \times Y) = \upsilon(X) \times \upsilon(Y)$. Submitted.
[17] S. Salbany, T. Todorov, *Nonstandard and standard compactifications of ordered topological spaces.* Top. Appl. **47** (1992) 35-52.
[18] K.D. Stroyan, W.A.J. Luxemburg, *Introduction to the Theory of Infinitesimals.* Academic Press, New York 1976.
[19] M. Weir, *Hewitt-Nachbin spaces.* North-Holland Publishing Company, Amsterdam 1975.
[20] K. Wicks, *Fractals and Hyperspaces.* Lect. Notes Math. **1492**, Springer-Verlag, Berlin 1991.
[21] S. Willard, *General topology* Addison-Wesley, Reading 1970.

Gerhard-Mercator Universität-GH Duisburg
Lotharstr. 65
D-47057 Duisburg
Federal Republic of Germany

E-mail render@math.uni-duisburg.de

STEVEN C LETH
Some nonstandard methods in geometric topology

0. Introduction

In this paper we will look at some ways in which nonstandard methods can be applied to problems in geometric topology. This is an area in which currently very little research has been done. Early work on applications to general topology was done by Robinson [15], Luxemburg [8], and Wattenberg [17]. The book by Stroyan and Luxemburg [16] provides a fairly comprehensive look at the basic applications to topology, as does the early book of Machover and Hirschfeld [9] and the more recent survey article of Lindstrøm [7] (this article is also a very good introduction to nonstandard analysis, especially for people with limited logic background).

Wattenberg considered some of the same types of questions as we look at here in [18], where some homotopy and homology theory was used in the nonstandard model to obtain a classification theorem for certain covering spaces. Chuaqui and Bertoglio [2] and Narens [13] used techniques somewhat similar to those in section 2 of this work to obtain a nonstandard proof of the Jordan Curve Theorem. Andersen obtained a very general result about fixed-point and related theorems in [1]. Nonstandard approaches to homology and cohomology are considered in [10], [14], and [19].

While many of the recent results using nonstandard methods make use of the now fundamental constructions of Loeb measure or nonstandard hulls, the main tool that we will use here is simply the existence of infinitesimals and the transfer principle. Nevertheless, since the nonstandard model "realizes all limits" these are often powerful techniques, even when considering such objects as sets in $\mathbb{R}^2$, which is what we are almost exclusively concerned with here.

In section 1 we consider some standard notions of connectedness which are similar to the classical "path-connected" and natural from the nonstandard point of view. Here the nonstandard model allows us to generalize the standard definition of path-connected in ways that are not available classically.

In section 2 we give an example of an approximation argument which is simplified by the existence of the infinitesimals. More specifically we show that any simple closed curve in $\mathbb{R}^2$ can be approximated to within any $\epsilon > 0$ by a simple closed polygon. The main benefit of the infinitesimals in this argument is that the proof can proceed more naturally than in the classical case.

In section 3 we take a brief look at a well-known open question about sets which have the fixed point property, and try to give some indication of why nonstandard methods may prove to be helpful in considering such a problem.

We will work in some nonstandard model $*V$ of a standard universe V as described, for example, in [7].

We assume the following elementary facts and definitions (see e.g. [7]):
- If a is an element of a standard topological space, the **monad** of a is defined to be the intersection of all *standard open sets containing a.
- A standard set A is open if and only if for all $a \in A$, the monad of A is contained in *A.
- A standard set A is compact if and only if every $x \in$ *A is in the monad of some standard point $a \in A$.
- A standard function f is uniformly continuous on the metric space X if and only if for any $x, y \in$ *X, whenever $|x - y|$ is infinitesimal, then $|f(x) - f(y)|$ is infinitesimal.

We will use the convention of not placing a "*"next to the nonstandard version of a standard function. Thus, in the above statement, the points x and y are in the nonstandard model and f is a standard function. While it would technically be more correct to write "*$f(x)$" to indicate that we are applying the nonstandard model's version of the function f to x, this notation leads to an overabundance of stars, and will be avoided here.

1. Nonstandard Notions of Connectedness

Sometimes the nonstandard model suggests new standard definitions. In this section we consider some notions of connectedness which are similar to path-connected and yet closer to capturing the notion of connectedness.

The concept of "path-connected" is intuitively very simple, whereas the more general notion of "connected" is defined in a cumbersome way. Our goal here is to mimic the way that path- connectedness is defined, and yet obtain a more general notion which is closer to true connectedness.

Definition 1
We call a subset A of $\mathbb{R}^n$ **SP-connected** *(for "Star-Polygonally Connected") if and only if for every two points $p, q \in A$ there exists an internal polygonal path P from p to q such that $st(P) \subseteq A$.*

A subset A of $\mathbb{R}^n$ will be called **HP-connected** *(for "Hyperfinite Path Connected") if and only if for every two points $p, q \in A$ there exists a *continuous function $f :$ *$[0, 1] \to$ *$\mathbb{R}^n$ such that:*
*$f(0) = p$, $f(1) = q$, and for all $t \in$ *[0,1], either $f(t) \in$ *A or $st(f(t)) \in A$*

Implicit in the above definition is that every point of the polygonal path P (for SP-connected) must have a standard part, i.e. P must contain only finite points. If we require only that the standard parts of points in P be in A when these standard parts exist, it is easy to see that we could have non-connected sets satisfy the definition. The graph of the function $f(x) = \frac{1}{x^2}$ would be a simple example. Similarly we interpret the requirement "$st(f(t)) \in A$" in the definition of HP-connected to mean that $st(f(t))$ exists and is in A.

Proposition 2
Path connected $\implies$ SP-Connected $\implies$ HP-Connected $\implies$ Connected

Proof. If A is path connected, then given p and q in A there exists a continuous function $f : [0,1] \to A$ with $f(0) = p$ and $f(1) = q$. Then if H is any element of *$\mathbb{N}$–$\mathbb{N}$ we can obtain the desired internal polygonal path from p to q by connecting successive points of the form $f(\frac{k}{H})$.

If A is SP-connected, then it follows quickly from the transfer property that A is HP- connected. This is because in the standard universe any polygonal path is the range of a continuous function with domain $[0,1]$, so that the existence of an internal polygon as is guaranteed by the definition of SP-connected implies that a *continuous function $f : {}^*[0,1] \to {}^*\mathbb{R}^n$ with the desired properties exists.

Now suppose that A is HP-connected but not connected, so that there exists open sets U and V which disconnect A. Choose $p \in U$ and $q \in V$. We now let z be defined internally by:

$$z = sup\{x \in {}^*[0,1] : \text{for all } y \in {}^*[0,x], f(y) \in {}^*U\}$$

Now $f(z)$ must be a boundary point of *U, and since *U is open, $f(z)$ is not in *U, which in turn implies that $st(f(z))$ is not in U (again since U is open). Similarly $st(f(z))$ is not in V. But the condition that $A \subseteq U \cup V$ and the corresponding transferred relation now imply that $f(z) \notin {}^*A$ and $st(f(z)) \notin A$, contradicting that A is HP-connected. $\square$

Proposition 3
The notions of Path Connected, SP- Connected, HP-Connected and Connected are all distinct, even for bounded sets in dimension 2. More Specifically:

1) The closure of the set $A = \{(x, sin(\frac{1}{x})) : x \in (0,1)\}$ is SP-Connected but not Path Connected.

2) The set $B = \{(x, sin(\frac{1}{x})) : x \in (0,1)\} \cup \{(0,0)\}$ is HP-Connected but not SP-Connected.

3) The set $C = ((\mathbb{Q} \cap [0,1]) \times [0,1]) \cup ((\mathbb{R}-\mathbb{Q}) \cap [0,1]) \times [-1,0))$ is connected but not HP-Connected.

Proof. 1) It is well-known that set A is not path- connected (see, e.g. chapter 3 of [12]). To see that it is SP-connected it suffices to show that if p is a point on the interval from -1 to 1 on the y–axis, and q is any point on the "sine" portion of the curve, then there is a polygonal curve of the desired type connecting p to q. We may begin with any straight-line segment which starts at p and connects to a point on the curve an infinitesimal distance from the y–axis. It is then easy to see that a sequence of line segments, each of infinitesimal length, with endpoints in the set A can be used to complete the connection to q. Since the set A is compact, every point in our connecting sequence of line segments is infinitely close to a point of A, so that the necessary condition is satisfied.

2) We first note that any set which is internally path- connected must be HP-connected. Thus, to show that the set B is HP-connected it suffices to show that the point (0,0) can be joined to some point on the rest of the curve by an appropriate

internal continuous function. Any straight line segment of infinitesimal length from (0,0) to a point on the rest of the curve can be the range of this function.

To see that B is not SP-Connected we note that any polygonal path P from (0,0) to a standard point q which stays infinitely close to $*B$ on the rest of the curve must contain some points which are in the monad of points on the y–axis other than (0,0), thus violating the condition that $st(P) \subseteq A$.

3) We begin by showing that C is connected. Suppose for contradiction that the open sets U and V separate C, and without loss of generality we assume $(0,0) \in U$. Let $r = sup\{x : x \in [0,1]$ and all $(y,0)$ in C such that $0 \le y \le x$ are also in $U\}$. Then if $r < 1$ and rational $(r,0)$ must be in C and yet a boundary point of both U and V, and thus in neither of these sets since they are open. If r is irrational then the point $(r, -\frac{1}{2})$ must be in C and a boundary point of both U and V (there are some points on the x-axis in V and arbitrarily close to $(r,0)$, and since V is open there must also be entire line segments with irrational x coordinates arbitrarily close to r in V.) In either case this contradicts that $C \subseteq U \cup V$.

It is easy to see that C cannot be HP-connected, since any two points with distinct x coordinates cannot be connected by a satisfactory internal function; any segment of the path along the x-axis fails both conditions in parts of the monad of any irrational point. $\square$

The set C above is an example of a connected set with $\beth_1$ HP-connected components.

It is well-known that if a set is both open and connected it must be path connected (see, e.g. chapter 5 of [3]). Thus, in that case all of these notions are equivalent. Our new notions have the advantage of being equivalent to connectedness in another important case, as the result below indicates.

Proposition 4
If A is compact and connected then A is SP-connected.

Proof. We first note that since A is connected, given any $\epsilon > 0$ and any points p, q in A there exists a polygonal path from p to q which stays within ϵ of A. Then by transfer any two points in $*A$ may be connected by an internal polygonal path which stays within an infinitesimal distance of $*A$. But since A is compact, the standard part of every element in $*A$ is in A, so that the condition for SP-connectedness holds. $\square$

We cannot replace "compact" with "closed" in the above proposition. For example the closure of the set $\{(x, \frac{1}{x}(sin\frac{1}{x}))\}$ is closed and connected but not SP-connected. In fact this example can be improved to show that closed and connected is not sufficient to imply HP-connected either. If small portions of the above curve are replaced by small "$sin(\frac{1}{x})$"-like pieces near all the maximums so that the entire curve remains connected but no segment from maximum to maximum is path connected then the resulting set is not HP-connected. This is essentially because the range of the internal connecting function in the definition of HP-connected must stay in $*A$ except in the monad of standard points of A. In particular, any infinite parts of the range must connect only points which are internally path connected.

The following result gives a standard characterization of the notion of SP-connected. A similar characterization of HP-connectedness is currently unknown.

Proposition 5
A is SP-connected if and only if for any two points p and q in A there exists a compact connected subset C of A such that both p and q are in C.

Proof. If A is SP-connected, then given any p and q in A the standard part of a path P as is given in the definition of SP-connected will be our desired compact set. The other direction follows immediately from proposition 4 above. □

We are interested in the degree to which these new notions are similar to full connectedness. Proposition 4 above gives a condition under which both new notions are equivalent to connectedness. Another result in this direction is given below.

Proposition 6
If A is connected and has only finitely many path-connected components, then A is HP-connected.

Proof. It suffices to assume that there are only two path connected components, for then the result follows easily by induction. Since A is connected there must be a point which is in the closure of both path connected components. Call such a point p and let B be the component which contains p and C be the other component. Then there exists a sequence of points $\langle p_n \rangle$ in C which converge to B. Since p_1 is path-connected to all the points in the sequence $\langle p_n \rangle$, for each n there is a standard continuous function $f_n : [0, 1 - \frac{1}{n}] \to A$ such that $f_n(0) = p_1$ and $f_n(1 - \frac{1}{n}) = p_n$. By transfer there exists an $H \in {}^*\mathbb{N}-\mathbb{N}$ and an internal function $f_H : [0, 1 - \frac{1}{H}] \to {}^*A$ such that $f_H(0) = p_1$, and $f_H(1 - \frac{1}{H})$ is in the monad of p. If we now let the function f be equal to f_H on $[0, 1 - \frac{1}{H}]$, and parameterize the straight-line segment from $f(1 - \frac{1}{H})$ to p on the interval $[1 - \frac{1}{H}, 1]$ we see that the points p_1 and p are HP-connected. As points within one path connected component are certainly HP-connected, this completes the proof. □

Proposition 7
Both HP-connectedness and SP-connectedness are preserved under continuous mappings.

Proof. The result in the SP-connected case follows immediately from proposition 5 and the classical facts that both connectedness and compactness are preserved under continuous mappings. The proof for HP-connected is exactly like the proof for path-connected: If g is a continuous function and A is HP-connected, then given points $g(p)$ and $g(q)$ in $g(A)$, if f is an internal HP-connecting function from p to q, then $g \circ f$ is the connecting function from $g(p)$ to $g(q)$. □

It is not known whether the full notion of connectedness can be captured by a definition similar to that of HP-connected above.

2. Approximation Techniques

An important example of a type of result for which nonstandard methods can be useful are those which involve approximations by simpler structures. Often the existence of infinitesimals provides a method for more simplistic approaches to be successful. A good example of this is given below, where we consider a classical result about approximation of simple closed curves by polygons.

Theorem 8
Let C be a simple closed curve in $\mathbb{R}^2$. Then, given $\epsilon > 0$, there exists a simple closed polygon P such that for all $p \in P$ $dist(p,C) < \epsilon$ and for all $c \in C$ $dist(c,P) < \epsilon$.

Intuitively, the most natural proof would be to start by subdividing the circle into small pieces, and form a polygon by connecting the points mapped from these endpoints. In a standard proof, however, it is difficult to see how to obtain a simple approximation from this first polygon. The use of infinitesimals avoids this problem.

Proof. Since C is a simple closed curve, there exists a homeomorphism f from S^1 onto C (where S^1 is the unit circle). Now let $\epsilon > 0$ be given. Inside the nonstandard universe *V we construct a polygonal approximation P_0 (which will not necessarily be simple) as follows. For H any element of *N$-$ N, let:

$$S = \{p_k : p_k = f(\cos(\frac{2\pi k}{H}), \sin(\frac{2\pi k}{H})), \quad k = 0, 1, \ldots, H-1\}$$

Form P_0 by connecting all the consecutive points of S with straight–line segments, and connecting p_{H-1} to p_0. We will write $\overline{p_i p_{i+1}}$ for the (closed) line segment from p_i to p_{i+1}. By the uniform continuity of f the polygon P_0 is a close approximation to C, but there is no reason to suppose that it is simple. We now wish to "throw away" various points in the polygon so as to make it simple. We will inductively define sets of the form T_n which will contain the points we want to remove.

We will first construct an internal polygon P with the desired properties and obtain the standard conclusion by transfer. To this end we make the following definitions inside the nonstandard model:

$$T_0 = \{p_i \in S : \text{there exists } p_k, p_j \text{ with } \frac{3}{4}H < k < H, \text{ and } 0 < j < \frac{1}{4}H,$$
$$\text{such that } \overline{p_k p_{k+1}} \text{ intersects } \overline{p_{j-1} p_j}\}$$

If $k = H-1$ we replace $k+1$ by 0 in the above definition. The numbers $\frac{3}{4}H$ and $\frac{1}{4}H$ are chosen so as to keep the definition internal. In fact, since f is one–to–one it is not

55

hard to see that any such p_k and p_j must be infinitely close. Now let:

$S_0 = S$
$S_{n+1} = S_n - T_n$
$T_n = \{p_i \in S_n : k_n < i < j,$ where p_{k_n} is the first point in S_n with index greater than all elements in T_{n-1} such that $\overline{p_{k_n} p_{k_n+1}}$ intersects a segment with indices $> k_n + 1$ and j is the largest index such that $\overline{p_{j-1} p_j} \cap \overline{p_{k_n} p_{k_n+1}} \neq \emptyset\}$
$q_{k_n+1} = $ an intersection point of $\overline{p_{k_n} p_{k_n+1}}$ and $\overline{p_{j_n-1} p_{j_n}}$
$U_{n+1} = S_{n+1} \cup \{q_{k_n+1}\}$

Let N be the largest index such that T_N is nonempty. (We note that if P_0 is already simple, then every T_n is empty. In this case we let $N = 0$)

We can now construct our simple polygonal approximation P by connecting all the consecutive points of U_{N+1} (whether labeled as "q" or "p" points, we connect successive indices; we note that by construction points with "q" labels have been given indices of "p" points which were in some T_n, so that we do not have any duplicate indices).

The diagram below shows an example of what some of the above sets and points and the final polygon P would look like in a specific case: $T_0 = \{p_0, p_1, p_{H-1}\}$, while $T_1 = \{p_4, p_5, \ldots, p_{12}\}$.

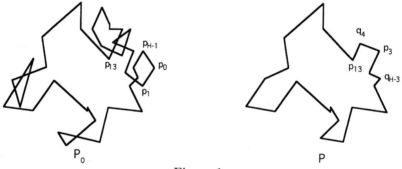

Figure 1

It is easy to see by the construction that P is now simple, for only successive line segments intersect each other. It remains to show that the polygon P is infinitely close to the curve C.

By the uniform continuity of f all the line segments in P_0 and P are of infinitesimal length. Thus, any endpoints of segments which intersect are infinitely close. But since the points in any one T_n are contained in the image of an arc of the circle from $f^{-1}(p_{k_n})$ to $f^{-1}(p_{j_n})$ they are all infinitely close. The points q_{k_n+1} have been included so as to make sure that U_{n+1} still contains points in every monad in which there are points of S. Thus every point of C is within an infinitesimal distance of P and every point of P is within an infinitesimal distance of C.

For any standard $\epsilon > 0$, the nonstandard universe now satisfies the property that there exists a simple closed polygon which is an ϵ approximation to C. By transfer, the standard universe also has this property, finishing the proof. □

In a standard proof in which the infinitesimals are merely replaced by small real numbers there is a major difficulty in knowing how small the initial partition of the circle needs to be in order to ensure that the portions removed are still small enough. In the proof above we do not know how the size of the removed portions compares to the size of an individual segment, but we know that both are infinitesimals, which is sufficient. For a classical proof of this theorem and a systematic development of the subject, see [11].

It would be interesting to see if the use of infinitesimals also simplifies the corresponding proof in 3 dimensions. The classical proof of this result, due to Bing, is quite difficult. In two dimensions it is possible to show that there are "inside" and "outside" approximations, while this is not true in three dimensions (the "horned sphere" is a counterexample).

3. Fixed Points

In this section we briefly consider a direction for possible additional research. There is some reason to believe that nonstandard methods may be helpful in the following open problem:

Open Question
Suppose that E is compact and connected in $\mathbb{R}^2$, and that $\mathbb{R}^2 - E$ is connected. Must E have the fixed point property?

A compact and connected set in the plane is sometimes called a **continuum**. Thus the question can be restated as: If E is a non-separating plane continuum and f is a continuous function from E to E is there always a point $a \in E$ such that $f(a) = a$? The book by Klee and Wagon [6] contains a nice discussion of this problem.

While we will not attempt a systematic look at this question here, we would like to give some indication of why nonstandard methods may be of use. Using a very similar argument to that in the last section it is easy to show that any non–separating plane continuum is an intersection of topological disks, although the classical proof of this fact is also straightforward. This implies that there are internal sets H which are internally homeomorphic to a disk which contain $*E$ and which satisfy $\mathrm{st}(H) = E$ (this in itself provides no advantage to simply using a good standard approximation).

With E, f and H as above, it is not possible, in general, to find an internal continuous function from H to H which stays infinitely close to f (if it were, we could make use of the Brouwer Fixed Point Theorem to prove the open question). However, it appears that approximating f by a function from such an H to H might be a viable method of approaching this problem if the following condition does not occur:

(#) G is a maximally connected subset of $*\mathbb{R}^2 - *E - B$, where B is some ball of infinitesimal radius, and for some $p \in \overline{G} \cap *E$, $\ f(p) \in \overline{G} \cap *E$.

With this in mind, we illustrate one example of a relevant proposition which uses a nonstandard proof. The hypothesis of the proposition below appears to be an important case in which the above condition could occur.

Proposition 9
Let E be a non–separating plane continuum, and f a continuous function from E to E. Suppose that there is a point $p \in {}^*E$, and a ball of infinitesimal radius B such that if we let

$$D = \text{the connected component of } {}^*E - B \text{ containing } f(p)$$

then $D \cap {}^*st(D) = \emptyset$, the connected component of ${}^*E$ in $\overline{B}$ containing p intersects D, and $B \cup D$ does not disconnect the plane. Then f has a fixed point in $st(D)$

The proof of this proposition is somewhat similar to that of the classical result that "snakelike" continua have the fixed point property (see e.g. [4] for definitions and a proof of this result).

Proof. Since both D and ${}^*st(D)$ are closed sets in ${}^*\mathbb{R}^2$, there exists an infinitesimal $\eta_1 > 0$ which is the distance between the two sets, and an infinitesimal η_2 which is the maximum distance that any point in one set is from the other set. We let p_0 be any standard reference point in the complement of E. We note that since the complement of E is open and connected it is polygonally connected. We also let q be a point on the connected component of ${}^*E$ in $\overline{B}$ containing p, as is given in the hypothesis.

Since $B \cup D$ does not separate the plane, and since open connected sets are always polygonally-connected the point p_0 can be connected to a point on B by a polygonal path P_1 in ${}^*\mathbb{R}^2 - {}^*E$ such that every point in ${}^*st(D)$ is within $\frac{\eta_1}{2}$ of P_1, i.e. by a path which stays "between" D and ${}^*st(D)$.

By transfer to the standard model there must exist polygonal connections in $\mathbb{R}^2 - E$ from p_0 to points infinitesimally close to p. By transfer again (back to the nonstandard model) there must exist an internal polygonal connection P_2 in ${}^*\mathbb{R}^2 - {}^*E$ which is infinitesimally close to D but which is always at least $2\eta_2$ from ${}^*st(D)$ from p_0 to B (except near p itself). These two polygonal sets are thus infinitely close to D and "enclose" D.

We can now partition the set enclosed by P_1, P_2 and the boundary of B into a sequence of solid (i.e. including the interior) quadrilaterals $A_0, A_1, \ldots, A_K$, where each A_j is of infinitesimal size, and $q \in A_0$ and $p_0 \in A_K$. As each A_j is infinitesimal, if any point in any $D \cap A_j$ is mapped by f into itself, then taking standard parts immediately yields the desired fixed point in $st(D)$. However, under the assumption that no point in any $D \cap A_j$ is mapped by f into itself, and using the fact that removing any A_j with $0 < j < K$ disconnects D, we see by induction that for all j there are points of $f(A_j \cap D)$ in $\cup_{i=j+1}^{K} A_i$. As K is finite (in the nonstandard model), this is a contradiction which completes the proof. □

The condition that $B \cup D$ does not disconnect the plane can be replaced with more complicated weaker conditions, but cannot be simply removed, as is illustrated in the diagram below. In this picture, the interior of the circle is part of the set E, and both

p and $f(p)$ are in $^*E$ and infinitely close to the circle (there must be a sequence of curves approaching the circle in the standard set E in order for this to occur). The set $\text{st}(D)$ could now be the circle itself (depending on the size of the ball B), and the function f need not have a fixed point on this circle.

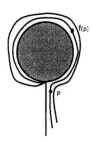

Figure 2

The only advantage of the nonstandard model in the above proof is that we were able to use transfer to find "enclosing" polygonal paths in the complement of the set. While this does not seem to be a powerful use of the nonstandard structure, it is worth noting that if, in the hypotheses of the proposition, we replace $^*\text{st}(D)$ by first taking an arbitrary ϵ and then finding a disjoint subset of E which is everywhere within ϵ of D, we are left with a very weak assumption, from which we can certainly not deduce a "near" fixed point. Similarly the condition (#), above, is a much stronger condition than any single "approximated" version. (Note, for example, that the set G in the statement cannot contain any balls of noninfinitesimal size, for if it did we would see by taking standard parts that E separates the plane). This provides some motivation for the use of nonstandard methods in considering this problem.

Hagopian [4] has shown that every path-connected non-separating plane continuum has the fixed point property. In view of proposition 4, above, it is possible that this problem could also be approached by generalizing Hagopian's proof to the SP or HP-connected case.

References

[1] R.M. Andersen, *"Almost" implies "near"*, Trans. Amer. Math. Soc. **296** (1986), 229-237.
[2] R.Chuaqui and N. Bertoglio, *An elementary Geometric Nonstandard Proof of the Jordan Curve Theorem*, Geometriae Dedicata **51** (1994), no. 1, 15.
[3] J. Dugundji, *Topology*, Allyn and Bacon, 1966.
[4] C.L. Hagopian, *Uniquely arcwise connected plane continua have the fixed-point property*, Trans. Amer. Math. Soc. **248** (1979), 85-104.
[5] O.H. Hamilton, *A fixed-point theorem for pseudo-arcs and certain other metric continua*, Proc. Amer. Math. Soc. **2** (1951), 173-174.
[6] V. Klee and S. Wagon, *Old and New Unsolved Problems In Plane Geometry and Number Theory*, MAA, Dolciani Mathematical Expositions– no. 11, 1991.

[7] T. Lindstrøm, *An invitation to nonstandard analysis*, Nonstandard Analysis and its Applications (N. Cutland, ed.), Cambridge University Press, 1988, 1-105.
[8] W.A.J. Luxemburg, *A general theory of monads*, Applications of Model Theory to Algebra, Analysis, and Probability (Luxemburg, ed.), Holt, 1969, 18-86.
[9] M. Machover and J. Hirschfeld, *Lectures on Nonstandard Analysis*, Lecture Notes in Mathematics, vol. 94, Springer-Verlag, 1969.
[10] M. C. McCord, *Nonstandard analysis and homology*, Fund. Math. **74** (1972), 21-28.
[11] E. Moise, *Geometric Topology in Dimensions 2 and 3*, Springer-Verlag, 1977.
[12] J. R. Munkres, *Topology, a First Course*, Prentice–Hall, 1975.
[13] L. Narens, *A nonstandard proof of the Jordan curve theorem*, Pacific J. Math. **36** (1971), 219-229.
[14] J.P. Reveilles, *Infinitesimaux et Topologie*, C.R. Acad. Sc. Paris **298** (1984).
[15] A. Robinson, *Nonstandard Analysis*, North-Holland, 1966.
[16] K. Stroyan and W.A.J. Luxemburg, *Introduction to the Theory of Infinitesimals*, Academic Press, 1976.
[17] F. Wattenberg, *Nonstandard Topology and extensions of monad systems to infinite points*, J. Symbolic Logic **36** (1971), 463-576.
[18] F. Wattenberg, *Nonstandard analysis and the theory of shape*, Fundamenta Mathematicae **98** (1978).
[19] R. Živaljevich, *On a cohomology theory based on hyperfinite sums of microsimplexes*, Pacific J. Math. **126** (1987).

Department of Mathematical Sciences
University of Northern Colorado
Greeley CO 80639
USA

E-mail leth@hopper.univnorthco.edu

Thomas Walter
Delayed bifurcations in perturbed systems analysis of slow passage of Suhl-threshold

0. Introduction

We are interested in ordinary differential equations of the following kind:

(1) $$\dot{x} = f(x, \mu) \quad , \quad x(t), \mu \in \mathbb{R} \text{ and } f \in C^2(\mathbb{R}) \quad ,$$

where μ is a parameter.

A value μ_0 for which the flow of (1) is not structurally stable is called a *bifurcation value* for (1), (see [2]). Simplifying we say that (1) undergoes a *bifurcation*, if the fixed point behaviour changes. In expansion to the basic definition we consider a change of a parameter-value, which stands for an external influence. In this paper we describe the *delayed dynamic bifurcation* that can occur when μ is replaced by a slowly varying function crossing the bifurcation point.

In order to study dynamic bifurcations, we add a second independent equation to (1)

$$\dot{\mu} = \varepsilon$$

where $\varepsilon > 0$.

Given initial values t_0 and μ_0, this is solved by

$$\mu(t) = \mu_0 + \varepsilon (t - t_0)$$

so it is advantageous to introduce a new time-variable $\tau = \mu(t)$ and to consider the equivalent nonautonomous equation

(2) $$\varepsilon \frac{d}{d\tau} x = f(x, \tau)$$

as a representation of (1) with the parameter replaced by τ varying at rate ε.

Because we are interested only in slowly variation rates, we use infinitesimals to give a technical meaning to "slowly varying" (section 1 and 2). We later use the Cauchy-Principle to extend our results to sufficiently slowly varying real parameters (section 3) and physical experiments (section 4). The Nonstandard Analysis conventions are from common literature, [3] or [4].

1. Delayed Bifurcations without Perturbation

Suppose x_F is a fixed point for all parameter values of (1) and hence it is a solution of (2), and suppose this fixed point is stable for $\tau < 0$ and unstable for $\tau > 0$. Then delayed bifurcations are given by the following definition:

Definition 1.1. (Delayed Bifurcation) *Consider a system of type (2) with $0 \approx \varepsilon > 0$ and suppose x_F as above. If there exist initial values (τ_0, x_0), where $\tau_0 < 0$, $\tau_0 \not\approx 0$, and $x_0 \not\approx x_F$, such that the solution of (2) with $x(\tau_0) = x_0$ is satisfying*

$$\exists \tau > 0,\ \tau \not\approx 0\ \forall \tau' \in (0, \tau) : x(\tau') \approx x_F$$

then it is a delayed bifurcation.

In addition to this first definition we would like to fix the terminology *delay*:

Definition 1.2. (Delay and Maximal Delay) *Consider a delayed bifurcation of definition 1.1. The delay D is given by*

$$D = \operatorname{st}\tau \in \mathbb{R} \cup \{\infty\}$$

if the initial conditions fulfill $x_0 \not\approx x_F$. The maximal delay is the supremum of all values of D (st: standard-part map, here st : $^*\mathbb{R} \to \mathbb{R} \cup \{\infty\}$).

The main result of this part is the following theorem:

Theorem 1.1. (Delayed Bifurcation without Perturbation) *Suppose f is of the form*

$$f(x, \mu) = \mu \cdot x - g(x)$$

and $g : {}^*\mathbb{R} \to {}^*\mathbb{R}$ satisfies

(a) $g \in C^2$
(b) $g(0) = g'(0) = 0$

then the following bifurcation phenomenons occur:

<u>Static case ($\varepsilon = 0$):</u> *The point 0 is a fixed point of (1) which is stable for $\mu < 0$ and unstable for $\mu > 0$.*

<u>Dynamic case ($0 \approx \varepsilon > 0$):</u> *The slowly varying point 0 undergoes a delayed bifurcation and the delay is at least equal to the absolute value of the initial parameter value $|\mu_0|$.*

Proof. x_F is fixed because $f(\mu, x_F) = 0 - g(0) = 0$. This fixed point is stable for $\mu < 0$, because the Jacobian has only one eigenvalue equal to μ which is negative and it is unstable for $\mu > 0$ vice versa.

To show the delay of this bifurcation we have to consider the solution of (2) and to show:

$$\forall \tau : \tau < |\tau_0|\ \text{and}\ \tau \not\approx |\tau_0| \quad \Rightarrow \quad x(\tau; x_0, \mu_0) \approx 0\ .$$

We can restrict this proof to $x \geq 0$.

We choose initial values (τ_0, x_0) such that $\operatorname{st} x_0$ is an element of the domain of attraction of $x_F(\tau_0)$. It suffices to look for $\tau > 0$ and $|x_0| \leq 1$. For this τ we choose $\delta \in \mathbb{R}^+$ such that $\tau + \delta < |\tau_0|$.

Taylor-expansion of g with respect to $x_e = 0$ yields

$$g(x) = a_0 + a_1 x + R_1(x)\ .$$

Because of our assumptions of theorem 1.1 ($g(0) = g'(0) = 0$) a_0 and a_1 vanish and because of the stability of x_F we can assume $x_0 \leq 1$ ($x(\tau)$ reaches every standard neighborhood of x_F by an infinitesimal parameter-change). Furthermore we assume $x(\tau) \leq 1$. Hence

$$g(x) = R_1(x) = \frac{1}{2!} g''(\xi) x^2 \quad \text{and} \quad |g(x)| \leq \tilde{a}_2 x^2 \quad (\tilde{a}_2 \geq \max_{\xi \in [0,1]} \left|\frac{1}{2!} g''(\xi)\right| \in \mathbb{R}^+) \ .$$

With the same arguments we assume $x(\tau) \leq \frac{\delta}{3\tilde{a}_2} \in \mathbb{R}^+$ and get

$$|g(x)| \leq \max_{\xi \in [0,1]} \frac{1}{2!} |g''(\xi)| x^2 \leq \tilde{a}_2 x^2 \leq \frac{\delta}{3} x \ .$$

The solution of

$$\varepsilon \frac{d}{d\tau} \xi = \left(\tau + \frac{\delta}{3}\right) \xi$$

fulfills the relation

$$\xi(\tau) \geq x(\tau) > 0$$

if $|\xi(\tau)| \leq \frac{\delta}{3\tilde{a}_2} \in \mathbb{R}^+$. This solution is given by

$$\xi(\tau) = x_0 \, e^{\frac{1}{2\varepsilon}\left((\tau+\frac{\delta}{3})^2 - (\tau_0+\frac{\delta}{3})^2\right)} \ .$$

Hence for δ as chosen above ($\Rightarrow \ 0 < \delta < |\tau_0|$) it follows

$$\xi(|\tau_0| - \delta) = x_0 \, e^{-\frac{1}{2\varepsilon}\left(\frac{2}{3}\delta|\tau_0| - \frac{1}{3}\delta^2\right)} \approx 0 \ ,$$

and for all $\tau' \in [0, |\tau_0| - \delta] : \xi(\tau') \approx 0$, and hence $x(\tau') \approx 0$ (and so all our assumptions are all right). This is the delay of the bifurcation. □

The next question is to ask for a *longer* delay than that of theorem 1.1. This does in general not exist!

Theorem 1.2. (Maximal Delay) *Consider the same system as in theorem 1.1 with the same assumptions. Then the maximal delay is given by $|\tau_0|$.*

We prove theorem 1.2 only for the transcritical bifurcation, for this result gives the general case (see [5] or [6]). So in combination with theorem 1.1 the delay of these bifurcations is exactly given by the absolute value of initial parameter value.

These results can be applied to many well-known bifurcations:

Corollary 1.3. (Delayed Transcritical Bifurcation) *The transcritical bifurcation*

$$\varepsilon \frac{d}{d\tau} x = \tau x - x^2$$

is delayed and the maximal delay is given by the absolute value of the initial parameter value.

Corollary 1.4. (Delayed Pitchfork Bifurcation) *The pitchfork bifurcation*

$$\varepsilon \frac{d}{d\tau} x = \tau x - x^3$$

is delayed and the maximal delay is given by the absolute value of the initial parameter value.

Proof. These are the cases $g(x) = x^2$ and $g(x) = x^3$ of theorems 1.1 and 1.2. □

How to Get Results in the Reals?

The above results make use of the nonstandard-setting $0 \approx \varepsilon > 0$. We use the permanence-principles to get results in the reals:

Theorem 1.5. (The Cauchy-Principle) *Let $\Phi(x)$ be an internal formula with x the only free variable. Then:*
If $\Phi(x)$ holds for each infinitesimal x then there is a standard $e > 0$ in $\mathbb{R}$ so that $\Phi(y)$ holds for all y with $|y| \leq e$ in $^*\mathbb{R}$.

(See for instance [3], II.7.1.(iii) or [4], p. 109).

So this means we should get delayed bifurcations for sufficiently small *real* values of ε.

To apply theorem 1.5 we have to give proper internal formulations of the nonstandard-results. These do not exist for external elements such as "$\approx$". So first of all we have to make a standard-definition of delayed bifurcations, which should be equivalent to definition 1.1. This works using an equivalent "$\varepsilon - \delta$−criterion".

We consider (2) with standard $\varepsilon = e \in \mathbb{R}^+$. Suppose (1) has fixed point x_F for all parameter values, which is stable for $\tau < 0$ and unstable for $\tau > 0$. Let $x_e(\tau)$ be the solution of corresponding equation (2). The standard form is given by the following theorem:

Theorem 1.6. (Delayed Bifurcation–Standard Form) *The following condition is equivalent to Definition 1.1: If there exist initial values $(\tau_c < 0, x_0 \neq x_F)$ such that the solution $x_e(\tau)$ with $x_e(\tau_0) = x_0$ satisfies*

$$\exists \mu \in \mathbb{R}^+ \; \forall \delta \in \mathbb{R}^+ \; \exists E \in \mathbb{R}^+ \; \forall e \in \mathbb{R}^+ \; \forall t \in (0, \frac{\mu}{e}) :$$

$$0 < e < E \quad \Rightarrow \quad |x_e(t) - x_F| < \delta$$

then $x_e(\tau)$ is a delayed bifurcation.

Proof. The equivalence of definitions 1.1 and the condition on theorem 1.6 is shown using the Cauchy-principle 1.5 and the transfer-principle. □

Theorem 1.6 shows the advantage of Nonstandard-Analysis, where the corresponding definition 1.1 is much more straight forward and clear! Now we can apply theorem 1.5 to our first results and we get the corresponding ones for sufficiently small *real* ε. This works in the same way for the results of section 2.

2. Delayed Bifurcations with Perturbation

In some applications we have to consider the influence of *perturbation* in our calculations. Two different types of perturbation are investigated: constant perturbation and noise. For a constant perturbation a it is possible to give a condition for the existence of delay: if $\exists \varrho \approx 0 : |a| \leq \alpha(\tau, \varepsilon)$, then the bifurcation is perturbed for all $\tau' < \tau < |\tau_0| \; \tau \not\approx |\tau_0|$. There $\alpha(\tau, \varepsilon)$ is the *first limit value*, see [5] or [6].

Noise

We are looking for a simple, but fruitful model of noise $\sigma(t)$, which yields to different results:
- our noise has constant absolute value $\pm\sigma$.
- our noise is constant for a duration Dt which is small: $Dt \approx 0$. Corresponding $D\tau = \varepsilon\, Dt$ is small in respect to ε:
$$\frac{D\tau}{\varepsilon} \approx 0 \ .$$
- both signs of noise have the same probability.

Furthermore we begin by investigating the linear system

(3) $$\varepsilon \frac{d}{d\tau}\xi = \tau \cdot \xi + \sigma(t) \ .$$

The solution of (3) is the connection of the solution for time-intervals Dt, where $\sigma(t)$ is constant.

Then the following theorem holds:

Theorem 2.1. (Noise) *Consider the linear system (3). The solution will most probably remain near by the unstable fixed point $x_F \equiv 0$ for all $\tau' \in (0, \tau)$, $\tau < |\tau_0|$ and $\tau \not\approx |\tau_0|$, if it exists $\varrho \approx 0$ such that*
$$|\sigma| \le \varrho \cdot \beta(\tau, Dt, \varepsilon) \ .$$

$\beta(\tau, Dt, \varepsilon)$ *is called* <u>second limit value</u> *and is given by*

(4) $$\beta(\tau, Dt, \varepsilon) = \frac{1}{\sqrt[4]{2\pi}} \frac{\sqrt[4]{\varepsilon}}{\sqrt{Dt}} e^{-\frac{1}{2\varepsilon}\tau^2} \ .$$

Proof. We discuss the behaviour of solutions of (3).

Discrete System: Between $\tau_0 + nD\tau$ and $\tau_0 + (n+1)D\tau$ the noise is constant. So we define discrete quantities for $n \in {}^*\mathbb{Z}$:

$$\tau_n = \tau_0 + n \cdot D\tau \qquad \xi_n = \xi(\tau_n) \qquad \sigma_n = \sigma \cdot r_n = \lim_{\substack{\tau \to \tau_n \\ \tau > \tau_n}} \sigma(t(\tau)) \quad (\text{where } r_n \in \{-1, +1\}) \ .$$

Solving $\varepsilon \frac{d}{d\tau}\xi = \tau\xi + \sigma \cdot r_n$ gives

$$\xi_{n+1} = e^{\frac{1}{2\varepsilon}(\tau_{n+1}^2 - \tau_n^2)} \xi_n + \frac{\sigma_n}{\varepsilon} e^{+\frac{1}{2\varepsilon}\tau_{n+1}^2} \int_{\tau_n}^{\tau_n + D\tau} e^{-\frac{1}{2\varepsilon}\tau'^2} d\tau' \ .$$

Because of $D\tau/\varepsilon \approx 0$ we get the difference equation

$$\xi_{n+1} = e^{\frac{1}{2\varepsilon}(\tau_{n+1}^2 - \tau_n^2)} \xi_n + \frac{\sigma\, r_n\, D\tau}{\varepsilon}$$

with solution

$$\xi_n = \xi_0\, e^{\frac{1}{2\varepsilon}(\tau_n^2 - \tau_0^2)} \frac{D\tau}{\varepsilon} \sigma e^{+\frac{1}{2\varepsilon}\tau_n^2} \sum_{\nu=0}^{n-1} r_\nu\, e^{-\frac{1}{2\varepsilon}\tau_\nu^2} \ .$$

This expression is — beside factors — a random variable of type

$$R(n) = \sum_{\nu=0}^{n-1} r_\nu e^{-\frac{1}{2\varepsilon}(\tau_0+\nu D\tau)^2} .$$

By substitutions

$$\gamma = \frac{D\tau^2}{2\varepsilon} \qquad \mu = \nu + \frac{\tau_0}{D\tau} \qquad m = n - 1 + \frac{\tau_0}{D\tau}$$

we get

$$R(m) = \sum_{\mu=\frac{\tau_0}{D\tau}}^{\mu=m} r_\mu e^{-\gamma\mu^2} .$$

Delay means we have to look for

$$\forall m' \in {}^*\mathbb{Z} : \frac{\tau_0}{D\tau} < m' \leq m \quad \Rightarrow \quad R(m') \approx 0 ,$$

where m is chosen such that we pass the bifurcation point: $m \cdot D\tau > 0$ and $m \cdot D\tau \not\approx 0$. Instead of this we regard

$$R = \sum_{\mu=-\infty}^{\mu=+\infty} r_\mu e^{-\gamma\mu^2}$$

because it's possible to neglect the rest of summation (see 5] or [6]). To continue we use that both signs of r_μ have the same probability. Then (because of symmetry): *The expected value of R vanishes.* To get a delayed bifurcation, the fluctuations of the solution of (3) must be small. So we look for the standard deviation s (second central moment). It is given by the summation of $p_r (R_r - E\,R)^2$ over all possibilities r of the r_μ, where p_r is the probability of a possibility r of the r_μ's and R_r is the value of R for this r:

$$s^2(R) = \mathrm{var}\,R = \sum_r p_r (R_r - E\,R)^2 = \sum_r p_r \left(\sum_\mu r_\mu e^{-\gamma\mu^2}\right)^2 .$$

We simplify this problem by regarding

$$S(N) = \sum_{\mu=1}^{N} r_\mu \cdot 1 .$$

This is the well-known *Bernoulli Distribution*.

We choose N such that

$$\int_{-\infty}^{+\infty} e^{-\gamma x^2} dx = \frac{\sqrt{\pi}}{\sqrt{\gamma}} = \sqrt{2\pi}\,\frac{\sqrt{\varepsilon}}{D\tau} = N .$$

Then we get the following two lemmas:

Lemma 2.2. (Comparison to Bernoulli Distribution)

$$\mathrm{var}\,R \leq \mathrm{var}\,S(N) .$$

Proof. (lemma 2.2): We compare two Bernoulli Distributions

$$S_A = \sum_{\nu=1}^{M} r_\nu H \quad \text{and} \quad S_B = \sum_{\nu=1}^{m} r_\nu h$$

and assume $M \cdot H = m \cdot h$. Then

$$H > h \quad \Rightarrow \quad \text{var} S_A > \text{var} S_B \quad .$$

This is obviously because

$$\text{var} S_A = \frac{1}{4}(2H)^2 M > \frac{1}{4}(2h)^2 m = \text{var} S_B \quad .$$

Lemma 2.2 follows by applying this to Riemann oversums of $e^{-\gamma \mu^2}$. □

Lemma 2.3. (**Standard Deviation**)

$$s^2(R) \leq \sqrt{2\pi} \frac{\sqrt{\varepsilon}}{D\tau} \quad .$$

The proof follows by straight calculation. □

Regarding the neglected factors, we get the condition of theorem 2.1 for a delayed bifurcation. □

To complete these results we consider

(5) $$\varepsilon \frac{d}{d\tau} x = \tau \cdot x - g(x) + \sigma(t(\tau)) \quad .$$

Lemma 2.4. (**General Case with Noise**) *Consider (5) where* $g : {}^*\mathbb{R} \to {}^*\mathbb{R}$ *fulfills the conditions of theorem 1.1 ($g \in C^2$ and $g(0) = g'(0) = 0$) and in addition*

(6) $$g(x > 0) > 0 \quad \text{and} \quad g(x < 0) < 0 \quad .$$

Then the same holds as in theorem 2.1 for all initial values which are elements of basin of attraction of $x_F \equiv 0$.

Proof. The conditions of theorem 2.1 give delayed bifurcations for $\sigma = 0$.
If $\sigma \neq 0$ it suffices to look for $x(\tau_0) = 0$. Because of symmetry-relation (6) we have

$$x(\tau) \not\approx 0 \quad \Rightarrow \quad \xi(\tau) \not\approx 0$$

where $\xi(\tau)$ is solution of the linear system (3) to initial condition $\xi(\tau_0) = 0$. If the conditions of theorem 2.1 are fulfilled, then ξ is small and hence x is small. □

Application of lemma 2.4 gives delayed pitchfork bifurcations with noise:

Corollary 2.5. (**Delayed Pitchfork Bifurcation with Noise**) *The dynamic pitchfork bifurcation with noise*

$$\varepsilon \frac{d}{d\tau} x = \tau x - x^3 + \sigma(t(\tau))$$

has a delay, if, for $\beta(\tau, Dt, \varepsilon)$ as in (4), the amplitude of noise σ fulfills

$$\exists \varrho \approx 0 : |\sigma| \leq \varrho \cdot \beta(\tau, Dt, \varepsilon).$$

Proof. This is the case $g(x) = x^3$ of lemma 2.4 □

In the presence of noise we have in some sense the opposite behaviour than without noise. The unperturbed system shows the "proper delay" if ε is as small as possible. But then the limit values are becoming (by an exponential rule) small, so there is no chance to fulfill the condition of theorem 2.1.

For more details about the limit values α and β see [5] or [6].

3. Numerical Simulations

Application of the Cauchy-principle 1.5 gives real versions of these results for sufficiently small ε. So we can do numerical simulations. We found in every case $\varepsilon = \frac{1}{30}$ to be sufficiently small!

Different numerical simulations were done by **Mathematica** (version 2.1) to show the meaning of these results. We use frequently the routine NDSolve, an adaptive Runge-Kutta algorithm.

Here we want to present only two simulations, the delayed pitchfork bifurcation and the meaning of noise. Figure 1 shows the delayed Pitchfork bifurcation for different initial values.

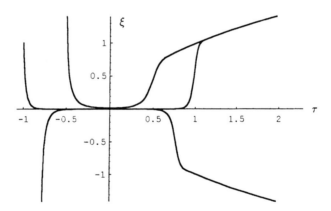

Figure 1: Delayed Pitchfork bifurcation, different initial values, $\varepsilon = 0.035$

The influence of noise is shown in figure 2. The noise was simulated using the random number generator of **Mathematica**. We choose two different cases, first $\sigma_1 = 3\beta$ and second $\sigma_2 = \frac{1}{3}\beta$. In the first case there is only a short delay whereas in the second we reached in fact the maximal delay. This is the meaning of theorem 2.1. We show two different curves for each case.

Numerical values:

$$\varepsilon = 0.1 \quad \tau_0 = -1 \quad Dt = 0.01 \quad (\Rightarrow 2000 \text{ iterations})$$
$$\beta = 0.0239 \quad \sigma_1 = 0.0718 \quad \sigma_2 = 0.008 \quad .$$

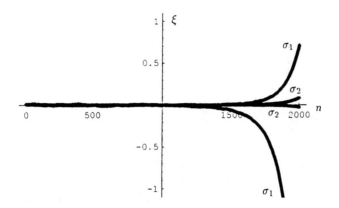

Figure 2: Delayed bifurcation with noise, $\sigma_1 = 3\beta$ and $\sigma_2 = \frac{1}{3}\beta$

4. Physical Example: High-Power Magnetic Resonance Experiments

Ferromagnetic samples excited by strong microwave fields show a variety of nonlinear phenomena, ending up in chaos. We investigate a ferromagnetic sample (Yttrium-Iron-spheres YIG) in an external field consisting of a static field **H** and a transverse (driving) component $\mathbf{h} \cdot \cos\omega t$. A review of recent results is given in [1].

We assume to have the following complex magnetic field inside the sphere (only its real part is of physical meaning):

$$\mathbf{m}(\mathbf{r},t) = a_0(t)\mathbf{m}_0 e^{-i\omega_0 t} + a_1(t)\mathbf{m}_1(\mathbf{r}) e^{-i\omega_1 t} \ .$$

This is the *two modes model*, because we only consider a uniform (non loci-dependent) and one nonuniform mode. It shows to be very successful in many applications (see for example [7]). This yields to our equations of motion

(7)
$$\begin{aligned}
\dot{a}_0 &= -\underbrace{\left(i(\omega_0-\omega)+\Gamma_0\right)}_{\alpha_1} a_0 - \varrho^* a_1^2 - \underbrace{i\gamma h}_{\mu} = -\alpha_1 a_0 - \varrho^* a_1^2 - \mu \\
\dot{a}_1 &= -\underbrace{\left(i(\omega_1-\frac{\omega}{2})+\Gamma_1\right)}_{\alpha_2} a_1 + \varrho a_0 a_1^* = -\alpha_2 a_1 + \varrho a_0 a_1^*\ .
\end{aligned}$$

Here Γ is a phenomenological damping constant ($\Gamma \in \mathbb{R}^+$), γ the gyromagnetic ratio, and ϱ the dipolar coupling coefficient. Then the first-order Suhl threshold is given by

(8)
$$\begin{aligned}
h_s &= +\sqrt{\frac{1}{\gamma^2 \varrho\varrho^*} \alpha_1 \alpha_1^* \alpha_2 \alpha_2^*} \\
&= +\sqrt{\frac{1}{\gamma^2 \varrho^*}\left(\Gamma_0^2+(\omega_0-\omega)^2\right)\cdot\left(\Gamma_1^2+(\omega_1-\frac{\omega}{2})^2\right)}\ .
\end{aligned}$$

We found two different fixed points. The first (simple) fixed point is given by a vanishing nonuniform mode

$$a_0 = -\frac{\mu}{\alpha_1} \qquad a_1 = 0 \quad . \tag{9}$$

The second fixed point exists only for $h \geq h_s$ and is given by

$$a_0 = -\frac{\mu_s}{\alpha_1} = -\frac{i\gamma}{\alpha_1} h_s \qquad a_1^2 = -\frac{1}{\varrho^*}(\mu - \mu_s) = -\frac{i\gamma}{\varrho^*}(h - h_s) \quad , \tag{10}$$

if $h \geq h_s$ and $\alpha_i = \Gamma_i \in \mathbb{R}^+$ (resonant pumping $\omega = \omega_0 = 2\omega_1$). Because of the square of a_1, (10) describes two fixed points which can not distinguished by experiment.

To understand the stability of these fixed points we look for the eigenvalues of the Jacobian. We get the following bifurcation behaviour at the Suhl threshold (8): There exists one fixed point (9), which is stable for $h < h_s$ and which becomes unstable for $h > h_s$. Furthermore two new stable fixed points (10) exist only for $h > h_s$, which are equal to the first fixed point at $h = h_s$. This is exactly the behaviour of the pitchfork bifurcation. For this, section 2 shows the existence of delay even in the presence of noise (theorem 2.1/corollary 2.5). So we can expect to have a delayed bifurcation when passing the Suhl threshold sufficiently slow.

4.1 Numerical Simulation

We use {Mathematica (version 2.1) to simulate the slow passage of first-order Suhl threshold without noise. Here we choose the following numerical data:

$$\gamma = 1 \qquad \alpha_1 = \alpha_2 = 1 \qquad \Gamma_1 = \Gamma_2 = 1 \qquad \varrho = 1 + i \quad .$$

Then we get the Suhl threshold at $h_s = 1/\sqrt{2}$. Using $\varepsilon = 0.0005$ we get the following characteristic delay of the bifurcation, depending only on initial-values of parameter h. There the equations of motion (7) are solved using the routine NDSolve.

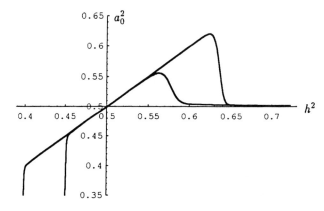

Figure 3: Slow passage of Suhl threshold, numerical result

4.2. Experimental Results

The numerical results are in agreement to our measurements. We want to present our results for an intermediate parameter-change speed $\varepsilon_0 = 1.5 W/s$. We show two different initial values:

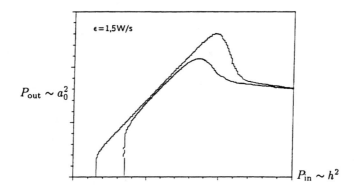

Figure 4: Slow passage of Suhl threshold, experimental result

These results correspond to the numerical simulation (figure 3) and to the theoretical results, because there exists a delay and its duration is given by the initial parameter value.

As discussed in [5] the presence of noise makes it impossible to find a delay of bifurcation when the parameter speed ε becomes to small (because the second limit value grows up by an exponential rule). So when we choose smaller values of ε we see the break-down of delay. We found this characteristic break-down in our measurements, the results are shown in [6].

Finally we want to point out that the easy way to get these real results shows the advantage of the nonstandard method.

The author would like to thank the organization committee of CIMNS 94 for the fruitful conference.

References

[1] H. Benner, F. Rödelsperger, and G. Wiese. *Chaotic dynamics in spin-wave instabilities*, Nonlinear Dynamics in Solids (H. Thomas, ed.), 129–155, Springer 1992.
[2] John Guckenheimer and Philip Holmes. *Nonlinear Oscillations, Dynamical Systems, and Bifurcations of Vector Fields.* Springer-Verlag, New York, 2nd edition, 1986.
[3] Albert E. Hurd and Peter A. Loeb. *An Introduction to Nonstandard Real Analysis.* Academic Press, Inc., Orlando, 1985.
[4] Detlef Laugwitz. *Zahlen und Kontinuum.* Bibliographisches Institut, Mannheim, 1986.
[5] Thomas Walter. *Delayed Bifurcations in Perturbed Systems*, Preprint Fachbereich Mathematik vol. 1685, Technische Hochschule Darmstadt, Darmstadt, 1994.

[6] Thomas Walter. *Verzögerte Bifurkationen als Effekt langsamer Parameteränderung.* Verlag Shaker, Aachen, 1994, Technische Hochschule Darmstadt, Dissertation.

[7] X. Y. Zhang and H. Suhl. *Spin-wave-related period doublings and chaos under transverse pumping*, Physical Review A, vol. 32, 2530-2533, 1985.

Fachbereich Mathematik
TH Darmstadt
D–64289 Darmstadt

E-mail walter@mathematik.th-darmstadt.de

ALAIN ROBERT

Functional analysis and NSA

1. Brief review of naive NSA — Internal Set Theory

Most mathematicians use set theory in a very naive way, and they would be hard put to define *which* set theory they use. Most students only have a very limited knowledge of the axioms of this theory. In fact, they get used to its machinery through practice, thanks to the naive interpretation of sets (bags containing mathematical entities) and the famous "$\in$" as "belonging" relation.

It is possible to deal with NSA in a quite similar fashion and we propose to interpret the new term "standard" as "classically and uniquely defined". This interpretation makes obvious the following statements

$$0,\ 1,\ 10^3,\ \pi,\ e\ ,\ldots \text{ are standard numbers.}$$

This new term can be applied to *any mathematical entity*. For example

$\mathbb{N},\ \mathbb{R},\ \mathbb{C},\ [0,1],\ \mathcal{F}([0,1],\mathbb{C}),\ L^2(\mathbb{R}^4),\ L^1(0,1),\ldots$ are standard sets,

$\sin,\ exp,\ \ldots$ are standard functions,

$T(\mathbb{S}^4)$ (tangent space to the 4-dim. sphere)

is a standard fibered space,...

With this interpretation of the term standard, it is obvious that if f is a standard function and x a standard element in the domain of f, then $f(x)$ is standard. In particular, all constructions of set theory which start from standard sets and produce well defined sets furnish standard ones

E and F standard sets $\Longrightarrow$ $E \times F$, E^F, $\mathcal{P}(E)$ are standard sets.

But we admit that in any infinite set, there are elements which cannot be defined classically uniquely (we have only a limited way of defining explicitly objects, because we have limitations of time, language,...)

In any infinite set E, there are some non standard elements.

For example, if we are dealing with the Lebesgue spaces $L^p(\mathbb{R}^n)$, since there are infinitely many of them ($1 \leq p \leq \infty, n \geq 1$), some must be non standard. It is true that the function $p \longmapsto L^p(\mathbb{R}^3)$ is classically and uniquely defined so is a standard function. If $L^p(\mathbb{R}^3)$ is not standard, this is because p is not standard. Similarly, $(p,n) \longmapsto L^p(\mathbb{R}^n)$ is well defined uniquely classically, so that this function is standard.

Let us see what happens in the particular case of the set $\mathbb{N}$ of integers. We have seen that $1, 2, 3, 4,\ldots\ 10^6,\ldots\ 1000^{100}$ are standard. But since this set is infinite, there must exist a non standard integer $\nu \in \mathbb{N}$. Such an integer must be different from all quoted integers. Hence, for example

$$\nu > 1000^{100}.$$

This shows that the non standard natural integers are big! Let us introduce another term for this property: we say that n is an *illimited* integer. In other words, standard integers are also called *limited* integers.

More generally, if $z \in \mathbb{C}$ is a complex number, we *define*
$$z \text{ limited} \iff \exists \text{ a limited integer } n \in \mathbb{N} \text{ with } |z| \leq n.$$
This notion is solely concerned with the absolute value of the complex number z under consideration. One could define similarly this notion in any valued field (p-adic field, field of quaternions, etc.). If E is any normed vector space over a valued field (typically over the field $\mathbb{R}$ or $\mathbb{C}$) and $x \in E$, let us define similarly
$$x \text{ limited} \iff \|x\| \text{ limited} \iff \exists \text{ a limited integer } n \in \mathbb{N} \text{ with } \|x\| \leq n.$$
We shall use the notation $\|x\| \ll \infty$ for limited elements.

By definition, for natural integers n we have
$$n \text{ limited} \iff n \text{ standard}.$$
This relation also holds for rational integers $n \in \mathbb{Z}$, but it fails in a more general context. For example, if $\nu \in \mathbb{N}$ is an illimited integer, we have in particular $\nu > 1$ hence $1/n < 1$ proving that the rational number $1/\nu$ is limited. But $1/\nu$ is not standard since
$$x \text{ standard} \iff 1/x \text{ standard}$$
(if x can be defined classically uniquely, $1/x$ can also be defined classically uniquely and conversely!). If ν is as above, the rational number $\varepsilon = 1/\nu > 0$ satisfies
$$\varepsilon < 1/n \text{ for all limited integers } n \in \mathbb{N}.$$

We call *infinitesimal* any number x (or vector in a normed space) which has the property
$$|x| < 1/n \text{ for all limited integers } n \in \mathbb{N}$$
(resp. $\|x\| < 1/n$ ...) and denote this property by $x \simeq 0$. Elements which are neither infinitesimal nor illimited are also called *appreciable*. Let us emphasize at this point that the only new *attribute* that we add to traditional set theory is the predicate *standard*. It is *undefined* in set theory and extends its power. From it, we introduce new terms

limited, illimited, infinitesimal, etc.

which are defined (or derived) from the term *standard*. Any statement, theorem... which does not contain the new term (or its derived) is called *classical* (or traditional, internal). When $\nu \in \mathbb{N}$ is an illimited integer, we have $\nu > n$ for all standard integers n. In other words, *all standard integers belong to the finite interval* $[0, \nu] = S$ *of* $\mathbb{N}$.

This property is true more generally: in each set E, there is a *finite part* S containing all the standard elements (of E). But there is no set (no part of the set E) that contains precisely its standard elements. At most, we can speak of the *class* of the standard elements in a set. These classes do not share all properties of sets. For example, the class of illimited integers does not have a smallest element. The class of limited integers is bounded but has no biggest element. The set $\mathbb{N}$ is still *well ordered:* any non empty *part* of $\mathbb{N}$ has a smallest element. In particular, the **induction**

principle holds as before, provided that we take care to apply it to **subsets** and not classes. In other words, induction applies to classical statements (but not necessarily to NSA statements). There is a limited NSA version of induction that works for *all statements P*:

$$P(0) \text{ is true and}$$
$$(P(n) \implies P(n+1)) \text{ for standard } n \in \mathbb{N}$$
$$\text{implies } P(n) \text{ true for all standard } n \in \mathbb{N}.$$

(this is *non standard —or external— induction*).

For reference, we list a few **principles** that we are going to apply

Idealization

   a) *In each infinite set E, there are some non standard elements.*
   b) *In each set E, there is a finite part S containing all its standard elements.*

Logical continuation (or extension)

   Let E and F be two standard sets. If for each standard $x \in E$, a certain construction furnishes a standard element $y \in F$, then there exists a unique standard function $f : E \longrightarrow F$ extending the construction.

Of course, all principles that we are going to state and use can be *proved* from the axioms (I), (S), (T) given by Nelson (1977). This is explained in several easily available books (cf. references at the end).

2. Topology and NSA

Let (X, d) be a standard metric space. If s and t are two standard elements of X, $d(s, t)$ is standard hence limited. Consequently, if an element $x \in X$ is such that $d(x, s)$ is limited for one standard element s, it will be limited for all standard elements t (by the triangle inequality) and we define

$$x \text{ limited in } X \iff d(x, s) \text{ limited for all standard elements } s \in X.$$

If A is a standard part of X, its closure is

$$\overline{A} = {}^S\{x \in X : \exists a \in A, a \simeq x\}.$$

This means that the standard elements of $\overline{A}$ are precisely those for which there exists an $a \in A$ infinitely close to x ($a \simeq x$). The map

$$A \longmapsto \overline{A} : \mathcal{P}(X) \longrightarrow \mathcal{P}(X)$$

is the standard map extending (by logical continuation) the preceding definition.

Definition. *A point x in a metric space (X, d) is called near standard (and we abbreviate this situation by "x is ns.") if there exists a standard point s with $d(x, s) \simeq 0$ (i.e. any standard open ball containing s also contains x).*

Definition. *A metric space X is called S-compact if each point in X is ns.*

Theorem. *Let X be a standard metric space. Then*
$$X \text{ compact} \iff X \text{ is S-compact}.$$

Proof. If a standard metric space contains a non ns. point x, it is not compact. Indeed, for each standard s, take a standard open ball $B = B_s$ centered at s and not containing x. By logical continuation, there exists a unique standard family of open balls $y \longmapsto B_y$ extending the preceding construction. Since $\bigcup B_y$ is a standard set containing all standard points of X, (B_y) is a covering of X. However, for all standard finite subcoverings F, we have
$$x \notin \bigcup_F B_y \text{ and in particular } \bigcup_F B_y \neq X.$$
Transfer can be applied to the second (classical) inequality:

the open covering (B_y) of X has no finite subcovering.

(To visualize this part of the proof, the reader may like to look at the following particular example: $X = \mathbb{R}$, x illimited and for s standard let us choose the standard open interval $]s - 1, s + 1[$ which does not contain x; the unique standard extension is given by $y \longmapsto]y - 1, y + 1[$ for *all* y and the open intervals $]y - 1, y + 1[$ cover $\mathbb{R}$ but no standard finite subfamily contains x, no finite subfamily covers $\mathbb{R}$ entirely.) Conversely, if all points $x \in X$ are ns. then X is compact. Let indeed $(U_i)_{i \in I}$ be a standard open covering of X. Take a finite set S containing all standard points of X and a corresponding finite part J of the index set I such that $\bigcup_J U_i \supset S$. If x is an arbitrary point, it is ns. by assumption so that $x \simeq s$. Since there is a standard U_i containing s, this U_i contains x by definition. This proves that $\bigcup_J U_i$ is a finite subcovering (by transfer, there is also a standard finite subcovering).

Lemma. *Let (x_n) be a standard sequence in a (standard) metric space X. Then this sequence admits a convergent subsequence precisely when there exists an illimited integer ν such that x_ν is near standard.*

Proof. If the given sequence has a convergent subsequence, it also has a standard convergent subsequence (x_{n_i}). The limit of this standard subsequence is a standard element y and $x_{n_i} \simeq y$ for all illimited i. Conversely, assume that ν is illimited such that x_ν is near standard, say $x_\nu \simeq y$ where y is a standard point. For each standard m, the ball $B_{1/n}(y) = \{x \in X : d(x, y) \leq 1/n\}$ has a non empty intersection with the set of points of the sequence $\{x_n\}$ (indeed, the point x_ν belongs to this intersection). Since this intersection $B_{1/n}(y) \cap \{x_n\}$ is standard and not empty, it must contain a standard element x_{n_m}. By logical continuation, there is a (unique) standard sequence extending $m \longmapsto x_{n_m}$ as desired.

Proposition. *Let X be a standard metric space and $A \subset X$ be S-compact. Define*
$$A^S = {}^S\{x \in X : \exists a \in A, a \simeq x\}.$$
Then A^S is compact.

Proof. It is known (and easy to see) that A^S is closed. It is enough to show that each standard sequence of A^S has a convergent subsequence. The preceding lemma shows that we have to find a near standard element x_ν (ν illimited) in any such sequence. But the sequence $\varepsilon_n = d(x_n, A)$ is such that $\varepsilon_n \simeq 0$ for all limited integer n. By Robinson's lemma, there must be an illimited integer ν such that $\varepsilon_n \simeq 0$ for all $n \leq \nu$. In particular, $x_\nu \simeq a$ for some $a \in A$. By hypothesis, this point a is ns. hence x_ν is also ns. (This proof has been shown to me by Y. Perrin.)

Definition. *A function f between two metric spaces is called* S-continuous *when*

$$x \simeq y \text{ (in } X) \Longrightarrow f(x) \simeq f(y) \text{ (in } Y).$$

In other words, f is S-continuous when

$$d(x,y) \simeq 0 \Longrightarrow d'(f(x), f(y)) \simeq 0.$$

S-continuity is a notion of *experimental continuity*. It is independent from continuity in general but equivalent to it for standard maps. A good relation between continuity and S-continuity is given by the following result.

The continuous shade theorem. *Let X be a standard metric space and g be any S-continuous function $X \longrightarrow Y$. Assume $g(x)$ ns. in Y for all standard $x \in X$. Then the standard function f defined by logical continuation from $f(x) = \operatorname{st} g(x)$ (for standard $x \in X$) is continuous.*

The closed graph theorem (topological version). *If $f : X \longrightarrow Y$ is a mapping between metric spaces, where Y is compact, then*

$$\operatorname{Graph}(f) \text{ closed in } X \times Y \Longleftrightarrow f \text{ continuous.}$$

Proof. Without loss of generality, we may assume f standard (hence X and Y are standard spaces) and prove f continuous at all standard points $x \in X$. We have to show

$$x' \simeq x \Longrightarrow f(x') \simeq f(x).$$

But since Y is compact, $f(x')$ is nearly standard, say $f(x')^* = \operatorname{st} f(x') = y$. The standard point (x, y) has its coordinates infinitely close to $(x', f(x'))$ which belongs to the graph of f

$$(x, y) \simeq (x', f(x')) \in \operatorname{Graph}(f) \Longrightarrow (x, y) \in \text{closure of } \operatorname{Graph}(f).$$

Since we assume that the graph of f is closed, we infer that (x, y) belongs to this graph, namely $y = f(x)$, and by definition $f(x') \simeq f(x)$.

Counterexample when Y is not compact. Let $f : \mathbb{R} \longrightarrow \mathbb{R}$ be defined by $f(0) = 0$ and $f(x) = 1/x$ if $x \neq 0$. Then the graph of f is

$$\operatorname{Graph}(f) = (0,0) \cup \{(x,y) \in \mathbb{R}^2 : xy = 1\}$$

hence is closed, but f is not continuous.

3. Operators between normed spaces

Let us begin with a few classical (and elementary) results concerning linear operators $T : E \longrightarrow F$ between normed spaces. Recall that T is *continuous* if it is continuous at the origin and this is the case if it is *bounded*. In this case we define

$$\|T\| = Sup_{\|x\|=1} \|Tx\| = Sup_{\|x\|\neq 0} \|Tx\|/\|x\| < \infty.$$

Proposition-definition. *The operator T is S-continuous if any of the following equivalent conditions is satisfied*

i) $x \simeq y \Longrightarrow Tx \simeq Ty$,
ii) $x \simeq 0 \Longrightarrow Tx \simeq 0$,
iii) x *limited* $\Longrightarrow Tx$ *limited*,
iv) *There exists a limited* $M > 0$ *such that* $\|Tx\| \leq M \cdot \|x\|$,
v) $\|T\| = Sup_{\|x\|=1}\|Tx\|$ *is limited.*

For standard operators, this definition is the same as continuity, but it may differ from it even in finite dimension. For example, let $E = F = \mathbb{R}$ and look at the operator $x \longmapsto y = \nu x$ where ν is illimited. This operator is continuous but not S-continuous. However, quite generally

$$\text{S-continuous and linear} \Longrightarrow \text{continuous}.$$

The first interesting case is $F = \mathbb{C}$. The set of linear forms φ with $\|\varphi\| \leq 1$ is the unit ball of the dual E' of E. Although we have only defined formally the notion of near standard point in a metric space, let us say that a linear form φ is ns. when there exists a standard linear form ψ such that $\varphi(x) \simeq \psi(x)$ for all standard $x \in E$. We abbreviate this weak proximity relation by $\varphi \simeq_\sigma \psi$ (in fact, although the weak topology on E' is not usually metrizable, its restriction to the unit ball usually is...).

Lemma. *Every limited linear form is near standard in the weak sense.*

Proof. Let φ be limited in E'. Then

$$x \text{ standard} \Longrightarrow x \text{ limited} \Longrightarrow \varphi(x) \text{ limited in } \mathbb{C} \Longrightarrow \varphi(x) \text{ ns.}$$

There is a unique standard ψ such that

$$\psi(x) = st\,\varphi(x) \text{ for all standard } x \in E$$

(logical continuation principle). This function ψ is linear. Moreover, if M is a standard integer greater than or equal to $\|\varphi\|$, we have

$$|\psi(x)| = st\,|\varphi(x)| \leq M \text{ for all standard } x \text{ such that } \|x\| \leq 1.$$

By transfer we conclude that $\|\psi\| \leq M$, i.e. ψ is limited and continuous q.e.d.

In particular this lemma shows that the unit ball in the dual of a standard normed space is weakly compact. Since this property is classical, transfer shows that it is still true in *any* normed space.

Definition. *A linear map T between two normed spaces E and F is S-compact when any of the following equivalent conditions is satisfied*
  i) *x normed in $E \Longrightarrow Tx$ near standard in F,*
  ii) *x limited in $E \Longrightarrow Tx$ near standard in F.*

In particular if $F = \mathbb{C}$ we see that a linear form φ is S-compact whenever it is limited (we have used this observation in the proof of the lemma). Since near standard elements are limited, we also see that

$$T \text{ S-compact} \Longrightarrow T \text{ S-continuous}$$
$$(\Longrightarrow T \text{ continuous})$$

If T_1 and T_2 are S-continuous (resp. S-compact) operators and λ_i are limited scalars, then $\lambda_1 T_1 + \lambda_2 T_2$ is S-continuous (resp. S-compact). But the *class* of S-continuous (resp. S-compact) operators is not a *set* (and in particular not a vector space!). The sum of a finite family of S-continuous operators is not necessarily S-continuous. However, standard finite (i.e. limited) sums of S-continuous (resp. S-compact) operators are S-continuous (resp. S-compact).

Recall that a compact linear $T : E \longrightarrow F$ is one for which the image of the unit ball is a relatively compact subset of F. All continuous finite rank linear maps are compact. The non standard characterization of standard compact spaces shows that for standard operators T

$$T \text{ S-compact} \Longleftrightarrow T \text{ compact.}$$

Examples.

1/ Let E be any normed space and choose $0 \simeq \varepsilon \in \mathbb{R}$. Then $\varepsilon I = \varepsilon 1_E$ is an S-compact operator. But such an operator εI can only be compact if $\dim E < \infty$ or $\varepsilon = 0$.

2/ Let $E = \ell^2$ be the space of complex sequences $(x_i)_{i \in \mathbb{N}}$ with $\sum |x_i|^2 < \infty$. This is a Hilbert space with orthonormal basis $\mathbf{e}_i = (0, ..., 1, 0, 0, 0, ...)$ (the non zero element 1 being situated at the ith place). Select an illimited integer $\nu \in \mathbb{N}$ and for $x \in \ell^2$ define

$$Tx = T(x_i) = x_\nu \cdot \mathbf{e}_1 = (x_\nu, 0, 0, 0, \dots) \in \ell^2.$$

Then T is compact (since $\operatorname{Im} T = \mathbb{C} \cdot \mathbf{e}_1$, it is an operator of finite rank) and S-compact.

3/ Let $(\lambda_n)_{n \in \mathbb{N}}$ be a sequence of scalars with $(\lambda_n) \to 0$. Then the operator T in ℓ^2 defined by $T(\mathbf{e}_n) = \lambda_n \cdot \mathbf{e}_n$ is a compact operator. Its eigenvalues are the λ_n with corresponding eigenvectors $\mathbf{e}_n$.

4/ As before, let T denote the operator in ℓ^2 defined by $T(\mathbf{e}_n) = \lambda_n \cdot \mathbf{e}_{n+1}$. Then T is compact as soon as the sequence $\lambda_n \to 0$. Let us list some elementary properties of (S-)compact operators.

Proposition 1. *Let $K : E \longrightarrow F$ be an S-compact linear map between two normed spaces E and F. If $T_1 : E_1 \longrightarrow E$ is S-continuous and $T_2 : F \longrightarrow F_2$ is standard continuous, then the following composite is an S-compact linear map*

$$T_2 \circ K \circ T_1 : E_1 \longrightarrow E \xrightarrow{K} F \longrightarrow F_2.$$

Proof. If x_1 is limited in E_1, then $x = T_1(x_1)$ is limited in E (recall that by definition, the norm of T_1 is limited). By hypothesis, $Kx \simeq y$ standard in F. Since the norm of T_2 is also limited $T_2 Kx \simeq T_2 y$ and the element T_2 is standard by assumption. This proves $T_2 \circ K \circ T_1 x_1$ ns.

Proposition 2. *Let $K : E \longrightarrow F$ be an S-compact linear map between two normed spaces E and F. If $T : E \longrightarrow F$ is a linear map such that $\|K - T\| \simeq 0$, then T is also an S-compact map.*

Proof. If x is a normed vector (in E), Kx is ns. say $Kx \simeq y$ standard (in F). Then $Tx \simeq Kx \simeq y$ proves Tx is ns. This proposition generalizes example 1 above (take $K = 0$ and $T = \varepsilon I$).

Proposition 3. *Let $K : E \longrightarrow F$ be a standard compact linear map between two normed spaces E and F. Then the transpose $K' : F' \longrightarrow E'$ is also a compact linear map.*

Proof. Let φ be a continuous linear form on F with $\|\varphi\| = 1$. If x is limited in E, $y = Kx$ is limited in F and $\varphi(Kx)$ is limited in $\mathbb{C}$. For y standard in F we can define $\psi(y) = st\ \varphi(y)$. By standardization, this defines a standard linear form $\psi \in F'$. I claim that $\psi \circ K \simeq \varphi \circ K = K'(\varphi)$ thus showing that $K'(\varphi)$ is ns. Take *any* x in the unit ball of E. By assumption $y = Kx$ is ns. and we can write $y = Kx \simeq z$ standard in F with successively

$$\varphi(Kx) \simeq \varphi(z) \text{ since } \|\varphi\| \leq 1$$
(because φ is a contracting map)
$$\varphi(z) \simeq \psi(z) \text{ since } z \text{ is standard}$$
(by definition of ψ)
$$\psi(z) \simeq \psi(Kx) \text{ since } \psi \text{ standard}$$
(ψ is S-continuous at the standard point z)

This proves
$$\varphi(Kx) - \psi(Kx) \simeq 0 \quad (\|x\| \leq 1)$$

hence $\|\varphi \circ K - \psi \circ K\| \simeq 0$, $\varphi \circ K \simeq \psi \circ K$ standard (i.e. $\varphi \circ K$ is ns.)

Observe that the first part of the proof is an NSA formulation of the classical result

the unit ball in F' is weakly compact

and the definition of ψ is simply $\psi \simeq_\sigma \varphi$ for the weak topology. The second part of the construction by-passes Ascoli's theorem.

Proposition 4. *Let $K : E \longrightarrow F$ be a standard compact linear map between two normed spaces E and F. Let $\hat{K} : \hat{E} \longrightarrow \hat{F}$ be the canonical extension of K to the completed Banach spaces. Then*

$$\hat{K}(\hat{E}) \subset F \text{ and } \hat{K} : \hat{E} \longrightarrow F \text{ is a compact linear map.}$$

Proof. Recall that E can be identified to a subspace of E'' (normed dual of E'). With this identification $\hat{E}$ is simply the closure of E in E''. From Prop.3 we infer that K' and then $K'' : E'' \longrightarrow F''$ are compact linear maps. Restricting K'' to $\hat{E}$ = closure of E, we get $\hat{K}(\hat{E}) \subset \overline{F} \subset F''$. Now let x be a standard vector in $\hat{E}$, and assume $x \simeq y \in E$. Then y is limited and since K is compact, Ky is ns., say $Ky \simeq z$ st. in F. We have
$$\hat{K}x \simeq Ky \simeq z.$$
Since $\hat{K}$ is standard, $\hat{K}x$ is standard and we must have $\hat{K}x = z \in F$. By transfer we deduce $\hat{K}(\hat{E}) \subset F$.

Proposition 5. *Let $T : E \longrightarrow F$ be an S-compact operator where E and F are standard normed spaces. The standard operator K defined by*
$$Kx = st\, Tx \text{ for all standard } x \in E$$
is a compact operator.

Proof. For standard $x \in E$, Tx is ns. and its standard part st Tx is well (and uniquely) defined. Logical continuation takes care of the definition of the standard operator K. Let B denote the closed unit ball in E. Then $T(B)$ is an S-compact subset of F and $K(B)$ is contained in the compact set $T(B)^S$ (cf. section 2 for the notation and relation between compactness and S-compactness). (This proposition has been communicated to me by Y. Perrin; its proof is due to L. Haddad.)

Important example of compact operator.

Let $k : I \times J \longrightarrow \mathbb{C}$ be a standard continuous function on the product of two compact intervals of $\mathbb{R}$ and define
$$Kf(s) = \int_I k(u,s)f(u)du.$$
It is obvious that
$$|Kf(s)| \leq \|k\|_\infty \cdot \|f\|_1 \quad (f \in \mathcal{L}^1(I)).$$
One proves that Kf is continuous (first when f is standard) and then that K is a standard continuous linear map
$$\mathcal{L}^1(I) \longrightarrow \mathcal{C}(J).$$

Let us prove that it is (S-)compact (both notions coincide since K is standard). For $\|f\|_1 \leq 1$ and a standard s, let us define
$$h(s) = st\, g(s) = \int_I k(u,s)f(u)du$$
(this is possible since $g(s)$ is limited, in fact $|g(s)| \leq \|k\|_\infty$). By standardization, there is a unique standard function h taking the precedingly prescribed values at standard points. For s and t standard
$$|h(s) - h(t)| \leq |g(s) - g(t)| \leq Max_u |k(u,s) - k(u,t)|.$$

For any standard $\varepsilon > 0$, we can use transfer on
$$|h(s) - h(t)| \leq Max_u |k(u,s) - k(u,t)| + \varepsilon$$
and deduce that this is valid even if s and t are not standard. Take in particular $s \simeq t$ so that $k(u,s) \simeq k(u,t)$ for all u and
$$Max_u |k(u,s) - k(u,t)| \simeq 0.$$
This leads to
$$|h(s) - h(t)| \leq \varepsilon, \ |h(s) - h(t)| < 2\varepsilon$$
and since ε standard is arbitrary positive, we infer
$$h(s) \simeq h(t) \text{ whenever } s \simeq t.$$
Now if s is arbitrary in J and s^* is its standard part we can write successively (this by-passes Ascoli's theorem)
$$h(s) \simeq h(s^*) \simeq g(s^*) \simeq g(s),$$
$$\|h - g\|_\infty = Max_s |h(s) - g(s)| \simeq 0.$$
This proves that h is the standard part of $g = Kf$ hence Kf is ns.

Exercise. Prove that if the function $k \in \mathcal{L}^2(I \times J)$, then the operator K defined by the same formula as before is a compact operator $\mathcal{L}^2(I) \longrightarrow \mathcal{L}^2(J)$.

4. Spectral properties of operators — Quasi-eigenvectors

Let E be a complex normed space. We shall call *operator* in E any continuous linear map $E \longrightarrow E$. [We shall eventually be mainly interested in *standard* operators in E, in which case E is assumed to be standard too.] Let us recall that in infinite dimension, an operator may have no eigenvector (and hence no eigenvalue). The *spectrum* σ_T of a continuous operator T is the set of *spectral values*, namely the set of $\lambda \in \mathbb{C}$ such that the operator $T - \lambda I$ is *not continuously invertible*. It is known quite generally that
$$\sigma_T \neq \emptyset \text{ is a compact set contained in the disc } \|\lambda\| \leq \|T\| \text{ in } \mathbb{C}.$$
The set ρ_T of *regular values* is the complement of the spectrum σ_T of T: $\mu \in \mathbb{C}$ is regular when $T - \mu I$ is continuously invertible.

Definition. *We say that a vector v is a* quasi-eigenvector *of an operator T, relatively to a* standard quasi-eigenvalue λ *when*
$$\|v\| \not\simeq 0 \text{ and } T(v) \simeq \lambda v.$$

Observe that if a normed vector v satisfies $Tv = \mu v$ for some $\mu \in \mathbb{C}$, then v is a quasi-eigenvector of T with respect to $\lambda = st\, \mu$. The *existence* of quasi-eigenvectors in general is assured by the following general result.

Theorem 1. *Let T be a standard continuous operator in a Banach space E. Then T has some quasi-eigenvectors. More precisely, if λ is any standard point in the boundary of the spectrum σ_T of T, then λ is a quasi-eigenvalue of T.*

Proof. Let us first prove the following *lemma:* If μ is a regular value for T such that $\|(T - \mu I)^{-1}\|$ is limited, then $\lambda = st\,\mu$ is also a regular value. Write indeed

$$T - \lambda I = (T - \mu I) + (\mu - \lambda)I = (T - \mu I) \cdot [I - (\lambda - \mu)(T - \mu I)^{-1}].$$

Since $\lambda - \mu \simeq 0$ and $\|(T - \mu I)^{-1}\|$ is limited, $(\lambda - \mu)(T - \mu I)^{-1} \simeq 0$ and

$$I - (\lambda - \mu)(T - \mu I)^{-1} \text{ is invertible (by the geometric series!).}$$

Hence $T - \lambda I$ is the product of two continuously invertible operators: λ is a regular value as asserted in the *lemma*. Let us return to the proof of the theorem. Let λ be a standard point of the boundary of the spectrum σ_T. Hence there exists a regular value $\mu \simeq \lambda$ and by the lemma $\|(T - \mu I)^{-1}\|$ must be illimited. Namely, there exists a normed vector y with $(T - \mu I)^{-1} y = z$ illimited. Divide by the illimited scalar $\|z\|$ and put $z_1 = z/\|z\|$. Then

$$(T - \lambda I)z_1 \simeq (T - \mu I)z_1 = y/\|z\| \simeq 0.$$

This proves that $Tz_1 \simeq \lambda z_1$ and z_1 is a quasi-eigenvector of T relative to the quasi-eigenvalue λ q.e.d.

For compact operators, quasi-eigenvectors are nearly eigenvectors. This is made quite precise in the next statement which will play a crucial role in this section.

Theorem 2. *Let K be a standard compact operator in a Banach space E. If v is a quasi-eigenvector of K relatively to a standard $\lambda \neq 0$, then v is ns. and the standard part of v is an eigenvector of T relatively to the eigenvalue λ.*

Proof. Since K is compact and v limited, Kv is ns., say $Kv \simeq w$ st. and

$$w \simeq Kv \simeq \lambda v, \quad w/\lambda \simeq v, \quad K(w/\lambda) \simeq Kv \simeq \lambda v \simeq w.$$

Since the extreme terms are standard, they must be equal

$$K(w/\lambda) = w$$

thereby proving the theorem.

The special case $\lambda = 1$ is typical and will be systematically studied first (it concerns *fixed points* of K on the unit sphere). This is why we shall generally put $T = I - K$ and study its kernel N (eigenspace of K if 1 is an eigenvalue), its invertibility etc. We shall also say that T is a *compact perturbation of the identity* in this case. Observe that with these notations $T + K = I$ and symmetrically $K = I - T$.

Theorem 3. *Let T be a standard operator in a Banach space E. Define $K = I - T$ and $N = Ker(T)$. If there exists an element $a \in E$ with*

$$dist(a, N) \text{ illimited and } T(a) \text{ limited}$$

then K is not a compact operator.

Proof. Put indeed $a_1 = a/dist(a, N)$ at a distance 1 from N by construction. We can find $n \in N$ with $\|a_1 - n\| \simeq 1$. Furthermore

$$T(a_1 - n) = T(a_1) = T(a)/dist(a, N) = \lim./\text{illim.} \simeq 0.$$

By definition

$$K(a_1 - n) = a_1 - n - T(a_1 - n) \simeq a_1 - n.$$

This shows that the limited element $a_1 - n$ is a quasi-eigenvector for K (rel. to the quasi-eigenvalue $\lambda = 1$) although situated at distance 1 from the corresponding eigenspace N of K. From the preceding proposition, K is not a compact operator. One can interpret the statement of the theorem by saying that a compact perturbation T of the identity does not shrink wildly (or preserves the orders of magnitude). In the same vein, if $dist(a, N) \not\simeq 0$ and $T(a) \simeq 0$ then T is not a compact perturbation of the identity.

Particular case. When the operator T is injective (i.e. 1 is not an eigenvalue of K), we have $N = 0$, $dist(a, N) = \|a\|$ and the theorem proves the following three implications (in sequence)

$$\|a\| \text{ illimited and } T(a) \text{ limited} \implies K \text{ not compact},$$
$$K \text{ compact and } T(a) \text{ limited} \implies \|a\| \text{ limited},$$
$$K \text{ compact} \implies T^{-1} \text{ is S-continuous at } 0 \in T(E).$$

When K is a compact operator not having the eigenvalue 1, $T = I - K$ is injective and $T^{-1} : T(E) \longrightarrow E$ is continuous. A fortiori $T(E)$ is complete hence closed in E (more generally, for any closed subset A of E, $T(A)$ is closed in $T(E)$ and in E). The continuity of T^{-1} on $T(E)$ (closed) would also follow from the Banach homomorphism theorem. When T is not injective, one can still prove that $T(E)$ is closed.

Proposition. *Let K be a compact operator in a Banach space E, $T = I - K$. Then $T(E)$ is a closed subspace of E.*

Proof. When T is standard (transfer takes care of the generalization). In this case, it is also enough to show that a standard b close to an element of $T(E)$ belongs to the image of T. More precisely, assume $b \simeq T(a)$ for some $a \in E$. Let us show how we can modify a in order to have $b \simeq T(s)$ and s standard (this will immediately give $b = T(s)$ because both sides are standard). Since b is standard, $T(a)$ is limited and by theorem 1

$$dist(a, N) \text{ is limited} \qquad (N = KerT).$$

Take an element $n \in N$ with $\|a - n\|$ limited. Since K is compact $K(a - n)$ is ns., say $K(a - n) \simeq v$ standard. We then have

$$b \simeq Ta = T(a-n) = T(K(a-n) + T(a-n)) \simeq T(v+b)$$

with $v + b$ standard. Hence $b = T(v+b) \in Im\,T$.

Motivation. Suppose that a practical situation (e.g. physical experiment) leads to a continuous operator T in an infinite dimensional Banach space E. It may be interesting to find the eigenvalues of T. However, one can safely say that such a situation certainly arises in an idealized way. Certainly, numerical computations have a finite character and only approximations of the ideal situation can be measured and evaluated. For example, instead of E, only a finite dimensional subspace V of E will be treated by a computer (V could be determined by a finite element method or a finite difference scheme). Instead of the operator T in E, only an approximation A in V can be handled. This finite dimensional approximation will always have eigenvalues (in $\mathbb{C}$) and corresponding eigenvectors. For practical purposes a normed eigenvector $v \in V$ of A will be sufficient for all purposes. From $Av = \lambda v$, we infer $Tv \approx \lambda v$ and approximate eigenvectors may be as good as genuine eigenvectors for practical purposes... It is easy to give a description of this situation in NSA.: it is quite natural to take a finite set S containing all standard elements of E and define V as the linear span of S.

Quite generally, the notion of *finite approximation scheme* according to F. Trêves is the following. Let E and H be two Banach (or Hilbert) spaces connected by a standard homomorphism

$$J : E \longrightarrow H \text{ injective, continuous with closed image}$$

(hence J is a topological isomorphism $E \longrightarrow J(E) \subset H$ by the Banach theorem). The (non standard) finite dimensional space V together with two linear maps

$$R : E \longrightarrow V \text{ (restriction)}, \quad T : V \longrightarrow H \text{ (extension)}$$

is an approximation (or numerical model) of E and J provided that

a) $Jx = st\,T(Rx)$ for every standard $x \in E$,
b) If $Tv \simeq_\sigma y$ st. $\in H$, then $y = Jx$ for some standard $x \in E$.

This approximation scheme is called *stable* if moreover

c) R and T are S-continuous (i.e. $\|R\|$ and $\|T\|$ limited).

5. Finiteness theorems for compact operators

In this section, we are going to exploit systematically the NSA principle:

*In any set E, there is a finite part S
containing all standard elements of E.*

First, we start with a more classical observation due to F. Riesz.

Proposition. *Let F be a closed subspace of a Banach space E. If $F \neq E$, there exists a normed element $x \in E$ such that $dist(x, F) \simeq 1$.*

Proof. We can choose an element $z \in E$ not in F, and then $y \in F$ with
$$\text{dist}(z, F)/\|z - y\| \simeq 1$$
(in fact, one can replace z by a non zero multiple $z' = \lambda z$ so that $\text{dist}(z', F)$ is limited and *not* $\simeq 0$, then choose $y' \in F$ so that $\text{dist}(z', F) \simeq \|z' - y'\|$). Since $y \in F$ we have $\text{dist}(z - y, F) = \text{dist}(z, F)$ and
$$\text{dist}((z - y)/\|z - y\|, F) = \text{dist}(z, F)/\|z - y\| \simeq 1.$$
We can take $x = (z - y)/\|z - y\|$ to prove the proposition.

Remark. When E is a Hilbert space, one can choose $x \perp F$ and in this case, we can find a normed x with $\text{dist}(x, F) = 1$.

Corollary 1. *Let E be a Banach space such that every limited element of E is ns. Then E is finite dimensional.*

Proof. Let S be a finite part containing all standard elements of E and let F be the subspace generated by S. Then F has finite dimension, hence is closed in E. If $F \neq E$, there exists a normed $x \in E$ with $\text{dist}(x, F) \simeq 1$ and in particular
$$\|x - y\| \not\simeq 0 \text{ for all standard elements } y$$
(the standard elements in question are all in F). This proves that x has no standard part.

Corollary 2. *Let E be a Banach space in which the identity operator $I = \text{id}_E$ is an S-compact operator. Then E is finite dimensional.*

Riesz' basic observation. With the above notations, assume $(I - K)(E) \subset F$. For every $x \in E$, we have then $Kx \equiv x \mod F$ hence
$$\text{dist}(Kx, F) = \text{dist}(x, F).$$
Put $T = I - K$ so that $T(E) \subset F$ by hypothesis. Then for any $y \in F$
$$Kx - Ky = x - Tx - (y - Ty) = x - (y + Tx - Ty) = x - y' \equiv x \mod F$$
and in particular if x is normed, chosen as in the proposition
$$\|Kx - Ky\| \geq \text{dist}(Kx, F) = \text{dist}(x, F) \simeq 1$$

Applications.

1/ Let K be an S-compact operator in a Banach space E, $T = I - K$ and $N = \text{Ker } T$ (= eigenspace of K with respect to 1 if this is an eigenvalue of K). There is a finite part S containing all standard elements of N. Consider the finite dimensional space F generated by S. It is closed in N. Moreover
$$T = I - K \text{ sends } N \text{ into } \{0\} \in F.$$
Take x normed in N. Since K is S-compact, Kx is ns. so that
$$\text{dist}(x, F) = \text{dist}(Kx, F) \simeq 0.$$

By the proposition the inclusion $F \subset N$ must be an equality. In particular, the space N is finite dimensional. We have thus obtained a first essential finiteness theorem concerning compact operators

any non zero eigenvalue of a compact operator has a finite geometric multiplicity.

2/ The second finiteness result concerns the finiteness of the codimension of the image of a compact perturbation $T = I - K$ of the identity (recall that we have proved that the image $T(E)$ is indeed closed in this case). Since the kernel of the transpose of K precisely consists of the linear forms vanishing on $T(E)$ (or on the closure of $T(E)$), the finiteness of the codimension of $T(E) = \overline{T(E)}$ results from the finiteness of the dimension of the kernel of $T' = I - K'$ (which is also a compact perturbation of the identity). Alternatively, we can again take a finite part S containing all standard elements of E and let F be the vector space generated by S and (the closure of) $T(E)$. Since S is finite, F is closed. Let $x \in E$ be limited so that Kx is ns., say $Kx \simeq y$ standard (hence in $S \subset F$). Then

$$0 \simeq \text{dist}(Kx, F) = \text{dist}(x, F)$$

(recall $Kx = x - T(x) \equiv x \mod T(E) \subset F$). This proves $F = E$ and $\overline{T(E)}(= T(E))$ has finite codimension in E.

3/ Together, the preceding two finiteness results show that compact perturbations of the identity are *index operators*. For such operators, one defines

$$\text{index } T = \text{codim } T(E) - \dim \text{Ker}(E) \in \mathbb{Z}.$$

We shall see that the index of any compact perturbation of the identity vanishes: operators having this property are usually called *Fredholm operators*. In this respect, compact perturbations of the identity behave like operators in finite dimensional spaces.

Proposition. *Let K be a compact operator in the Banach space E. Then*

$$\bigcup_{j \geq 0} \text{Ker}(I - K)^j \text{ is finite dimensional.}$$

Proof. For $j \geq 1, (I - K)^j = I - K(*) = I - K_j$ with a compact operator K_j. This proves that each $\text{Ker}(I - K)^j$ is a finite dimensional subspace of E. The proposition asserts moreover that these subspaces have bounded dimension. Let $T = I - K$ (or $T + K = I$) as before and put $N_j = \text{Ker } T^j = \text{Ker}(I - K)^j$. As long as N_j properly contains N_{j-1} choose a normed vector $x_j \in N_j$ such that $\text{dist}(x_j, N_{j-1}) \geq 1/2$. For j standard, one can even choose x_j standard. There is a unique (finite or infinite) standard sequence (x_j) extending the construction. We still have

$$\|x_j\| = 1, \ x_j \in N_j \ \text{dist}(x_j, N_{j-1}) \geq 1/2$$

for all indices j. Write now

$$x_j = K(x_j) + T(x_j) \equiv K(x_j) \mod N_{j-1}.$$

For all $i \neq j$, say $i < j$ to fix ideas,

$$\|Kx_j - Kx_i\| \geq \text{dist}(Kx_j, V_{j-1}) = dist(x_j, V_{j-1}) \geq 1/2.$$

Since all elements of the standard set $\{Kx_j\}$ are near standard, this set is finite (it is closed, discrete and compact). This proves stationarity.

One shows similarly (by duality) that the sequence of iterated images stops decreasing
$$E \supset T(E) \supset T^2(E) \supset ... \supset T^m(E) = T^{m+1}(E) \text{ (m minimal)}.$$
If T is injective, the last equality gives $E = T(E)$: this is the first case of the index zero property. In other words

if 1 is not an eigenvalue of K, $I - K$ is both injective and surjective and 1 is not a quasi eigenvalue of K either:
$T = I - K$ *is a topological isomorphism, i.e. 1 is not a spectral value of K.*

One can replace the number 1 in the preceding statement by any standard $\lambda \neq 0$ which is not an eigenvalue (replace K by K/λ)
$$\lambda \text{ standard,} \neq 0, \text{ not eigenvalue}$$
$$\Longrightarrow \lambda \notin \sigma_K.$$
By contraposition, we deduce the following first essential spectral properties of compact operators.

Theorem 1. *Let K be a standard compact operator in a Banach space E. Then, if λ is a spectral value of K, we have*

either $\lambda = 0$,
or λ not standard,
or λ is an eigenvalue.

To elucidate completely the structure of the spectrum of compact operators, we need to study more closely the case of eigenvalues. For this purpose, a finer analysis of the sequence of iterated kernels and images has to be made (observe that this *analysis* is in fact very *algebraic!*).

From the fact that the two sequences
$$N_i = \operatorname{Ker}(I - K)^i, \text{ and } F_j = \operatorname{Im}(I - K)^j$$
are stationary, we can deduce by a purely algebraic argument that they stop moving at the same rank. Take
$$n \text{ minimal with } N_n = N_{n+1} = N_{n+2} = ...$$
$$m \text{ minimal with } F_m = F_{m+1} = F_{m+2} = ...$$

a) Observe $F_n \cap N_n = \{0\}$. Indeed $x \in N_n \Longrightarrow T^n x = 0$ and $x \in F_n \Longrightarrow \exists y$ such that $x = T^n y$ hence $T^{2n} y = 0$, $y \in N_{2n} = N_n$ and $x = T^n y = 0$.
b) One also proves $F_m = F_n$ (whence $m \leq n$). Certainly $F_m \subset F_n$ by minimality. Let us show the converse inclusion. Take $z \in F_n$. For some positive integer k we shall have $T^k z \in F_m = T^k(F_m)$ and there exists a $t \in F_m \subset F_n$ with $T^k z = T^k t$. Hence $T^k(z - t) = 0$, $z - t \in N_k \subset N_n$. But z and t belong to F_n, so that $z - t \in F_n \cap N_n = 0$. This proves $z = t \in F_m$ as was to be shown.
c) The direct sum decomposition now follows from the fact that $E = F_n + N_n$. Indeed, if $x \in E$, $T^n x \in F_n = F_{2n} = T^n F_n$ and there is an $y \in F_n$ such that

$$T^n x = T^{2n} y, \text{ whence } T^n(x - T^n y) = 0, \ x - T^n y \in N_n$$
$$x = (x - T^n y) + T^n y \in N_n + F_n.$$

We shall write $N_\infty = N_n$ and $F_\infty = F_n = F_m = \text{Im } T^\infty$. Thus $E = N_\infty \oplus F_\infty$ and $\text{Im } T^\infty$ isomorphic to E/N_∞. The stationarity of the increasing sequence of iterated kernels proves that T is injective on E/N_∞ where $N_\infty = \bigcup \text{Ker } T^n = \text{Ker } T^\nu$ and nilpotent on N_∞. Moreover $\text{Im } T^\infty = \bigcap \text{Im } T^n = \text{Im } T^\nu$ is a topological complement of N_∞. The two factors are stable under T which appears as a sum of a nilpotent operator $T|_{N_\infty}$ in a finite dimensional space ($K|_{N_\infty}$ is a unipotent operator) and a topological isomorphism $T|_{F_\infty}$. Restricting K to F_∞ simply removes $\lambda = 1$ from its spectrum.

We see now (algebraically) why compact perturbations of the identity are Fredholm operators

$$\text{Ker } T \in N_\infty \Longrightarrow \text{Ker } T = \text{Ker } T|_{N_\infty},$$
$$\text{Im } T \supset F_\infty \Longrightarrow \dim E/\text{Im } T = \dim N_\infty/T(N_\infty)$$

implies

$$\dim \text{Ker } T = \dim \text{Ker } T|_{N_\infty} = \text{codim }(T(N_\infty), N_\infty) = \text{codim } T(E).$$

Theorem 2. *Let K be a standard compact operator in a Banach space E. Its spectrum σ_K is a non empty compact set having the following properties*

a) $1 \simeq \lambda \in \sigma_K \Longrightarrow \lambda = 1$,
b) $\lambda \in \sigma_K$ and $\lambda \not\simeq 0 \Longrightarrow \lambda$ is standard,
c) for any standard $\varepsilon > 0$, $\sigma_K \cap \{|\lambda| \geq \varepsilon\}$ only has standard elements,
d) for any $\varepsilon > 0$, $\sigma_K \cap \{|\lambda| \geq \varepsilon\}$ is a finite set,
e) any sequence of distinct elements of σ_K tends to 0.

Proof. Proof of a). Let w be a normed eigenvector corresponding to $\lambda \simeq 1$. Then w is a quasi-eigenvector with respect to 1 and its standard part $v = w^*$ exists and is an eigenvector corresponding to the eigenvalue 1

$$v = w^* \in \text{Ker } T = N \subseteq N_\infty.$$

Obviously v is also normed. Consider the standard decomposition

$$E = N_\infty \oplus F_\infty.$$

There is a standard (continuous) linear form φ which vanishes on F_∞ and such that $\varphi(v) = 1$. Hence

$$\varphi(w) \simeq \varphi(w^*) = \varphi(v) = 1$$

($\Longrightarrow \varphi(w) \neq 0!$). Since $F_\infty = T^n E$ (for some $n \geq 1$), $T^n w \in F_\infty$ and

$$0 = \varphi(T^n w) = \varphi((1-\lambda)^n w) = (1-\lambda)^n \varphi(w)$$
$$\Longrightarrow (1-\lambda)^n = 0 \Longrightarrow \lambda = 1.$$

Having proved a), we observe that b) follows from a) applied to K/λ^* (λ^* denoting the standard part of λ). Then b) implies c) which in turn implies d) when ε is standard first (principle: in any infinite set, there are some non standard elements) and for all

$\varepsilon > 0$ by transfer (legitimate by the fact that d) is a classical property). The last implication d)$\Longrightarrow$e) is classical.

Comments.

1/ The assertions d) and e) of the proposition are classical. Hence they also hold by transfer for all compact operators. The assumption that K is standard is needed only for the first three assertions (the reader will easily construct counterexamples to a), b) and c) for non standard compact operators).

2/ Observe that 0 belongs to the spectrum of all compact operators in an infinite dimensional space E. For example, the operators

$$K : \mathbf{e}_n \longmapsto \mathbf{e}_{n+1}/(n+1) \ (n \geq 0)$$

in (c_o), or ℓ^2, or ℓ^1... have spectrum $\sigma_K = \{0\}$ reduced to this single point. These operators can be viewed as integral operators

$$Kf(x) = \int_0^x f(t)dt = \text{ primitive of } f \text{ vanishing at } x = 0$$

on suitable Banach spaces (completions of the space of polynomials with basis $\mathbf{e}_n = x^n$, $n \geq 0$).

About the bibliography

[1], [2] and [3] are references for the internal set theory of NSA; [4] is a reference for compact operators.

References

[1] André Deledicq, Marc Diener: *Leçons de calcul infinitésimal*, Armand Colin (1989).
[2] Francine Diener, Georges Reeb: *Analyse Non Standard*, Hermann (1989).
[3] Alain Robert: *Nonstandard Analysis*, John Wiley & Sons (1988).
[4] Jean Dieudonné: *Foundations of Modern Analysis (Chap.12)*, Academic Press (1960).

Institut de Mathematiques
Emile-Argand 11
CH-2007 Neuchatel
Switzerland

E-mail alain.robert@maths.unine.ch

W. A. J. LUXEMBURG
Near-standard compact internal linear operators

1. Introduction

In order to apply Robinson's celebrated result (see [2], Theorem 4.1.13 and Corollary 4.1.14) that a subset of a topological space is compact if and only if all its entities in an enlargement are near-standard it is often necessary to obtain an intrinsic characterization of the near-standard entities that are involved. For instance, in the case of Ascoli's theorem (see [1], Theorem 3.8.1 and Theorem 3.8.2) its nonstandard proof is based on the characterization of the near-standard internal continuous functions.

Similarly, it is useful for applications of the theory of bounded linear operators to have criteria for an internal linear operator to be near-standard in some sense. Since there are to my knowledge no published results around of this nature, the purpose of the present article is to make a beginning with such an investigation. As a first step we shall limit our investigation to the solution of the problem to give an intrinsic characterization for an internal bounded linear operator defined on a Hilbert space enlargement to be a near-standard element of the two-sided ideal of the compact standard linear operators. Such internal linear operators will be called near-standard compact. For special classes of internal linear operators such as the self-adjoint and normal operators the characterization of being near-standard compact turns out to be similar to Robinson's characterization of compactness for standard bounded linear operators (see [2] Theorem 7.1.4).

2. Notation and preliminaries

Let $\mathcal{M}$ be a superstructure in the sense of [3] and let $^*\mathcal{M}$ be a κ-saturated model of $\mathcal{M}$, where the cardinal number κ is larger than $\text{card}(\mathcal{M})$. We assume that $\mathcal{M}$ contains R, the field of real numbers and so also the field of complex numbers C. The model of $\mathbb{R}$ with respect to $^*\mathcal{M}$ will be denoted as usual by $^*\mathbb{R}$. If we denote the set of natural numbers by $N := \{1, 2, \cdots\}$, then we shall denote the external set of the infinitely large natural numbers by $^*N^\infty$. By $\inf(^*\mathbb{R})$ we shall denote the monad of the infinitesimals of $^*\mathbb{R}$ and instead of $a \in \inf(^*\mathbb{R})$ we shall often write $a \simeq 0$. The external subset of the near-standard numbers of $^*\mathbb{R}$, the so-called limited or finite numbers of $^*\mathbb{R}$, will be denoted by $\text{fin}(^*\mathbb{R})$ rather than by $\text{ns}(^*\mathbb{R})$. The algebraic and order preserving homomorphism of $\text{fin}(^*\mathbb{R})$ onto $\mathbb{R}$ with kernel $\inf(^*\mathbb{R})$ will be called as usual the standard part operator of $^*\mathbb{R}$ and will be denoted by "st". If no confusion can arise we will often use the same notation "st" for the standard part of a near-standard element of a topological structure.

Let $H \in \mathcal{M}$ be a Hilbert space and let $(\cdot,\cdot)$ denote its inner product and if $h \in H$ its norm by $||h|| = (h,h)^{\frac{1}{2}}$. Its κ-enlargement with respect to *$\mathcal{M}$ will be denoted by *H. For the sake of simplicity in the notation we shall denote the inner product of *H as well as its norm (deleting the asterisk) again by $(\cdot,\cdot)$ and $||\cdot||$ respectively. The elements of H in *H will again be denoted by the same letter without the upper left asterisk.

In the results to follow an important role will be played by the *$\mathcal{M}$-non-standard hull $\widehat{H}$ of H (see [1], Sections 15 and 16). For the sake of completeness we recall that $\widehat{H} := \text{fin}(*H)/\inf(*H)$ is defined as the quotient space of the external inner product space $\text{fin}(*H)$ over $\mathbb{C}$ of all the elements of *H of limited norm and its linear subspace $\inf(*H)$ of all the infinitesimal elements of *H, i.e., the elements of *H with infinitesimal norms. The linear homomorphism of $\text{fin}(*H)$ onto $\widehat{H}$ with kernel ($=$ null space) $\inf(*H)$ will be denoted by π. The images $\pi(x)$ of the elements $x \in \text{fin}(*H)$ in $\widehat{H}$ will be denoted by $\hat{x}$. From the κ-saturation property it follows that $\widehat{H}$ is a Hilbert space with respect to the inner product $(\hat{x},\hat{y}) := \text{st}(x,y)$, where $x \in \text{fin}(*H)$ and $y \in \text{fin}(*H)$ and $\hat{x} = \pi(x)$ and $\hat{y} = \pi(y)$ (see [1], Theorem 3.16.1). Since $H \subseteq \text{fin}(*H)$ (externally) and $H \cap \inf(*H) = \{0\}$, $\pi(H)$ is a closed linear subspace of $\widehat{H}$ which is isomorphic and isometric with H. It is customary to denote the copy $\pi(H)$ of H in $\widehat{H}$ again by H. Then H being identified with a closed linear subspace of $\widehat{H}$ we can decompose $\widehat{H}$ as the topological direct linear sum of H and its orthogonal complement H_0 of H in $\widehat{H}$. In symbols, $\widehat{H} = H \oplus H_0$.

By $B(H)$ we denote the Banach algebra of all the bounded linear operators defined on H and mapping H into H. The two-sided ideal of $B(H)$ consisting of the compact linear operators is denoted by $\text{Cpt}(H)$.

The κ-enlargement *$B(H)$ of $B(H)$ with respect to *$\mathcal{M}$ can be identified as the internal Banach algebra of all internal "bounded" linear operators on *H. The corresponding κ-enlargement *$\text{Cpt}(H)$ is the two-sided ideal of the internal compact linear operators defined on *H.

If $T \in \text{fin}(*B(H))$, i.e., $T \in *B(H)$ and its norm $||T|| \in \text{fin}(*\mathbb{R})$ is a limited number, (observe that we follow here the same convention of dropping the asterisk of the norm symbol), then its $\widehat{H}$-hull $\widehat{T}$, or shortly the hull of T, is defined by, for all $\hat{x} \in \widehat{H}$, $\widehat{T}(\hat{x}) := \pi(T(x))$, where $x \in \text{fin}(*H)$ and $\hat{x} = \pi(x)$. The "hull-mapping" $T \to \widehat{T}$ of $\text{fin}(*B(H))$ into $B(\widehat{H})$ with kernel $\inf(*B(H))$, the space of internal operators of *H with infinitesimal norm, is an algebraic homomorphism, i.e., $\widehat{(T+S)} = \widehat{T} + \widehat{S}$ and $\widehat{(TS)} = \widehat{T}\widehat{S}$. Because of the special role of the asterisk notation in nonstandard analysis we shall denote the adjoint operation for linear operators by the number symbol #. We recall that the hull-mapping also preserves adjoints, i.e., for all $T \in \text{fin}(*B(H))$ we have $\widehat{(T^\#)} = (\widehat{T})^\#$.

Observe that if $T \in \text{fin}(*B(H))$, then $||\widehat{T}|| = \text{st}(||T||)$. Furthermore, if $T \in B(H)$ is a (standard) bounded linear operator, then its $\widehat{H}$-hull $\widehat{T}$ is defined as the $\widehat{H}$-hull $\widehat{(*T)}$ of its *-image of T in *$B(H)$. For a (standard) bounded linear operator $T \in B(H)$ we have $||\widehat{T}|| = ||T||$ and so $\widehat{T}$ is a norm preserving extension of T defined on H to $\widehat{T}$

defined on $\widehat{H}$. The algebra of the $\widehat{H}$-hulls of the bounded linear operators defined on H into H is isomorphic and isometric with $B(H)$. The nonstandard hull $(B(\widehat{H}))\widehat{}$ of $^*B(H)$ is a closed subalgebra of $B(\widehat{H})$. Finally, we recall that for standard operators T the hull-operation preserves its spectrum $\sigma(T)$ and the approximate spectrum $\sigma_a(T)$ of T is equal to the point spectrum $\sigma_p(\widehat{T})$ of $\widehat{T}$.

3. Near-standard compact operators

Let $H \in \mathcal{M}$ be a Hilbert space. By ns($^*H$) we shall denote the space of the near-standard points of $^*H$, i.e., $a \in$ ns($^*H$) whenever there exists an element $x \in H$ such that $||a-x|| \simeq 0$ (observe that we have simplified the notation by dropping the asterisk and stated $||a - x|| \simeq 0$ rather than $^*||a -^* x|| \simeq 0$). If $a \in$ ns($^*H$), then we shall denote, as in the case of $\mathbb{R}$, by st(a) its standard part $x \in H$ which is infinitely close to a in the norm of $^*H$. Observe that ns($^*H$) $\subseteq$ fin ($^*H$) and that *if $a \in$ fin($^*H$), then $a \in$ ns($^*H$) if and only if $\pi(a) \in H$ and that in this case* st(a) = $\pi(a)$.

We recall that a linear operator T of H into H is a bounded (= norm continuous) linear operator if and only if $^*T$ maps near-standard points into near-standard points. Using his characterization of compact sets, referred to in the Introduction, Robinson (see [2], Theorem 7.1.4) derived at the following characterization of compact linear operators.

Theorem 1 (A. Robinson). *If $T \in B(H)$, then T is a compact linear operator if and only if $^*T$ (fin($^*H$)) $\subseteq$ ns ($^*H$), i.e., $^*T$ maps finite points of $^*H$ into near-standard points of $^*H$.*

Using Robinson's result we shall first provide a nonstandard proof of the following well-known (standard) result.

Theorem 2. *If $T \in B(H)$, then T is compact if and only if its adjoint $T^\#$ is compact.*

Proof. Assume that T is compact and that $a \in$ fin($^*H$). Since the norm $||a||$ is finite there exists an element $y \in H$ such that for all $x \in H$, $(y - a, x) \simeq 0$. We shall show now that $^*(T^\#)a$ is near-standard and that its standard part is $T^\# y$. If this is not the case, then there exists a positive standard number $\epsilon \in \mathbb{R}$ such that $||T^\# y -^*(T^\#)a|| > \epsilon > 0$. Then there exists an element $z \in {}^*H_1 := \{a \in {}^*H : ||a|| \leq 1\}$ such that $(z, T^\# y -^* (T^\#)a) > \epsilon/2$. Hence, $(^*Tz, y - a) > \epsilon/2$. Since T is compact, however, we have by Robinson's theorem that $^*Tz$ is near-standard, and hence, by the definition of y, $(^*Tz, y - a) \simeq 0$ contradicting $(^*Tz, y - a) > \epsilon/2 > 0$. This completes the proof.

In the following theorem we present a few other characterizations for a linear operator to be compact.

Theorem 3. *If $T \in B(H)$, then the following conditions are equivalent.*
  (i) T is a compact linear operator, i.e., $T \in Cpt(H)$.
  (ii) $\widehat{T}$ is a compact linear operator, i.e., $\widehat{T} \in Cpt(\widehat{H})$.
  (iii) $\widehat{T}(\widehat{H}) \subseteq H$.
  (iv) $\widehat{T}(H_0) = \{0\}$, where H_0 is the orthogonal complement of H in $\widehat{H}$.

Proof. (i) $\Rightarrow$ (ii). If T is compact, then by Robinson's theorem, $^*T(^*H_1)$ is an internal subset of ns($^*H$), where $^*H_1 := \{x \in {}^*H : ||x|| \le 1\}$. Since, by [1], Theorem 3.6.1, the standard part of an internal set of near-standard points in a regular topological space is compact and the fact that if $a \in $ ns($^*H$), $\pi(a) = $ st(a), it follows from $\widehat{T}(\widehat{H}_1) \subseteq $ st($^*T(^*H)$), that $\widehat{T}$ maps the unit ball $\widehat{H}_1$ of $\widehat{H}$ into a compact subset of $\widehat{H}$, and so is compact.

(ii) $\Rightarrow$ (iii). If $\widehat{T}$ is compact, then T is compact and for all $x \in$ fin($^*H$) $\widehat{T}(\pi(x)) = \pi(^*T(x)) = $ st($^*Tx$) $\in H$.

(iii) $\Rightarrow$ (iv). If $\widehat{T}(\widehat{H}) \subseteq H$, then it follows immediately that $^*T$ maps finite point of $^*H$ in near-standard points. Then, by Theorem 2, $^*(T^{\#})$ also maps finite points in near-standard points, and so we have that $(T^{\#})\widehat{)}(\widehat{H}) \subseteq H$ also holds. Consequently, the orthogonal complement H_0 of H in $\widehat{H}$ is orthogonal to the range of $(T^{\#})\widehat{)}$ as well as to the range of $\widehat{T}$. Then it follows from a well-known theorem of bounded linear operators on a Hilbert space that the closure of the range is equal to the orthogonal complement of its kernel (= null space) that $\widehat{T}(H_0) = \{0\}$ as well as $(T^{\#})\widehat{)}(H_0) = \{0\}$; and the proof is finished.

(iv) $\Rightarrow$ (i) If $\widehat{T}(H_0) = \{0\}$, then, by the definition of $\widehat{T}$, we have that $^*T($ fin$(^*H)) \subseteq$ ns($^*H$). Hence, by Robinson's Theorem 1, we conclude that T is compact. This completes the proof the theorem.

Remark. It is of interest to observe that (iv) of Theorem 3 implies that if $T \in B(H)$ is compact, then H is a reducing subspace of $\widehat{T}$ as well as of $(T^{\#})\widehat{)}$. Furthermore, it also shows that the Calkin algebra $B(H)/Cpt(H)$ can be represented by a Banach algebra of bounded linear operators on the Hilbert space H_0.

We shall now turn to the problem, mentioned in the Introduction, of characterizing the near-standard elements of $^*(Cpt(H))$. For the sake of simplicity we shall introduce the following terminology.

Definition 1. *An internal operator $S \in {}^*B(H)$ is called* near-standard compact *or shortly* ns-compact *if there exists a standard compact operator $T \in Cpt(H)$ such that $||S - {}^*T|| \simeq 0$, i.e., $S \in $ ns($^*Cpt(H)$).*

Recalling that two internal linear operators $S_1, S_2 \in {}^*B(H)$ satisfy $||S_1 - S_2|| \simeq 0$ if and only if $\widehat{S}_1 = \widehat{S}_2$, it follows immediately from the definition that if $S \in {}^*B(H)$ is ns-compact, then $S \in $ fin($^*B(H)$).

In the following result we shall collect some of the immediate consequences of ns-compactness for future reference.

Theorem 4. *If $S \in {}^*B(H)$ is an internal linear operator on ${}^*H$, then the following conditions are equivalent.*
  (i) S is ns-compact.
  (ii) $S^\#$ is ns-compact.
  (iii) $S \in \text{fin}({}^*B(H))$ and the $\widehat{H}$-hull $\widehat{S}$ of S is the $\widehat{H}$-hull of a standard compact operator, and so in particular, $\widehat{S}$ and $(S^\#)\widehat{}$ are compact.
  (iv) $\widehat{S}$ is compact, $\widehat{S}(H) \subseteq H$ and $\widehat{S}(H_0) = \{0\}$.

Proof.   (i) $\Rightarrow$ (ii) There exists a linear operator $T \in \text{Cpt}(H)$ such that $\|S - {}^*T\| \simeq 0$. $T \in \text{Cpt}(H)$ implies by Theorem 2 that $T^\# \in \text{Cpt}(H)$. Then $\|S - {}^*T\| = \|S^\# - {}^*(T^\#)\|$ show that $S^\#$ is ns-compact.

  (ii) $\Rightarrow$ (iii) From $\|S^\# - {}^*T\| \simeq 0$ with $T \in \text{Cpt}(H)$ it follows that $\widehat{T} = (S^\#)\widehat{}$ and $(T^\#)\widehat{} = \widehat{S}$ and since T and $T^\# \in \text{Cpt}(H)$, the result follows from Theorem 3.

  (iv) $\widehat{S}$ being the $\widehat{H}$-hull of its standard part $T \in \text{Cpt}(H)$, the result follows again from Theorem 3.

  (iv) $\Rightarrow$ (i) Denote the restriction of $\widehat{S}$ to H by T. Then since $\widehat{S}$ is compact and $\widehat{S}(H) \subseteq H$, it follows that T is a compact operator of H into H. From the hypothesis $\widehat{S}(H_0) = \{0\}$ it follows, using Theorem 3, that S and ${}^*T$ have the same $\widehat{H}$-hull. Hence, for all $x \in {}^*H$, we have $\|{}^*Tx - Sx\| \simeq 0$ which implies that $\|{}^*T - S\| \simeq 0$, i.e., S is ns-compact; and the proof of the theorem is finished.

In the next section we shall present a more intrinsic characterization of ns-compactness.

4. Internal operators compact in the sense of Robinson

If $S \in {}^*B(H)$ is an internal ns-compact operator, then there exists a compact operator $T \in \text{Cpt}(H)$ such that $\|S - {}^*T\| \simeq 0$.

The Theorem of Robinson (Theorem 1) implies that if $x \in \text{fin}({}^*H)$, then ${}^*Tx \in \text{ns}({}^*H)$. From $\|Sx - {}^*Tx\| \leq \|S - {}^*T\| \|x\| \simeq 0$ it then follows that also S maps finite points in near-standard points.

Definition 2. *An internal bounded linear operator $S \in {}^*B(H)$ is called compact in the sense of Robinson or shortly R-compact if S maps finite points of ${}^*H$ into near-standard points of ${}^*H$.*

In view of this definition, Robinson's characterization of compact operators can now be stated in the form: *A bounded linear operator $T \in B(H)$ is compact if and only if ${}^*T$ is R-compact.*

Theorem 5.
  (i) If $S \in {}^*B(H)$ is R-compact, then $S \in \text{fin}({}^*B(H))$ and $\widehat{S}$ is compact and $\widehat{S}(\widehat{H}) \subseteq H$.
  (ii) If $S \in {}^*B(H)$ and is ns-compact, then S is R-compact.
  (iii) $S \in \text{ns}({}^*B(H))$ if and only if S is R-compact and S is near-standard.

Proof. (i) If S is R-compact, then $S(^*H_1)$ is an internal subset of $\text{ns}(^*H)$, and so, by Theorem 3.1 of [1], its standard part is a compact subset of H. Hence, there exists a positive real number m such that for all $x \in {}^*H_1$, we have $||Sx|| \leq m$, and so $||S||$ is finite. From $\text{st}(S(^*H))$ is compact it follows, as in the proof of Theorem 3, that $\hat{S}$ is compact. Finally, $S(\text{fin}(^*H)) \subseteq \text{ns}(^*H)$ implies that $\hat{S}(\widehat{H}) \subseteq H$.

(ii) See remark following Definition 2.

(iii) If $S \in \text{ns}(^*B(H))$, then there exists a linear operator $T \in B(H)$ such that $||S - {}^*T|| \simeq 0$. If now also S is R-compact, then it follows immediately that $^*T$ is R-compact as well. Then, by Robinson's Theorem, T is compact and the proof is finished.

Part (ii) and (iii) of the preceding theorem seem to suggest that ns-compactness and R-compactness may be equivalent properties. In the following example we will show that this is not the case and it will show further light on the relationship between these two properties.

Example. Let H be a separable Hilbert space and let $\{e_m : m = 1, 2, \cdots\}$ be an orthonormal basis of H. For each infinitely large natural number $\omega \in {}^*\mathbb{N}^\infty$ we define the following internal rank 1 operator S_ω as follows: For all $x \in {}^*H$, $S_\omega(x) := (x, e_\omega)e_1$. Then S_ω is an internal rank 1 operator and so also compact internally as an operator on $^*H$. First we observe that S_ω is R-compact. Indeed, if $x \in \text{fin}(^*H)$, then $S_\omega(x) \simeq \text{st}(x, e_\omega)e_1 \in H$. We shall now show that S_ω is not ns-compact, and even stronger that $S_\omega \notin \text{ns}(^*B(H))$. To this end, observe that for all $\hat{x} \in \widehat{H}$ we have $\hat{S}_\omega(\hat{x}) = \text{st}(x, e_\omega)e_1$, where $\pi(x) = \hat{x}$, $x \in \text{fin}(^*H)$ and so $\hat{S}_\omega \in B(\widehat{H})$ and $\hat{S}_\omega$ is a rank 1 operator on $\widehat{H}$. It is easy to see that for all $x \in H$, $\hat{S}_\omega(x) = \{0\}$. Indeed, if $x \in H$, then $(x, e_\omega) \simeq 0$. Hence, there is no non-zero standard operator $T \in B(H)$ such that $\hat{S}_\omega = \hat{T}$. This proves that S_ω is not near-standard and so certainly not ns-compact. Summarizing, the internal rank 1 operator S_ω is R-compact, but not ns-compact, showing that an R-compact internal operator need not be a near-standard operator. The adjoint $S_\omega^\#$ of S_ω is the internal operator $S_\omega^\#(x) = (x, e_1)e_\omega$, $x \in {}^*H$ and also of rank 1.

Since $e_\omega \notin \text{ns}(^*H)$ we have an example of a rank 1 operator on $^*H$ that is not R-compact, but it is an internally compact operator of $^*H$. This also shows that the property of an internal operator to be R-compact is not preserved under the adjoint operation.

Theorems 4 and 5 suggest the following result.

Theorem 6. *If $S \in {}^*B(H)$ is an internal linear operator on $^*H$, then S is ns-compact if and only if S is R-compact and $\hat{S}(H_0) = \{0\}$, where H_0 is the orthogonal complement of H in $\widehat{H}$.*

Proof. By Theorem 5 if S is ns-compact, then S is R-compact. By Theorem 4 (iv), $\hat{S}(H_0) = \{0\}$.

Conversely, if S is R-compact, then by (i) of Theorem 5, we have that $\hat{S}$ is compact. If we denote the restriction of $\hat{S}$ to H by T, then by the hypothesis that $\hat{S}(H_0) = \{0\}$, T is a compact operator from H into H, i.e., $T \in \mathrm{Cpt}(H)$. Since T is compact we have by Theorem 3 that $(^*T)\hat{}$ is T on H and 0 on H_0, i.e., S and $^*T$ have the same $\widehat{H}$-hull. Hence, for all $x \in {^*H}$, $||^*Tx - Sx|| \simeq 0$. So $||^*T - S|| \simeq 0$; and the proof is complete.

The property of ns-compactness is preserved under the adjoint operation, whereas as we have seen in the Example, the property of being R-compact is not. For this reason the following theorem is not without interest.

Theorem 7. *If $S \in {^*B(H)}$ is an internal operator, then S is ns- compact if and only if S and its adjoint $S^\#$ are R-compact.*

Proof. By Theorem 4 (ii) if S is ns-compact, then $S^\#$ is ns-compact, and so by Theorem 5, S and $S^\#$ are R-compact.

Conversely, assume that S and $S^\#$ are R-compact. Then in order to show that S is ns-compact we have to show, by Theorem 4 (iv), that $\hat{S}$ is compact, $\hat{S}(\widehat{H}) \subseteq H$ and $\hat{S}(H_0) = \{0\}$. To this end, observe first that by Theorem 5 (i), $\hat{S}$ and $(S^\#)\hat{}$ are both R-compact and $\hat{S}(\widehat{H}) \subseteq H$ and $(S^\#)\hat{}(\widehat{H}) \subseteq H$. To finish the proof we have to show that $\hat{S}(H_0) = \{0\}$. To this end, let $\hat{y}_0 \in H_0$ and let $y_0 \in \mathrm{fin}(^*H)$ such that $\hat{y}_0 = \pi(y_0)$. Then for all $x \in H$ we have that $(x, y_0) \simeq 0$. Since $S^\#$ is R-compact we have that for all $z \in \mathrm{fin}(^*H)$, $S^\# z \in \mathrm{ns}(^*H)$. Hence, for all $z \in \mathrm{fin}(^*H)$ we have that $(S^\# z, y_0) \simeq 0$. Then $(S^\# z, y_0) = (z, Sy_0)$ implies that for all $z \in \mathrm{fin}(^*H)$ $(z, Sy_0) \simeq 0$. Since $y_0 \in \mathrm{fin}(^*H)$ and S is R-compact it follows that $Sy_0 \in \mathrm{ns}(^*H)$. Taking $z := Sy_0$ in $(z, Sy_0) \simeq 0$ we obtain $||Sy_0||^2 = (Sy_0, Sy_0) \simeq 0$. Hence, $\hat{S}\hat{y}_0 = 0$, which completes the proof.

As an immediate corollary we have the following result.

Theorem 8. *If $S \in {^*B(H)}$, then S is ns-compact with a self-adjoint (normal) compact standard part operator if and only if S is R-compact and $||S - S^\#|| \simeq 0$ $(||SS^\# - S^\# S|| \simeq 0)$.*

Remark. If $S \in {^*B(H)}$ and S is R-compact and self-adjoint, then the spectrum of S is an internal sequence and a near-standard element of the κ-enlargement $^*c_0$ of the Banach space c_0 of the null-sequences.

References

[1] W. A. J. Luxemburg, *A general theory of monads*, Applications of Model Theory to Algebra, Analysis and Probability, (W. A. J. Luxemburg ed.), Holt, Rinehart and Winston, New York, 1969.

[2] A. Robinson, *Non-standard Analysis*, North-Holland, Amsterdam, 1966.
[3] K. Stroyan and W. A. J. Luxemburg, *Introduction to the Theory of Infinitesmals*, Academic Press, New York, 1976.

Caltech
Pasadena CA 91125-0001
USA

E-mail lux@cco.caltech.edu

Yvette Feneyrol-Perrin
Discrete Fredholm's equations

1. Introduction

The theory of integral equations, especially of linear integral equations, may be considered as an extension of linear algebra as well as precursory of functional analysis.

«The analogies between Algebra and Analysis and the idea to consider the functional equations - i.e. equations where the unknown is a function - as borderline cases of algebraic equations goes back to the beginning of infinitesimal calculus, which, in a certain way, fulfills a need of generalization from finite to infinite» (my translation from [1][1]).

This point of view explains that Pierre Cartier and myself, pursuing our ambitious project of «rebuilding mathematics on finitary bases» [3] have been interested by this theory and have attempted to apply the modern infinitesimal methods to it.

Let

(1) $$f(x) - \lambda \int_0^1 K(x,y) f(y) dy = g(x)$$

be the classical Fredholm equation, where
- K is a complex continuous function on $[0,1] \times [0,1]$,
- g is a complex continuous function on $[0,1]$.

A solution to this equation is a complex continuous function f on $[0,1]$ which satisfies (1) for any $x \in [0,1]$.

Infinitesimal methods seem to be very well adapted to solve such an equation in the following way. We consider a near interval $T = \{0 = x_0 < x_1 < \ldots < x_N = 1\}$ where $x_{i+1} - x_i$ is infinitesimal for any integer $i < N$, and we substitute to the equation (1) the following system of N linear equations:

(2) $$f(x_i) - \lambda \sum_{j=0}^{N-1} K(x_i, x_j) f(x_j) (x_{j+1} - x_j) = g(x_i), \ 0 \leq i < N.$$

Denote

$$A = \Big(K(x_i, x_j) (x_{j+1} - x_j) \Big)_{\substack{0 \leq i < N \\ 0 \leq j < N}}$$

the matrix of the system (2) and Y the vector $(g(x_i))_{0 \leq i < N}$ of $\mathbb{C}^N$.

Thus, system (2) has the following matricial form:

(3) $$(I - \lambda A) X = Y.$$

---

[1] «Les analogies entre Algèbre et Analyse et l'idée de considérer les équations fonctionnelles, c'est à dire où l'inconnue est une fonction, comme cas limites d'équations algébriques remontent aux débuts du calcul infinitésimal, qui en un sens répond à un besoin de généralisation du fini à l'infini».

First, we will solve equation (3) in a more general context, where :
- The $N \times N$ matrix A has a limited nuclear norm,
- The vector Y of $\mathbb{C}^N$ has a limited supremum norm,
- A solution of (3) is a vector X of $\mathbb{C}^N$ which has a limited supremum norm.

Then, we will assume that A is an integral matrix, i.e. A is defined by a S-continuous kernel $K(x,y)$ on a near interval, and we will return to Fredholm equation.

2. Nuclear norm of a matrix

Let $A = (a_{ij})_{1 \leq i,j \leq N}$ be a $N \times N$ matrix, (N can be limited or not).
Denote by A_j the jth column vector of A and by $\|A_j\|_\infty$ its supremum norm, i.e.

$$\|A_j\|_\infty = \text{Max}\{|a_{ij}| : 1 \leq i \leq N\}.$$

The nuclear norm of the matrix A is defined by

$$\|A\| = \sum_{j=1}^{N} \|A_j\|_\infty.$$

It is easy to verify that the nuclear norm satisfies the following properties :

2.1 Proposition
Let A and B be two $N \times N$ matrix, let X be a vector of $\mathbb{R}^N$ and $\|X\|_\infty$ its supremum norm. We have :
$$\|AX\|_\infty \leq \|A\| \|X\|_\infty$$
$$\|AB\| \leq \|A\| \|B\|$$
If $\tau(A)$ denotes the trace of the matrix A, then
$$|\tau(A)| \leq \|A\|.$$

2.2 Proposition
The matrix A has a limited nuclear norm iff its coefficients a_{ij} can be written as
$a_{ij} = c_{ij} \gamma_j$, where $|c_{ij}| \leq 1$, $\gamma_j \geq 0$ and $\sum_{j=1}^{N} \gamma_j$ is limited.

Proof. Assume that $\|A\|$ is limited. Put $\gamma_j = \|A_j\|_\infty$. Then $\gamma_j \geq 0$, $|c_{ij}| = \dfrac{|a_{ij}|}{\|A_j\|} \leq 1$ and $\sum_{j=1}^{N} \gamma_j = \|A\|$ is limited.

Reciprocally if A satisfies the proposition, then $\|A\| \leq \sum_{j=1}^{N} \gamma_j$ is limited.

3. Fredholm's determinant

Let $A = (a_{ij})_{1 \leq i,j \leq N}$ be a $N \times N$ matrix. We denote by

$$\Delta_A \begin{pmatrix} i_1 & \cdots & i_p \\ j_1 & \cdots & j_p \end{pmatrix} \quad \text{the determinant} \quad \begin{vmatrix} a_{i_1 j_1} & \cdots & a_{i_1 j_p} \\ \vdots & & \\ a_{i_p j_1} & \cdots & a_{i_p j_p} \end{vmatrix}.$$

The determinant of the matrix $I + \lambda A$ has the following series development :

$$\det(I + \lambda A) = \sum_{p=0}^{N} \sigma_p(A) \, \lambda^p$$

where

$$\sigma_p(A) = \frac{1}{p!} \sum_{i_1, \cdots, i_p} \Delta_A \begin{pmatrix} i_1, & \cdots, & i_p \\ i_1, & \cdots, & i_p \end{pmatrix}, (i_1, \cdots, i_p) \in \{1, \cdots, N\}^p.$$

By Hadamard's inequality, we show that

(4) $$|\sigma_p(A)| \leq \frac{1}{p!} p^{\frac{p}{2}} \|A\|^p.$$

3.1 Proposition
If the nuclear norm of A is limited, then $\det(I + A)$ is limited and the series $\sum_{p=0}^{N} \frac{1}{p!} \sigma_p(A)$ is S-convergent, i.e. for all unlimited integers $\mu \leq N$, $\sum_{p=\mu}^{N} \frac{1}{p!} \sigma_p(A)$ is infinitesimal.

Proof. It is a direct consequence of inequality (4).

3.2 Proposition
The function $A \to \det(I + A)$ is S-continuous at any point A whose nuclear norm is limited, in other words :
If $\|A\|$ is limited and $\|A - B\|$ is infinitesimal then $\det(I + A) - \det(I + B)$ is infinitesimal.

Proof. Suppose $\|A\|$ limited.

i) Let us prove that, for any limited integer n, the function σ_n is S-continuous at A. For this, we use Waring's formula which expresses $\sigma_n(A)$ by means of the trace of the powers of A :

$$\sigma_n(A) = n! \sum_{\alpha_1, \cdots, \alpha_n} (-1)^{n + \alpha_1 + \cdots + \alpha_n} \frac{\tau_1(A) 1^{\alpha_1} \cdots \tau_n(A)^{\alpha_n}}{\alpha_1! \alpha_2! 2^{\alpha_2} \cdots \alpha_n! n^{\alpha_n}}.$$

$\tau_k(A)$ denotes the trace of the matrix A^k and the sum $\sum$ is taken over the set of all the families of integers $\alpha_1, \ldots, \alpha_n$ such that $\alpha_1 + 2\alpha_2 + \ldots + n\alpha_n = n$. If n is limited, there is a limited number of terms in the sum $\sum$; moreover if k is limited it follows from proposition 2.1 that the mapping $B \to \tau_k(B)$ is S-continuous at any point B whose nuclear norm is limited. Therefore, the mapping σ_n is S-continuous at A.

ii) Let B be a matrix such that $\|B - A\|$ is infinitesimal. For any limited integer n, we have
$$\sum_{p=0}^{n} \frac{\sigma_p(A)}{p!} - \sum_{p=0}^{n} \frac{\sigma_p(B)}{p!} \simeq 0.$$
Suppose N unlimited. By Robinson's lemma, this relation is true for an unlimited integer $\nu \leq N$. The series $\sum_{p=0}^{N} \frac{\sigma_p(A)}{p!}$ and $\sum_{p=0}^{N} \frac{\sigma_p(B)}{p!}$ are S-convergent, then
$$\sum_{p=\nu}^{N} \frac{\sigma_p(A)}{p!} \simeq 0 \text{ and } \sum_{p=\nu}^{N} \frac{\sigma_p(B)}{p!} \simeq 0, \text{ so } \sum_{p=0}^{N} \frac{\sigma_p(A)}{p!} - \sum_{p=0}^{N} \frac{\sigma_p(B)}{p!} \simeq 0.$$

4. Fredholm's alternative

Let A be a $N \times N$ matrix with a limited nuclear norm.
4.1 Proposition
The two following properties are equivalent :
i)   *$det\,(I + A)$ is not infinitesimal,*
ii)   *There exists a matrix B, with a limited norm, such that*
$$(I + A)(I + B) = (I + B)(I + A) = I.$$

Proof.   ii) $\Longrightarrow$ i). From ii) it follows that
$$det\,(I + A) = [det\,(I + B)]^{-1}$$
and $det\,(I + B)$ is limited. Therefore $det\,(I + A)$ is not infinitesimal.
     i) $\Longrightarrow$ ii). Since the norm of A is limited, by proposition 2.2 we can express its coefficients a_{ij} in the following way :
$$a_{ij} = c_{ij}\gamma_j \text{ with } |c_{ij}| \leq 1,\ \gamma_j \geq 0 \text{ and } a = \sum_{j=1}^{N} \gamma_j \text{ limited}.$$

The matrix $I + A$ is invertible. Let B be the matrix such that $(I + A)(I + B) = I$. Then
$$B = -\frac{1}{\Delta}(b_{ij})_{1 \leq i,j \leq N} \text{ with } \Delta = det\,(I + A)$$
where
$$b_{ij} = \sum_{p=0}^{N-1} \frac{1}{p!} \sum_{i_1,\ldots,i_p} \Delta_A \begin{pmatrix} i & i_1 & \cdots & i_p \\ j & i_1 & \cdots & i_p \end{pmatrix},$$

$$\Delta_A \begin{pmatrix} i & i_1 & \cdots & i_p \\ j & i_1 & \cdots & i_p \end{pmatrix} = \gamma_j \gamma_{i_1} \cdots \gamma_{i_p} \begin{pmatrix} c_{ij} & c_{ii_1} & \cdots & c_{ii_p} \\ \vdots & & & \vdots \\ c_{i_pj} & c_{i_pi_1} & \cdots & c_{i_pi_p} \end{pmatrix}$$

$$= \gamma_j \gamma_{i_1} \cdots \gamma_{i_p} \Delta_C \begin{pmatrix} i & i_1 & \cdots & i_p \\ j & i_1 & \cdots & i_p \end{pmatrix}.$$

From Hadamard's inequality it follows that

$$\left|\Delta_C \begin{pmatrix} i & i_1 & \cdots & i_p \\ j & i_1 & \cdots & i_p \end{pmatrix}\right| \leq (p+1)^{\frac{p+2}{2}}.$$

Let us write
$$b_{ij} = d_{ij}\gamma_j.$$

Then we get

$$|d_{ij}| \leq \sum_{p=0}^{N} \frac{1}{p!}(p+1)^{\frac{p+2}{2}} \sum_{i_1\ldots i_p} \gamma_{i_1}\cdots\gamma_{i_p} = \sum_{p=0}^{N}\frac{1}{p!}(p+1)^{\frac{p+2}{2}} a^p = d.$$

Since $a = \sum_{j=1}^{N} \gamma_j$ is limited, so is d. Suppose now that Δ is not infinitesimal, and put

$$\frac{b_{ij}}{\Delta} = \left(\frac{|\Delta|}{\Delta}\frac{d_{ij}}{d}\right)\left(\frac{d}{|\Delta|}\gamma_j\right).$$

Then $\left|\frac{|\Delta|}{\Delta}\frac{d_{ij}}{d}\right|$ is less than 1 and $\sum_{j=1}^{N}\frac{d}{|\Delta|}\gamma_j$ is limited. By proposition 2.2 we conclude that the matrix B has a limited nuclear norm.

4.2 Proposition
1) If $\det(I+A)$ is not infinitesimal, the equation
$$(I+A)X = C$$
has a unique solution, and its supremum norm is limited.
2) If $\det(I+A)$ is infinitesimal, there exists a vector $X \in \mathbb{C}^N$ such that:
  i) $\|X\|_\infty$ is not infinitesimal,
  ii) $\|(I+A)X\|_\infty$ is infinitesimal.
Conversely, if there exists a vector X which satisfies the properties i) and ii), then $\det(I+A)$ is infinitesimal.

Proof.
1) Let B be a matrix such that $(I+B)(I+A) = I$ and $\|B\|$ is limited. The unique solution of $(I+A)X = C$ is :
$$X = (I+B)C = C + BC.$$
Then
$$\|X\|_\infty \leq \|C\|_\infty + \|B\|\|C\|_\infty = (1+\|B\|)\|C\|_\infty.$$
So $\|X\|_\infty$ is limited.
2) In order to prove ii) we consider the complex function : $z \to \det(I+zA)$.
It is analytic over $\mathbb{C}$ and takes limited value at any limited point. So, if $\det(I+A)$ is infinitesimal there exists $z \simeq 1$ such that $\det(I+zA) = 0$.
Let X be a vector of $\mathbb{C}^N$ such that :
$$(I+zA)X = 0 \quad \text{and} \quad \|X\|_\infty = 1.$$

Then
$$(I+A)X = (I+zA)X + (1-z)AX = (1-z)AX$$
$$\|(I+A)X\|_\infty \leq |1-z|\,\|A\|\,\|X\|_\infty = |1-z|\,\|A\| \simeq 0.$$

Reciprocally, let X be a vector whose supremum norm is appreciable and such that $\|(I+A)X\|_\infty$ is infinitesimal. Assume that $\det(I+A)$ is not infinitesimal. There exists a matrix B such that $\|B\|$ is limited and
$$X = (I+B)(I+A)X = (I+A)X + B(I+A)X.$$

It follows that
$$\|X\|_\infty \leq \|(I+A)X\|_\infty (1+\|B\|) \simeq 0.$$

This is contradictory with our hypothesis. Hence $\det(I+A)$ is infinitesimal. We revert now to the study of the classical Fredholm's equation.

5. Integral operators

5.1 Definitions

Let $T = \{0 = x_0 < x_1 < \ldots < x_N = 1\}$ be a near interval and $T^- = T \setminus \{1\}$. If x is a point of T^- we denote $x + dx$ its **successor** in T. Let $K : T \times T \to \mathbb{C}$ be a S-continuous function which takes limited values.

Denote $\mathcal{F}$ the set of all the internal mappings from T to $\mathbb{C}$. We say that the mapping $A_K : \mathcal{F} \to \mathcal{F}$ defined by

$$A_K f(x) = \int_T K(x,y)\, f(y)\, dy$$

$$= \sum_{y \in T^-} K(x,y)\, f(y)\, dy$$

is an **integral operator with kernel** K. We will identify this operator with its matrix

$$A_K = (K(x,y)dy)_{(x,y) \in T^- \times T^-}$$

Remark : If K takes limited values on $T \times T$, the nuclear norm of the matrix A_K is limited. Indeed

$$\|A_K\| = \sum_{y \in T^-} Max\{|K(x,y)|dy,\ x \in T^-\}$$

$$\leq |K|_\infty \sum_{y \in T^-} dy = \|K\|_\infty.$$

104

5.2 Fredholm's determinant of an integral operator

For $x_1, \ldots, x_p, y_1, \ldots, y_p$ in T^-, denote

$$K\begin{pmatrix} x_1, \ldots, x_p \\ y_1, \ldots, y_p \end{pmatrix} = \det \begin{pmatrix} K(x_1, y_1) & \cdots & K(x_1, y_p) \\ \vdots & & \vdots \\ K(x_p, y_1) & \cdots & K(x_p, y_p) \end{pmatrix}.$$

The determinant of the matrix $I + \lambda A_K$ is limited for any limited λ and admits the following S-convergent development:

$$\det(I + \lambda A_K) = \sum_{p=0}^{N} \sigma_p(K) \lambda^p$$

where

$$\sigma_p(K) = \frac{1}{p!} \sum_{x_1, \ldots, x_p} K\begin{pmatrix} x_1, \ldots, x_p \\ x_1, \ldots, x_p \end{pmatrix} dx_1 \ldots dx_p.$$

For each limited p, σ_p is infinitesimally near the integral

$$\frac{1}{p!} \int_T \cdots \int_T K\begin{pmatrix} x_1, \ldots, x_p \\ x_1, \ldots, x_p \end{pmatrix} dx_1 \ldots dx_p.$$

Hence $\det(I + \lambda A_K)$ is equal to the classical Fredholm's determinant of the operator A_K to within an infinitesimal.

It is easy to prove the following proposition:

5.3 Proposition

1) Let K and K' be two S-continuous kernels on $T \times T$, with limited values. If $\|K - K'\|_\infty \simeq 0$, then $\det(I + \lambda A_K) \simeq \det(I + \lambda A_{K'})$.

2) If T and T' are two near-intervals with 0 and 1 as extremities, T' being contained in T, if K is a S-continuous kernel on T and if K' is its restriction to T', then

$$\det(I + A_K) \simeq \det(I + A_{K'}).$$

6. Fredholm's equation

6.1 Proposition

Let K be a S-continuous Kernel on $T \times T$ with limited values. There exists a limited S-continuous kernel H on $T \times T$ such that

$$(I + A_K)(I + A_H) = (I + A_H)(I + A_K) = I$$

if and only if $\det(I + A_K)$ is not infinitesimal.

Proof. The matrix A_H is defined by

$$A_H = -\frac{1}{\Delta}(B(x,y)dy)_{(x,y) \in T^- \times T^-}$$

with
$$\Delta = \det(I + A_K),$$
$$B(x,y) = \sum_{p=0}^{N-1} \frac{1}{p!} \sum_{(x_1,\ldots,x_p)\in T^p} K\begin{pmatrix} x, x_1 \ldots x_p \\ y, x_1 \ldots x_p \end{pmatrix} dx_1 \ldots dx_p.$$

To prove that B is a limited, S-continuous function on $T \times T$, we use again the Hadamard's inequality :
$$\left| K\begin{pmatrix} x & x_1 \ldots x_p \\ y & x_1 \ldots x_p \end{pmatrix} \right| \leq \|K\|_\infty^{p+1} (p+1)^{\frac{p+1}{2}}.$$

So
$$|B(x,y)| \leq \sum_{p=0}^{N-1} \frac{1}{p!}(p+1)^{\frac{p+1}{2}} \|K\|_\infty^{p+1}$$
is limited.

For each limited integer p, the function $(x,y) \to K\begin{pmatrix} x & x_1 \ldots x_p \\ y & x_1 \ldots x_p \end{pmatrix}$ is S-continuous. Moreover the series which defines $B(x,y)$ is S-convergent for each $(x,y) \in T \times T$, therefore its sum $B(x,y)$ is S-continuous.

6.2 Proposition
If the function g is limited and S-continuous on T and if $\det(I + A_K)$ is not infinitesimal, the Fredholm's equation
$$(I + A_K)f = g$$
has a unique solution f which is S-continuous and limited.

Proof. Let H be the S-continuous kernel on $T \times T$ which is defined by proposition 6.1. The solution of the Fredholm's equation is
$$f = (I + A_H)g.$$

6.3 Proposition
If $\det(I + A_K)$ is infinitesimal, -1 is a near eigenvalue of A_K, i.e. there exists a function f which is S-continuous, with an appreciable supremum norm and such that $\|(I + A_K)f\|_\infty$ is infinitesimal.

Proof. It is a direct consequence of Proposition 4.2.

References

[1] Bourbaki, N. *Espaces vectoriels topologiques*, Hermann, 1967.
[2] Cartier, P. *A course on determinants*.
[3] Cartier, P. and Feneyrol-Perrin, Y. *Methodes infinitésimales appliquées au calcul des probabilités*, Lecture Notes, Springer (to appear).
[4] Goursat, E. *Cours d'Analyse Mathématiques*, Tome III, Gauthier-Villars, 1915.
[5] Grothendieck, A. *Produits tensoriels topologiques et espaces nucléaires*, Mémoire de l'A.M.S, n° 16.
[6] Grothendieck, A. *La théorie de Fredholm*, Bull. Soc. math. France, 84, 1956.
[7] Hochstadt, H. *Integral Equations*, Wiley classics Library, 1989.

[8] Ringrose, J. *Compact non-self-adjoint operators*, Van Nostrand Reinhold Company London, 1971.
[9] Robinson, A. *Non-standard Analysis*, North-Holland publishing company, Amsterdam, 1966.
[10] Simon, B. *Trace Ideals and their applications*, Camdbridge University Press, 1979.

Université Blaise Pascal—Clermont-Ferrand
63117 Aubiere Cedex
France

R. F. Hoskins
Nonstandard theory of generalized functions

1. Introduction

1.1.

NSA has been in the public domain for over a quarter of a century — in effect since the appearance of Abraham Robinson's seminal text [1] in the 1960's. It might have been expected that by this time it would have become generally accepted as a valuable and important tool, not only for the working mathematician but also for the general user of mathematics (in particular for the average engineer or physicist). As a matter of hard fact this has not happened. For example, as recently as March 1994, the Institute of Mathematics and its Applications (IMA) held a conference on "The Mathematical Education of Engineers" at Loughborough University, England. The central theme was the increasingly difficult problem of teaching mathematics to an engineering audience, and the SEFI (Société Européene pour la Formation des Ingénieurs) presented a possible "Core Curriculum in Mathematics for the European Engineer". Regrettably no mention of NSA appeared in the latter document and the overwhelming majority of the participants were almost wholly unaware of NSA and of its potential value in this context.

Much of the original interest in NSA focused on the possibility of using it to present a Leibnitzian approach to the calculus, and thereby to simplify the teaching of mathematics at an elementary level. Those participants at the IMA conference who had heard of NSA at all regarded it simply in this light. However this is not the only route by which NSA might become generally known, nor perhaps the most convenient or immediately useful. There is a vast potential audience of engineers, physicists, etc. who are already acquainted with conventional forms of elementary calculus, but who would still benefit from exposure to nonstandard methods. In particular, engineers and physicists both need an early introduction to generalized functions, more especially to the delta function and its derivatives. Standard approaches are still likely to be found too difficult for the comfort of most students. A nonstandard approach to generalized functions can provide both a solution to a real teaching problem and a salutary introduction to NSA as a general technique.

1.2. Standard Theories of Generalized Functions

A rigorous theory of the delta function has, of course, been in existence (at least in so far as mathematicians are concerned) since the publication of Schwartz's theory of distributions [2] in 1950/51. In the Schwartz theory a distribution on $\mathbb{R}$ is identified as a linear and continuous functional on the space $\mathcal{D}(\mathbb{R})$ of infinitely differentiable functions of compact support, that is as a member of the topological dual $\mathcal{D}'(\mathbb{R})$. Then each function f in $\mathcal{C}^0(\mathbb{R})$ defines a (regular) distribution $\mu_{(f)}$ as a functional through

$$<\mu_{(f)}, \varphi> := \int_{-\infty}^{+\infty} \varphi(x) f(x)\, dx, \quad \forall \varphi \in \mathcal{D}(\mathbb{R})$$

and the derivative of a distribution μ is defined by

$$<D\mu, \varphi> := <\mu, -\varphi'>.$$

In particular, the Dirac delta distribution δ is defined as the second derivative of the continuous function x_+: $\delta = D^2 x_+$. In the Schwartz model it appears as the functional on $\mathcal{D}(\mathbb{R})$ which carries each $\varphi \in \mathcal{D}(\mathbb{R})$ into the number $\varphi(0)$.

However the Schwartz theory was a fairly difficult subject for the non-mathematician when it first appeared and broadly speaking has remained so ever since, despite many attempts to make it more palatable to a wider audience. The most well known, and perhaps the most readily comprehensible, alternative definitions of distributions are those based on the definition of equivalence classes of sequences of ordinary functions. In 1957 Mikusinski and Sikorski published a sequential account of the theory of distributions in a book [3] which was intended to be a genuinely elementary introduction to the subject. It was subsequently extended in collaboration with Antosik [4] to include more advanced material. A little later, in 1958, Sir James Lighthill's famous elementary text [5] appeared. This offered a sequential treatment which was specifically designed "to be intelligible both to undergraduates and to working mathematicians whose study of analysis may not have gone much beyond the undergraduate level".

Lighthill's version of the theory of distributions was based on the ideas originally proposed by George Temple [6, 7]. It was taken up and expanded by D. S. Jones to form the magisterial and comprehensive account given in his 1966 text [8] (later extended in the edition [9] of 1982).

Rather less well known are those other attempts to simplify the Schwartz theory based on the idea of treating the distribution as a generalized derivative (of a continuous function). In many respects it is this aspect of distributions which might be claimed to be the most fundamental; it is surely the most immediately satisfying from a conceptual point of view. The basic idea was proposed in a paper by H. König [10], published in 1953. But it is in the much more extensive researches, [11, 12], of Sebastião e Silva in 1954/55 and 1960 that the full development of this approach is to be found. Silva was able to produce a genuine *axiomatic* theory of distributions, and to show that the original Schwartz account could be regarded as simply one of the possible models. For simplicity we consider here only finite-order distributions on $\mathbb{R}$ (more generally, on a non-degenerate interval $I \subset \mathbb{R}$). In the (standard) Silva approach we take "continuous function" and "derivative" as primitive notions and define (finite-order) distributions as elements of a linear space $\mathcal{C}_\infty(\mathbb{R})$ such that:

Axioms for (finite-order) distributions

S1.: If $f \in C^0(\mathbb{R})$ then f is a distribution; we write $f \equiv \mu_{(f)}$.
S2.: To each distribution μ there corresponds another distribution $D\mu$, called the *derivative* of μ, such that, if $\mu = \mu_{(f)}$ where $f \in C^1(\mathbb{R})$, then $D\mu = \mu_{(f')}$.
S3.: To each distribution μ there corresponds an integer $k \in \mathbb{N}_0$ such that $\mu = D^k \mu_{(f)}$, for some $f \in C^0(\mathbb{R})$.
S4.: For any $f, g \in C^0(\mathbb{R})$ we have $D^k \mu_{(f)} = D^k \mu_{(g)}$ if and only if $(f - g)$ is a polynomial of degree $< k$.

Silva himself put forward an entirely abstract model for these axioms in which distributions are defined as equivalence classes of ordered pairs (p, f), where $p \in \mathbb{N}_0$ and $f \in C^0(\mathbb{R})$. The most familiar model is of course the classical Schwartz one.

1.3. Naive "engineering" treatments

Even these alternative accounts of the theory of distributions have usually been considered to be too difficult for the comfort of many engineering students, and too time consuming. So-called "engineering" treatments which gloss over the mathematical difficulties are acknowledged to be not generally satisfactory, but they do have a conceptual basis which engineers and physicists find amenable. A rather uneasy compromise can be achieved by confining attention to delta functions and by accepting the fact that $\delta(x)$ represents a function only in some "symbolic" sense. It is then sufficient to devise an ad hoc set of formal rules for manipulating delta functions in algebraic expressions in combination with other, bona fide, functions.

In practice such an approach seems to work quite well for much of the time. Students are relieved to find that delta functions can be added to ordinary functions, multiplied by constants and subjected to the usual processes of the calculus; that is they can be differentiated, integrated and treated as solutions of differential equations. It is even found possible to compose them with (certain) other functions and to make sense of expressions like $\delta(\sin x)$. On the other hand it is necessary to recognize sooner or later that there are things one cannot do with delta functions. Unrestricted multiplication is the most obvious example; it is disconcerting to the student to find that the product $\delta(x).\delta(x) \equiv \delta^2(x)$ has no consistent meaning, (even in the context of distributions proper). Again, although $\delta(\sin x)$ seems to be well defined it is found not to be possible to assign a meaning to $\sin(\delta(x))$, nor to such expressions as $\exp(\delta(x))$, $\sqrt{\delta(x)}$, and so on. It is not easy to explain just why such anomalies should appear and yet to keep the discussion to an elementary level. Moreover to resolve the difficulties it seems necessary to turn to nonlinear theories of generalized functions such as those recently developed by Colombeau and others [13, 14].

These raise fresh teaching problems, especially for the broadly based audience of engineers and physicists which we presently have in mind.

2. Nonstandard treatments of distributions

2.1. Robinson's Theory of Distributions

Remarkably enough, nonstandard treatments of distributions were on the scene within a decade or so of the publication of Schwartz original text. Robinson's 1966 book on NSA contains a development of the Schwartz theory in nonstandard terms based on ideas already extant in a paper [15] published in 1961.

Robinson denotes by $\mathbf{Q}_0$ the set of all internal functions $F(x)$ on $^*\mathbb{R}$ which are *square-integrable over every finite interval, and writes

$$(F, \varphi) := {}^*\!\!\int_{-\infty}^{+\infty} F(x)\, {}^*\!\varphi(x)\, dx$$

for $F \in \mathbf{Q}_0$ and φ (standard) in $\mathcal{D}(\mathbb{R})$. He denotes by $\mathbf{Q}_1$ the set of all functions F in $\mathbf{Q}_0$ such that $(F, \varphi) \approx 0$ for every function $\varphi \in \mathbf{Q}_0$. Then each equivalence class $\alpha \equiv [F]$ in the quotient $\mathbf{Q}_0/\mathbf{Q}_1$ defines a linear functional on $\mathcal{D}(\mathbb{R})$ which Robinson refers to as a *pre-distribution* (unfortunately, as will be seen later, this term is used elsewhere in another sense). Those equivalence classes $[F]$ which define a *continuous* functional on $\mathcal{D}(\mathbb{R})$, and which may therefore be identified with Schwartz distributions, are referred to by Robinson as *Q-distributions*. The elements of the equivalence classes which represent distributions are always *functions* in the proper sense of the word (albeit functions defined on, and taking values in, $^*\mathbb{R}$); this, as Robinson remarks, ought to make the theory more intuitive. Moreover the difficulty over multiplication is seen to have a natural explanation. In general, if α and β are arbitrary pre-distributions there is no guarantee that all the products of the form $F(x)G(x)$, where F and G are internal functions belonging to α and β respectively, will belong to a uniquely defined pre-distribution which we could call the product of α and β.

Despite these advantages there has been little enthusiasm to adopt this approach as a means of introducing generalized functions to a broadly based audience. This may be partly because Robinson's theory is based on the original Schwartz conception of a distribution as a functional, but is mainly because the approach to NSA in Robinson's book is undeniably difficult. However, a simple ultrapower model of the hyperreals (with ad hoc definitions of internal sets and functions) is easy to introduce to an audience with a modest acquaintance with analysis, and the sequential definition of internal functions lends itself very naturally to a nonstandard formulation of the Mikusinski/Sikorski version of the theory of distributions. Such an account was presented by W.A.J. Luxemburg [16] in 1962 (only a year later than the first description of the nonstandard Robinson theory). Moreover in 1958 there had already appeared a remarkable anticipation of the essential ideas of this approach by Schmieden and Laugwitz [17]. Since then Detlef Laugwitz has developed an alternative introduction to a simple form of NSA which is closely related to an ultrapower approach [18]. This has been exploited to great effect to provide nonstandard treatments of delta functions and other distributions, with particular concern for the needs of physicists [19].

2.2. Naive "infinitesimal" treatments

Such nonstandard treatments of generalized functions do not seem to have had any great effect on the teaching of engineering mathematics, apart perhaps from a vague renewal of faith in the infinitesimal as such. The belief persists that NSA needs an explicit background of mathematical logic and model theory, which makes it difficult of access for all but those who are trained as mathematicians. It is instructive to note the following, not untypical, reaction from a paper by B. D. Craven [20], published as late as 1986:

> "This paper presents a simple, though rigorous approach to generalized functions in which the delta function is a genuine function whose domain and range are supersets of the real line.
>
> [...] The supersets of the real line used here do not require the full panoply of A. Robinson's nonstandard analysis, using symbolic logic; a simpler process suffices, which is constructible and computable. The class of generalized functions obtained here is wide enough for physics and engineering applications [...]"

It is not our main purpose here to discuss the merits or otherwise of Craven's theory, which is based on a simple field extension of $\mathbb{R}$ using formal power series. But the implications are important. It is obvious that more needs to be done to make it widely known that NSA is not difficult to teach and that it can be introduced in comparatively simple terms. What is more, the real advantages of nonstandard methods need to be made manifest. The advantages of a nonstandard approach to delta functions are not simply that they appear as functions, in the proper sense of the word, but that their properties can be established using (legitimately) the ordinary processes of elementary calculus in new, nonstandard situations. For example, the engineer would like to be able to use the integration-by-parts formula with which he is familiar to derive the sampling property attributed to the delta function in something like the following way: for $x > 0$ and $K > 0$

$$\begin{aligned}
\int_{-\infty}^{x} \delta(t)\varphi(t)\,dt &\equiv \int_{-K}^{x} H'(t)\varphi(t)\,dt \\
&= [H(t)\varphi(t)]_{-K}^{x} - \int_{-K}^{x} H(t)\varphi'(t)\,dt \\
&= \varphi(x) - \int_{0}^{x} \varphi'(t)\,dt \\
&= \varphi(0).
\end{aligned}$$

It is one of the real advantages of NSA that such an argument can be made entirely rigorous while remaining recognizably of the same form (see, for example, Laugwitz [19]).

Nonstandard representatives of $\delta(x)$ do not merely exhibit the kind of pointwise behaviour which the delta function is popularly supposed to display; in a very real sense they have all the properties which a derivative of the Heaviside unit step function "ought to" possess. In the same way nonstandard representatives of distributions in general have a powerful intuitive appeal when viewed as derivatives of S-continuous internal functions. The axiomatic formulation of the theory of distributions given by Sebastião e Silva seems to offer a particularly simple and straightforward basis for a nonstandard theory of distributions in general.

2.3. A Nonstandard Theory of Silva Distributions

A nonstandard treatment of the Silva theory is possible even with just a simple ultrapower model $^*\mathbb{R} = \mathbb{R}^{\mathbb{N}}/U$ of the hyperreals, and ad hoc definitions of internal sets and functions. Every internal function F in $^*\mathcal{C}^\infty(\mathbf{I})$, the algebra of all infinitely *-differentiable functions on $^*\mathbf{I}$, is *-continuous on $^*\mathbf{I}$ but not necessarily S-continuous. Accordingly we define, as in Hoskins and Pinto [21], the following (external) sets of functions in $^*\mathcal{C}^\infty(\mathbf{I})$:

Definition 1. $^S\mathcal{C}(\mathbf{I}) :=$ *the $^*\mathbb{R}_b$-submodule of all functions in $^*\mathcal{C}^\infty(\mathbf{I})$ which are finite-valued and S-continuous at each point of* $\mathbf{I}$.

Definition 2. $^S\mathcal{D}(\mathbf{I}) \equiv {}^*\mathcal{D}^\infty\left\{{}^S\mathcal{C}(\mathbf{I})\right\} :=$ *the space of all internal functions $F \in {}^*\mathcal{C}^\infty(\mathbf{I})$ such that F is a finite order *-derivative of some function in* $^S\mathcal{C}(\mathbf{I})$:

$$^S\mathcal{D}(\mathbf{I}) = \bigcup_{j=0}^{\infty} {}^*D^j\left\{{}^S\mathcal{C}(\mathbf{I})\right\} = \left\{F \in {}^*\mathcal{C}^\infty(\mathbf{I}) : \left[\exists j \in \mathbb{N}_0 \Phi \in {}^S\mathcal{C}(\mathbf{I})\ F = {}^*F^j\Phi\right]\right\}$$

The $^*\mathbb{R}_b$-submodule $^S\mathcal{D}(\mathbf{I})$ contains all those functions in $\mathcal{C}^\infty(\mathbf{I})$ which are nonstandard representatives of finite order distributions on $\mathbf{I}$. In contrast to Robinson's terminology Hoskins and Pinto refer to such internal functions as *(finite-order) pre-distributions*.

Definition 3. *Two internal functions $F_1, F_2 \in {}^S\mathcal{D}(\mathbf{I})$ are said to be distributionally equivalent if there is an integer $m \in \mathbb{N}_0$ and a polynomial p_m of degree $< m$, (with coefficients in $^*\mathbb{C}$) such that*

$$\forall x \in {}^*\mathbf{I}_b\ {}^*I_a^m(F_1 - F_2)(x) \approx p_m(x),$$

where $^*I_a$ denotes the *-indefinite integral operator for a given point $a \in {}^*\mathbb{R}$, and $\mathbf{I}_b \equiv \cup_{a,b \in \mathbf{I}, a<b} {}^*[a,b]$ is the bounded part of $^*\mathbf{I}$. We then write $F_1 \Xi F_2$.

Clearly this is equivalent to the Laugwitz criterion [19] that there exist internal functions Φ, Ψ in $^S\mathcal{C}(\mathbf{I})$ such that
(i): $\Phi(x) \approx \Psi(x)$ for all finite $x \in {}^*\mathbb{R}$, and
(ii): $F_1 = {}^*D^k\Phi$, and $F_2 = {}^*D^k\Psi$, for some $k \in \mathbb{N}_0$.

Definition 4. *If $F \in {}^S\mathcal{D}(\mathbf{I})$ then $\Xi[F]$ denotes the equivalence class containing F and*

$$\Xi\mathcal{C}_\infty(\mathbf{I}) := {}^S\mathcal{D}(\mathbf{I})/\Xi \equiv {}^*\mathcal{D}^\infty{}^S\mathcal{C}(\mathbf{I})/\Xi$$

We refer to $\Xi[F]$ as a Ξ- distribution.

Definition 5. *The (differential) order of a Ξ-distribution μ in ${}^\Xi C_\infty(\mathbf{I})$ is the smallest number $r \in \mathbb{N}_0$ for which there exists a Ξ-distribution $\nu = {}^\Xi[G] \in {}^\Xi C_\infty(\mathbf{I})$ with $G \in {}^S C(\mathbf{I})$, $\mu = {}^\Xi D^r \nu$ (Here, ${}^\Xi D\nu \equiv {}^\Xi[{}^*DG]$).*

${}^\Xi C_\infty(\mathbf{I})$ is a nonstandard model for the axiomatic definition of finite-order distributions on $\mathbf{I}$ as proposed by J. S. Silva. In this model the concept of standard continuous function on $\mathbf{I}$ has been replaced by that of Ξ-equivalence class of nonstandard ${}^S C(\mathbf{I})$ internal functions, and the concept of distributional derivative by that of Ξ-derivative. The addition of Ξ-distributions and multiplication by scalars can be readily defined within the model. Equipped with these operations ${}^\Xi C_\infty(\mathbf{I})$ is a ${}^*C_{bd}$-module which may be identified with $\mathcal{D}'_{fin}(\mathbf{I})$.

Remark: There is no difficulty in extending the treatment to include distributions of arbitrary order. Let $\{K_\alpha\}_{\alpha \in A}$ be the family of all compact subintervals of $\mathbf{I}$ with more than one point and let ${}^\pi\mathcal{D}(\mathbf{I})$ denote the ${}^*\mathbb{R}_{bd}$-submodule of ${}^*C^\infty(\mathbf{I})$ such that for each $\alpha \in A$ there exists $r_\alpha \in \mathbb{N}_0$ and Φ_α in ${}^S C(K_\alpha)$ for which we have

$$\forall x \in {}^*K_\alpha \; F(x) = {}^*D^{r(\alpha)}\Phi(X).$$

Extending the distributional equivalence relation Ξ to ${}^\pi\mathcal{D}(\mathbf{I})$ in the obvious way we write ${}^\Xi C_\pi(\mathbf{I}) := {}^\pi\mathcal{D}(\mathbf{I})/\Xi$.

2.4. The differential algebra ${}^\omega G(\mathbb{R})$

A nonstandard treatment of distributions can do much more than satisfy an immediate teaching need. Once nonstandard concepts and techniques have been introduced to, say, an engineering audience and *shown to be of real value*, there is obviously plenty of scope for their further exploitation. For example, it is clear that the space ${}^*C^\infty(\mathbb{R})$ contains many examples of internal functions which are not pre-distributions but rather nonstandard representatives of other types of generalized functions. A nonstandard theory of a differential algebra of such generalized functions is given in Hoskins and Pinto [22].

Once again it is possible to develop this within the simple model of NSA used above. Let ω denote the positive infinite hyperreal number $[n] \in {}^*\mathbb{R}$. Then we can introduce the following growth constraints for functions in ${}^*C^\infty(\mathbb{R})$:

ω-moderate functions: An internal function F in ${}^*C^\infty(\mathbb{R})$ is said to be of *ω-growth* on ${}^*\mathbb{R}_{bd}$ if, for each compact $K \subset \mathbb{R}$ there exists $m \in \mathbb{N}_0$ such that $|F(x)| \leq C\omega^m$ for all $x \in {}^*K$, where C is some number in ${}^*\mathbb{R}_{bd}$. If F is any internal function in ${}^*C^\infty(\mathbb{R})$ such that, for each $k \in \mathbb{N}_0$, the kth derivative ${}^*D^k F$ is a function of ω-growth on ${}^*\mathbb{R}_{bd}$ then F is said to be an *ω-moderate function*. The set ${}^\omega C^\infty(\mathbb{R})$ of all such functions is a differential sub-algebra of ${}^*C^\infty(\mathbb{R})$.

ω-small functions: An ω-moderate function $F \in {}^*C^\infty(\mathbb{R})$ is said to be *ω-small* on ${}^*\mathbb{R}_{bd}$ if $|F(x)| < \omega^{-m}$ for every $m \in \mathbb{N}_0$.

ω-null functions: If F is any internal function in ${}^\omega C^\infty(\mathbb{R})$ such that, for each $k \in \mathbb{N}_0$, the kth derivative ${}^*D^k F$ is ω-small on ${}^*\mathbb{R}_{bd}$ then F is said to be an

ω-*null function*. The set ${}^\omega\mathcal{U}(\mathbb{R})$ of all ω-null functions is an ideal of ${}^\omega\mathcal{C}^\infty(\mathbb{R})$.

We define the differential algebra

$${}^\omega\mathbf{G}(\mathbb{R}) := {}^\omega\mathcal{C}^\infty(\mathbb{R})/{}^\omega\mathcal{U}(\mathbb{R}),$$

and call the equivalence class ${}^\omega[F]$ containing $F \in {}^\omega\mathcal{C}^\infty(\mathbb{R})$ an ω-*generalized function* (with representative F). Note that the mapping which takes $f \in \mathcal{C}^\infty(\mathbb{R})$ into ${}^\omega[F] \in {}^\omega\mathbf{G}(\mathbb{R})$ defines an embedding $\mathcal{C}^\infty(\mathbb{R}) \subset {}^\omega\mathbf{G}(\mathbb{R})$ which is an embedding of (differential) algebras.

It is not difficult to show that the algebra of Colombeau's New Generalized Functions in the simplified form due to Biagioni [23] can be embedded in ${}^\omega\mathbf{G}(\mathbb{R})$. There are a number of disconcerting, and apparently paradoxical, features of Colombeau theory which become much easier to understand in the context of ${}^\omega\mathbf{G}(\mathbb{R})$, particularly if this can be given an appropriate physical interpretation. If, for example, we are dealing with electrical engineers then there is a natural *systems – theoretic* interpretation. As an indication of what might be done to introduce NSA to engineers we conclude with a sketch of such an interpretation.

3. A Systems-theoretic Interpretation

3.1. Systems-standard treatment

An alternative (standard) model of the Silva axioms with strong physical connotations can be presented in terms of the familiar engineering concept of a *signal processing system*: Each function $f \in \mathcal{C}^0(\mathbb{R})$ defines a (linear continuous) map of $\mathcal{D}(\mathbb{R})$ into $\mathcal{C}^\infty(\mathbb{R})$ through the convolution operator $T_{(f)}$:

$$T_{(f)}\varphi(x) := \int_{-\infty}^{+\infty} \varphi(y)f(x-y)\,dy = \int_{-\infty}^{+\infty} \varphi(x-y)f(y)\,dy$$

The operator $T_{(f)}$ admits a natural interpretation as a system which processes input signals φ in $\mathcal{D}(\mathbb{R})$ and turns them into output signals $T_{(f)}\varphi$ in $\mathcal{C}^\infty(\mathbb{R})$. More generally, if we denote by $\mathcal{E}[\mathcal{D}, \mathcal{C}^\infty]$ the set of all operators $T : \mathcal{D} \to \mathcal{C}^\infty$, then each such operator T can be considered as a *system*, with input signal space $\mathcal{D}(\mathbb{R})$ and output signal space $\mathcal{C}^\infty(\mathbb{R})$. There is a natural algebraic and differential structure for $\mathcal{E}[\mathcal{D}, \mathcal{C}^\infty]$ which corresponds to the possible ways in which systems may be interconnected:

$$\text{Sum}: (T_1 + T_2)\varphi := T_1\varphi + T_2\varphi \quad \text{Scalar multiple}: (aT)\varphi := a\{T\varphi\}$$
$$\text{Product}: (T_1.T_2)\varphi := (T_1\varphi)(T_2\varphi); \quad \text{Derivative}: (DT)\varphi := D(T\varphi)$$

With these definitions $\mathcal{E}[\mathcal{D}, \mathcal{C}^\infty]$ is a vector space and a differential algebra. Identifying each function f in $\mathcal{C}^\infty(R)$ with the system operator $T_{(f)}$ and using the system-theoretic definition of derivative gives an alternative model for the Silva axioms in which each distribution μ is interpreted as a (linear and translation-invariant) system $T_{(\mu)}$.

Example: The Dirac delta distribution now appears as the system $T_{(\delta)}$ which maps each input signal $\varphi \in \mathcal{D}(\mathbb{R})$ into itself:

$$T_{(\delta)}\varphi(x) : = D^2 \int_{-\infty}^{+\infty} x_+(y)\varphi(x-y)\,dy = \int_0^{+\infty} y\varphi^{(2)}(x-y)\,dy$$

$$= \int_0^{+\infty} -\varphi'(x-y)\,dy = \varphi(x).$$

In engineering terms δ represents (or may be identified with) the ideal amplifier of unit gain. In general the connection with the orthodox Schwartz definition of a distribution as a functional is easily established since we have

$$T_{(\mu)}\varphi(0) \equiv <\mu(y), \varphi(-y)>.$$

Such a systems-theoretic interpretation of distributions has an immediate conceptual appeal at an elementary level. Systems theory was originally concerned with electrical networks (or with mechanical analogues thereof), and has always had a strong intuitive physical basis. It is easy, for example, to give a plausible justification for the choice of $\mathcal{D}(\mathbb{R})$ as the (minimal) space of signal functions which ought to be admissible inputs for all systems of interest. Moreover the existence of such simple basic systems as the ideal time delay or the ideal differentiator makes it plain to the student at an early stage that the classical concept of function needs to be extended. In a typical naive engineering approach to systems theory this is achieved via the concept of "impulse response". Suppose for example that f is an infinitely differentiable function, $f \in C^{\infty}(\mathbb{R}) \subset C^0(\mathbb{R})$. Let $\theta \in \mathcal{D}(\mathbb{R})$ be such that $\int_{-\infty}^{+\infty} \theta(x)\,dx = 1$, and for $n = 1, 2, 3, ...$, define θ_n by

$$\theta_n(x) := n\theta(nx)$$

The system response $T_{(f)}\theta_n$ to the input signal θ_n is well defined for each $n \in \mathbb{N}$ and

$$\lim_{n \to \infty} T_{(f)}\theta_n(x) = \lim_{n \to \infty} \int_{-\infty}^{+\infty} f(x-y)\theta_n(y)\,dy = f(x), \quad \forall x \in \mathbb{R}.$$

Accordingly $f(x)$ is described loosely as the impulse response of the translation-invariant linear system $T_{(f)}$; that is, as the response to a hypothetical "unit impulse". The delta function thus makes its first appearance for the engineer as a generalized signal which is in some sense to be regarded as the limit of a sequence $(\theta_n)_{n \in \mathbb{N}}$ of "ordinary functions". Later on it is made to do duty as a generalized impulse response for the unit gain amplifier $T_{(\delta)}$, with δ' playing a corresponding role for the differentiator and so on. For a systems theoretic treatment of distributions therefore we require a rigorous definition of (generalized) impulse response which nevertheless allows it to retain its intuitive significance as a *signal*.

3.2. Sketch of a nonstandard systems theory

Following Colombeau we can introduce a special class of nonstandard delta functions or unit impulses. For $q = 0, 1, 2, ...$, let Φ_q denote the set of all functions $\theta \in \mathcal{D}(\mathbb{R})$ such that, for some $q \in \mathbb{N}$,

$$(A_1) \int_{-\infty}^{+\infty} \theta(x)\, dx = 1, \quad \text{and} \quad (A_2) \int_{-\infty}^{+\infty} x^k \theta(x)\, dx = 0, \quad 1 \leq k \leq q,$$

(condition (A_2) being vacuous for $q = 0$). Then for each $\theta \in \Phi_0$ there exists a nonstandard representative δ_θ of the δ function in $^*\mathcal{C}^\infty(\mathbb{R})$ given for $x = [x_n] \in {}^*\mathbb{R}$ by

$$\delta_\theta(x) := [(n\theta(nx_n))_{n \in \mathbb{N}}] \equiv \omega^* \theta(\omega x), \quad \text{where} \quad \omega \equiv [(n)] \in {}^*\mathbb{R}.$$

If $\theta \in \Phi_q$ then we shall say that δ_θ is a **Colombeau δ function of order q**.

Nonstandard impulse response: If T is any translation invariant operator in $\mathcal{E}[\mathcal{D}, \mathcal{C}^\infty]$ then there exists a nonstandard extension $^*T$ mapping $^*\mathcal{D}(\mathbb{R})$ into $^*\mathcal{C}^\infty(\mathbb{R})$. Thus for each $\theta \in \Phi$ there is a well defined internal function $^*T\delta_\theta \in {}^*\mathcal{C}^\infty(\mathbb{R})$ given by

$$^*T\delta_\theta(x) := \left[T(\theta_n(x_n))_{n \in \mathbb{N}}\right], \quad \text{where} \quad \theta_n \equiv n\theta(nx).$$

This we may call an impulse response function for the system T. The ω-generalized function $F_T \equiv {}^\omega[^*T\delta_\theta]$ itself we call the **ω-impulse response** of T.

Time invariant linear systems: For any system $T_{(f)}$, where $f \in \mathcal{C}^0(\mathbb{R})$, every impulse response function $T_{(f)}\delta_\theta$ is an ω-moderate function, and the same is true for any derivative $D^k T_{(f)}$ of the system. Moreover, although $^*f$ will not generally belong to $^\omega[F] \in \mathcal{C}^\infty(\mathbb{R})$, it is certainly the case that for any $\theta \in \Phi_0$ we have $f(x) \approx {}^*T_{(f)}\delta_\theta(x)$ on $^*\mathbb{R}_{bd}$. In particular this means that

$$f(x) = \operatorname{st}\left\{{}^*T_{(f)}\delta_\theta(x)\right\}, \quad \text{everywhere on} \quad \mathbb{R}.$$

This could be taken as a justification of the usual description of the function f as *the* "impulse response function" of the linear time-invariant system $T_{(f)}$. More precisely the ordinary function f, through its association with the linear, time-invariant system $T_{(f)}$, can be said to define a generalized function which admits nonstandard representatives of the form $^*T_{(f)}\delta_\theta$ in $^*\mathcal{C}^\infty(\mathbb{R})$. Each derivative $D^k T_{(f)}$ of the system will also be linear and time-invariant but will not necessarily be associated with an ordinary function since f need not be k times differentiable. Nevertheless a corresponding class of nonstandard ω-impulse response functions will exist, and the system can be said to be associated with, or even to define, a certain generalized function. This leads naturally and easily to a nonstandard model for the Silva axiomatic definition of (finite order) distributions in which each distribution can be identified with an equivalence class of ω-impulse functions or, if preferred, with the time-invariant linear system itself.

References

[1] Robinson, A.: *Nonstandard Analysis*, North Holland, (1966).
[2] Schwartz, L.: *Théorie des Distributions*, Hermann, (1950/51).

[3] Mikusinski, J. G. & Sikorski, R.: *The Elementary Theory of Distributions*, Rozprawy Matematyczne, Warszawa, (1957).

[4] Antosik, P.; Mikusinski, J. G. & Sikorski, R.: *Theory of Distributions: The sequential Approach*, Elsevier, (1973).

[5] Lighthill, M. J.: *Fourier Analysis and Generalized Functions*, CUP, (1958).

[6] Temple, G.: *Theories and Applications of Generalized Functions*, J. London Math. Soc., 28, (1953), pp.134-148.

[7] Temple, G.: *The Theory of Generalized Functions*, Proc. Roy. Soc. London, ser. A, 228, (1955), pp. 175-190.

[8] Jones, D. S.: *Generalized Functions*, McGraw-Hill, (1966).

[9] Jones, D. S.:*The Theory of Generalized Functions*, CUP, (1982).

[10] König, H.: *Neue Begründung der Theorie der "Distributionen" von L. Schwartz*, Math. Nachr., vol. 9, pp. 129-148, (1953).

[11] Silva J. S.: *Sur l'axiomatique de la théorie des distributions*, Rev. Faculdade de Ciências, Lisboa, 2^a série A 4, pp. 79-186, (1954/5)

[12] Silva J. S.: *Sur l'axiomatique des distributions et ses possibles modèles*, Centro Internazionale Matematico Estivo, Roma, Instituto Matematico, (1961).

[13] Colombeau, J. F.: *New Generalized Functions and Multiplication of Distributions*, North-Holland, (1984).

[14] Rosinger, E. E.: *Generalized Solutions of Nonlinear Partial Differential Equations*, Springer, (1987).

[15] Robinson, A.: *Nonstandard Analysis*, Proc. Roy. Acad. Sc., Amsterdam, ser. A, 64, 432-440, (1961).

[16] Luxemburg, W. A. J.:*Two applications of the method of ultrapowers to Analysis*, Bull. Amer. Math. Soc., ser. 2, 68, 416-419, (1962).

[17] Schmieden, C. & Laugwitz, D.: *Eine Erweiterung der Infinitesimalrechnung*, Math. Z., 69, pp. 1-39, (1958).

[18] Laugwitz, D.: *Ω-calculus as a generalization of field extension* In Nonstandard Analysis-Recent Developments, ed. A. Hurd, Springer, (1980).

[19] Laugwitz, D.: *Infinitesimals in physics*, Papers dedicated to Professor L. Iliev's 70th anniversary, Sofia, pp. 233-243, (1984).

[20] Craven, B. D.: *Generalized functions for applications*, J. Austral. Math. Soc., Ser. B, 26, pp. 362-374, (1985).

[21] Hoskins, R, F. & Sousa Pinto J.: *A nonstandard realisation of the J. S. Silva axiomatic theory of distributions*, Portugaliae Mathematica, 48(2), pp. 195-216.

[22] Hoskins, R, F. & Sousa Pinto J.: *Distributions, Ultradistributions, and other Generalized Functions*, Ellis Horwood, 1994.

[23] Biagioni, H. A.: *A Nonlinear Theory of Generalized Functions*, Springer, (1988).

Cranfield University
England

CHRIS IMPENS
Representing distributions by nonstandard polynomials

1. Introduction

One of the features of Nonstandard Analysis is, that it lowers the rank of mathematical concepts. A striking illustration of this is Robinson's famous characterization of compactness, with quantifiers running over numbers instead of over coverings. In this paper we reduce classical Schwartzian distributions to nonstandard functions, a simplification which might be useful in nonlinear operations with distributions.

Notation

Some standard notation: $\mathbb{R}$ for the set of real numbers, $\mathbb{N} = \{0, 1, \dots\}$ for the set of natural numbers, $U = [0, 1]$ and $W = [-1, 1]$ for real intervals, $\binom{n}{k} = n!/(k!(n-k)!)$ for binomial coefficients, $[x]$ for the largest integer less than or equal to x. For suitable functions, $f|A$ is restriction to A, $Df = f'$ the derivative, and If the primitive defined by $(If)(x) = \int_0^x f$.

Typical for Nonstandard Analysis in the superstructure approach (see [1], e.g.) are: the nonstandard extension $^*A$ of a set A, the relation $\approx$ of being infinitely close, and the standard part $^\circ a$ of a finite number a.

Concerning distributions (for which, see [3], e.g.) we use: $\hat{D}$ for distributional derivation, $\hat{f}$ for the regular distribution with continuous kernel function f, $\mathcal{D}$ for the set of all C^∞ functions with compact support, $\mathcal{D}_U$ for the set of all C^∞ functions with support in $]0, 1[$.

Finally, $B_n f$ is the Bernstein polynomial of order n for f. If the order is omitted from the notation, the fixed infinite hypernatural ω is understood. Properties not made explicit here can be found e.g. in [2].

2. Schwartzian distributions

We define a **distribution** as a continuous linear mapping $T : \mathcal{D} \to \mathbb{R}$, $\varphi \mapsto T(\varphi)$ from the space $\mathcal{D}$ of infinitely differentiable test functions with compact support into the reals. A distribution is either regular or singular.

(1) A **regular distribution** has the form $T(\varphi) = \int_\mathbb{R} f\varphi, \forall \varphi \in \mathcal{D}$ for some locally integrable kernel function f. Given a distribution T, we shall distinguish between a **kernel function** f for T and the **kernel** $\{f, g, \dots\}$ for T, which is the set of all possible kernel functions for T. The kernel $\{f, g, \dots\}$ of a regular distribution consists of functions which are equal almost everywhere, and among which there can be at most one continuous function. Hence one can identify a continuous function f with the regular distribution $\hat{f}$ defined by $\hat{f}(\varphi) = \int_\mathbb{R} f\varphi$.

(2) A **singular distribution** is not regular. Two examples will be considered in the sequel: the **delta distribution** defined by $\delta(\varphi) = \varphi(0), \forall \varphi \in \mathcal{D}$, and the **Principal Value Distribution** $P\frac{1}{x}$ defined by

$$\left(P\frac{1}{x}\right)(\varphi) = \text{P.V.} \int_{\mathbb{R}} \frac{\varphi(x)}{x} dx = \lim_{\varepsilon \to 0+} \int_{|x| \geq \varepsilon} \frac{\varphi(x)}{x} dx, \quad \forall \varphi \in \mathcal{D}.$$

Distributions are infinitely smooth objects: every distribution T has a distributional derivative, denoted by $\hat{D}T$, and defined by $\hat{D}T(\varphi) = -T(\varphi'), \forall \varphi \in \mathcal{D}$. In particular we have for a regular distribution $\hat{f}$ that $(\hat{D}\hat{f})(\varphi) = -\int_{\mathbb{R}} f\varphi', \forall \varphi \in \mathcal{D}$. The delta distribution is the second order derivative of a continuous function: $\delta = \hat{D}^2 \hat{f}$ if $f(x) = 0$ for $x \leq 0$ and $f(x) = x$ for $x \geq 0$.

3. Robinson's Representation

The idea to handle distribution theory with infinitesimal technology is as least as old as Nonstandard Analysis itself. (Taking into account the work of Laugwitz and Schmieden, it is older.) More than thirty years ago Robinson (*Non-Standard Analysis*, Th.5.3) proved that in an enlargement every distribution is regular. Actually, the theorem is more general and gives for *any* (even discontinuous) linear functional $T: \mathcal{D} \to \mathbb{R}$ a nonstandard polynomial P_T such that $T(\varphi) = \int_{*\mathbb{R}} P_T \varphi, \forall \varphi \in \mathcal{D}$.

Notice that this representation

(1) allows all standard test functions (i.e. $\mathcal{D}$), but not all nonstandard test functions (i.e. $^*\mathcal{D}$). Not surprisingly, the enlargement results in allowing test functions from some hyperfinite space $\mathcal{D}'$, strictly larger than $\mathcal{D}$ but strictly smaller than $^*\mathcal{D}$.

(2) has a strict equality and not merely equality up to infinitesimals.

This equality, nice as it is, has some drawbacks. To see this, let's apply Robinson's Theorem and represent the principal value distribution by a nonstandard polynomial P, so that P.V. $\int_{\mathbb{R}} \frac{\varphi(x)}{x} dx = \int_{*\mathbb{R}} P\varphi, \forall \varphi \in \mathcal{D}$, which leads to $\int_{*\mathbb{R}} (xP(x) - 1)\varphi = 0$, $\forall \varphi \in \mathcal{D}$.

We see that the polynomial $Q(x) = xP(x) - 1$ represents the zero distribution $\varphi \mapsto 0, \forall \varphi \in \mathcal{D}$. Yet $Q(x)$ is far from being the zero polynomial, having $Q(0) = -1$. What is worse, for any hyperreal constant w we also have (due to the strict equality) $\int_{*\mathbb{R}} (wQ)\varphi = 0$, so that $w(xP(x) - 1)$ also represents the zero distribution. The arbitrary (possibly infinite) constant w allows one to make the values of the representing polynomial as small or as large as desired. We can summarize these facts as follows:

> Robinson's Representation $T(\varphi) = \int_{*\mathbb{R}} P\varphi$ does not reduce a singular distribution T to a function P, but to a huge class $\{P,\dots\}$ of functions which are different beyond comparison.

Stated differently:

> In Robinson's Representation $T(\varphi) = \int_{*\mathbb{R}} \{P,\dots\}\varphi$ the kernel $\{P,\dots\}$ has no size and no values.

4. Another Representation

In this section we try to represent a distribution T by a "small" kernel $\{P_\alpha, P_\beta, \dots\}$ of nonstandard polynomials, preferably with $P_\alpha \approx P_\beta \approx \dots$. To this end we will use certain polynomials whose natural domain is the unit interval. Also, it soon appears (see Remark in Subsection 2 below) that strict equality is asking too much. Therefore we restrict ourselves to distributions $T : \mathcal{D}_U \to \mathbb{R}$ ($\mathcal{D}_U$ being the set of all test functions with support in $]0,1[$) and look for a representation $T(\varphi) = {}^\circ\!\int_{*U} \{P_\alpha, P_\beta, \dots\}\,\varphi$, $\forall \varphi \in \mathcal{D}_U$. We successively consider the space $\mathbb{B}$ of the kernel functions $P_\alpha, P_\beta, \dots$ and then the factor space $\mathbb{B}/\equiv$ of the kernels $\{P_\alpha, P_\beta, \dots\}$.

1. The space $\mathbb{B}$ of the kernel functions.

The starting point of our approach is the following well known fact: *On a compact interval, every distribution T is a finite-order distributional derivative of some continuous function*, i.e.: every $T : \mathcal{D}_U \to \mathbb{R}$ can be written as $T = \hat{D}^n \hat{f}$ for some continuous function f on $[0,1]$ and some natural n. Regarded this way, distribution theory amounts to differentiating nondifferentiable functions. This goal can be easily achieved directly by nonstandard means: the standard eye can not distinguish between a standard function f on $[0,1]$ and a nonstandard polynomial P for which ${}^\circ P = f$, and while f may not be differentiable, P is infinitely smooth. Such P's are readily obtained by applying Weierstrass' Theorem, which yields a sequence of polynomials converging uniformly to f. Generally these sequences do not have the property that P_n' converges to f' if the latter exists! So, if one wants to preserve the differentiability that f happens to have, very few choices for the P_n's are available. Borel (1905) had the idea to use polynomials of the form $\sum_{k=0}^n f(k/n) p_{nk}(x)$ as uniform approximations for a continuous function f, and Bernstein (1912) made the remarkable probabilistic choice $p_{nk}(x) = \binom{n}{k} x^k (1-x)^{n-k}$.

Let us fix an infinite hypernatural number ω and associate with each continuous function $f : U = [0,1] \to \mathbb{R}$ the nonstandard polynomial

$$Bf(x) = \sum_{k=0}^{\omega} \binom{\omega}{k} x^k (1-x)^{\omega-k} f(k/\omega),$$

which is its **Bernstein polynomial of order ω**.

Notice that Bf results from combining a probabilistic density function with the restriction of f to the hyperfinite time line $\{0, 1/\omega, 2/\omega, \dots, 1\}$. It is rather surprising that probability on the hyperfinite time line is linked here with distribution theory.

The basic properties of the linear operator $B : f \mapsto Bf$ are that it is a differentiability preserving imbedding, with a relative unicity property.

Property 1. (Imbedding): *Bf and f determine each other unambiguously, and $f = {}^\circ(Bf|^*U)$.*

Bernstein's proof of Weierstrass' Theorem shows that $B_n f \to f$ uniformly on U, which implies that f is the standard part of the restriction $(Bf)|^*U$. Notice the importance of using here a Bernstein polynomial $Bf = B_\omega f$ of infinite order ω. Reconstructing f from a Bernstein polynomial $B_n f$ of finite order is generally impossible, only the values $f(k/n)$ ($k = 0, 1, \ldots, n$) at the sampling points being determined by $B_n f$.

This property allows us to identify $f \in C[0,1]$ and $Bf \in {}^*C^\infty[0,1]$, and to view the linear operator $B : C[0,1] \to {}^*C^\infty[0,1]$ as imbedding the space $C[0,1]$ of continuous functions into the space of infinitely smooth functions ${}^*C^\infty[0,1]$.

Property 2. (Differentiability Preserving): *If f is differentiable at $a \in [0,1]$ (from the right if $a = 0$, from the left if $a = 1$), then $f'(a) = {}^\circ(DBf(a))$.*

This results from the property that $B'_n(a)$ converges to $f'(a)$.
We see that the imbedding $B : f \mapsto Bf$ preserves derivatives (up to infinitesimals).

Property 3. (Relative Unicity): *If, for some natural $k \in \mathbb{N}$ and $\forall \varphi \in \mathcal{D}_U$, we have $\int_{*U} (D^k Bf) \varphi \approx 0$, then $D^k Bf = 0$ on ${}^*U$.*

Since $\varphi(0) = \varphi(1) = 0$ for any $\varphi \in \mathcal{D}_U$, the assumption is first reduced to $\int_{*U}(Bf)(D^k\varphi) \approx 0$, $\forall \varphi \in \mathcal{D}_U$ after k integrations by parts and then to $\int_U f(D^k\varphi) = 0$, $\forall \varphi \in \mathcal{D}_U$ after taking standard parts. Hence $\hat{D}^k \hat{f} = \hat{0}$, and it is well known from distribution theory that this means that f is a polynomial of degree less than k. Bernstein polynomials have the property that every $B_n f$ is then also of degree less than k. Thus we see that $D^k Bf = 0$.

These combined properties suggest the use of the space of nonstandard polynomials $\mathbb{B} = \{D^k Bf \mid f \in C[0,1], k \in \mathbb{N}\}$, and to try the representation $T(\varphi) = {}^\circ\!\int_{*U}(D^k Bf)\varphi$, $\forall \varphi \in \mathcal{D}_U$. Notice that the third property does not say that the kernel function $D^k Bf$ in this representation is unique; it only shows that a given T can not have two different kernel functions $D^k Bf$ and $D^k Bg$ with the same order of derivation k. So we are still left with the task of grouping the different kernel functions of a single distribution into kernels.

2. The factor space $\mathbb{B}/\equiv$ of the kernels.

(0) *(Regular) Distributions of order 0.* Let us agree (there are other conventions) to call a continuous function f on U (identified with the regular distribution $\hat{f}$) a **distribution of order zero**. Trivially, a distribution T of order zero is regular. It is easy to find the functions in $\mathbb{B}$ that are kernel functions of T. First, $Bf \in \mathbb{B}$ is such a kernel function, as $(\hat{f})(\varphi) = \int_U f\varphi = {}^\circ\!\int_{*U}(Bf)\varphi, \forall \varphi \in \mathcal{D}_U$. Next, $(DBIf)$ is also a kernel function: for any $\varphi \in \mathcal{D}_U$ we have ${}^\circ\!\int_{*U}(DBIf)\varphi = $ (integration by parts) $-{}^\circ\!\int_{*U}(BIf)\varphi' = $ (standard parts) $-\int_U (If)\varphi' = $ (integration by parts) $\int_U f\varphi$. In the same manner one generates the whole sequence $\{Bf, DBIf, D^2 BI^2 f, \ldots, \}$ of kernel functions of $T = \hat{f}$. Besides these, there are no kernel functions in $\mathbb{B}$ for $\hat{f}$: if some $D^k Bg \in \mathbb{B}$ (g a continuous function, k a natural number) were a kernel function, $D^k Bg$ and $D^k BI^k f$ would be two kernel functions with the same order of differentiation k, hence $D^k Bg = D^k BI^k f$ by property 3. Therefore the representation for a continuous function $T = \hat{f}$ is fully described by

$$(\hat{f})(\varphi) = {}^\circ\!\!\int_{\bullet U} \{Bf, DBIf, D^2BI^2f, \ldots\}\varphi, \quad \forall \varphi \in \mathcal{D}_U.$$

Remark: *If strict equality were demanded, this representation breaks down for the distribution* $\widehat{x^2}$. *Its representation* $(\widehat{x^2})\varphi = \int_{\bullet U}(Bx^2)\varphi$, *written explicitly, is* $\int_0^1 x^2\varphi(x)\,dx = \int_{\bullet U}((\frac{1}{\omega}x + (1-\frac{1}{\omega})x^2)\varphi(x)\,dx$. *Simplifying (including multiplying by* ω, *which preserves the equality) we would find* $\int_{\bullet U}(x - x^2)\varphi(x)\,dx = 0, \forall \varphi \in \mathcal{D}_U$, *hence the contradiction that* $x - x^2 = 0$ *on* U.

(1) *(Regular or singular) Distributions of order 1.* We call $T : \mathcal{D}_U \to \mathbb{R}$ a **distribution of first order** if T is the distributional derivative of a continuous function, but is not itself a continuous function. This means that $T = \hat{D}\hat{f}$ with f continuous but not C^1 on U (if $f \in C^1[0,1]$, we have $T = \hat{f}'$ and T is a continuous function). If f' exists a.e. in U and is locally integrable, then $T = \hat{f}'$ and the distribution $T = \hat{D}\hat{f}$ is regular; otherwise, T is singular.

Combining integration by parts, the standard parts operator °(...), and the Relative Unicity property 3, we find that the representation is exhaustively described by

$$(\hat{D}\hat{f})(\varphi) = -\int_U f\varphi' = {}^\circ\!\!\int_{\bullet U} \{DBf, D^2BIf, D^3BI^2f, \ldots\}\varphi, \quad \forall \varphi \in \mathcal{D}_U.$$

There can not be a kernel function without derivation, because the representation $°\!\int_{\bullet U}(Bg)\varphi = \int_U g\varphi$ $(g \in C[0,1])$ defines a distribution $\hat{g}$ of order 0, which is not compatible with T being of order 1. Note that the kernel is independent of how T is written as a distributional derivative. If $T = \hat{D}\hat{f} = \hat{D}\hat{g}$, then f and g differ by a constant, and $DBf = DBg$.

(2) *(Singular) Distributions of order* $k \geq 2$. A **distribution of order k** $(k \geq 2)$ is of the form $T = \hat{D}^k\hat{f}$ with f continuous but not C^k on U. It is always singular, even if $f^{(k)}$ happens to be locally integrable on U. The representation is found to be

$$(\hat{D}^k\hat{f})(\varphi) = (-1)^k\int_U f\varphi^{(k)} = {}^\circ\!\!\int_{\bullet U}\{D^kBf, D^{k+1}BIf, D^{k+2}BI^2f, \ldots\}\varphi, \quad \forall \varphi \in \mathcal{D}_U.$$

If we write $[D^kBf]$ for the sequence $(D^{n+k}BI^nf)_{n\geq 0}$ of nonstandard polynomials from $\mathbb{B}$, we may summarize our findings as follows: *every distribution* $T : \mathcal{D}_U \to \mathbb{R}$ *has a sequence of kernel functions in* $\mathbb{B}$, *viz.* $T(\varphi) = {}°\!\int_{\bullet U}[D^kBf]\varphi, \forall \varphi \in \mathcal{D}_U$, *if* k *is the order of* $T = \hat{D}^k\hat{f}$ $(f \in C[0,1])$.

This description of distributions has several advantages: it is simple, both conceptually (reducing functionals to functions, thus considerably lowering the rank of the concept) and technically (it is very explicit, and does not require any saturation from the nonstandard model). Distributional derivation, expressed by way of the kernels, is almost a triviality: $\hat{D}[D^kBf] = [D^{k+1}Bf]$. The ideal would be, that the functions of a kernel $[D^kBf]$ satisfy $D^kBf \approx D^{k+1}BIf \approx D^{k+2}BI^2f \approx \ldots$. That this is not generally satisfied, is evidenced by the following example.

Example. Consider the regular distribution $T(x) = \widehat{1/(2\sqrt{x})} = \hat{D}\sqrt{x}$, which is of first order (the integrable function $1/(2\sqrt{x})$ is not continuous on $[0,1]$, but $\sqrt{x}$ is). The kernel of T is the sequence $[DB\sqrt{x}] = (DBx^{1/2}, D^2B(\frac{2}{3}x^{3/2}), \ldots)$. After calculation one finds that $DBx^{1/2} = \sqrt{\omega} + a_1 x + \ldots$ and $D^2B(\frac{2}{3}x^{3/2}) = \frac{4}{3}(\sqrt{2}-1)(1-\frac{1}{\omega})\sqrt{\omega} + b_1 x + \ldots$, so that the difference $DBx^{1/2} - D^2B(\frac{2}{3}x^{3/2})$ is infinite at $x=0$.

This example shows that the values of different functions in the same kernel may differ by infinite amounts. Yet we are far removed from the indeterminacy in Robinson's kernel, in two respects: (a) all the functions in a kernel share a common order of magnitude and (b) for some distributions (and not the least ones) we do have that all kernel functions are infinitely close to each other. More precisely, we have the following theorems.

Theorem A. *If T is a distribution of order $k \geq 1$, every kernel function F for T is of order $o(\omega^k)$, i.e. $F(x)/\omega^k \approx 0, \forall x \in {}^*U$.*

Theorem B. *Let $\{\delta_0, \delta_1, \ldots\}$ be the kernel of the delta-distribution $\delta : \varphi \mapsto \varphi(0)$, $\forall \varphi \in \mathcal{D}_W$. Then, taking ω even and equal to 2ν, we have $\delta_0(x) = C_\nu(1-x^2)^{\nu-1}$ with C_ν an infinite constant, and $\delta_0 \approx \delta_1 \approx \ldots$ on ${}^*W$.*

For a more detailed version of (B) and for both proofs, the reader is referred to the Appendices. Here we restrict ourselves to some comments.

As for (A), the estimate is better than could be expected. Generally, for a polynomial P of degree ω with $|P(x)| \leq M$ on $[0,1]$, we have V.A. Markov's optimal estimate $|D^k P(x)| \leq M2^k \omega^2(\omega^2-1^2)(\omega^2-2^2)\ldots(\omega^2-(k-1)^2)/(2k-1)!$. By Markov's inequality the kernel function $D^{n+k}BI^n f$ is of order $o(\omega^{2k+2n+1})$. Theorem A estimates the order of infinity much more efficiently, and uniformly for all the functions in the kernel. Applied to the example, we see that the first two kernel functions $DBx^{1/2}$ and $D^2B(\frac{2}{3}x^{3/2})$ differ at the origin by some multiple of $\sqrt{\omega}$, with (here $k=1$) $\sqrt{\omega}/\omega \approx 0$. This shows that the simple estimate in (A), while not optimal, can not be improved that much.

In (B) we leave the unit interval which is the canonical domain of Bernstein polynomials. An elementary affine transformation shows that the Bernstein polynomials to consider for a continuous function f on a compact interval $[a,b]$ are given by

$$B_n f(x) = (b-a)^{-n} \sum_{j=0}^{n} \binom{n}{j}(x-a)^j(b-x)^{n-j} f(a+j(b-a)/n).$$

The delta-distribution on W has the kernel $\{D^2 Bf, D^3 BIf, \ldots,\}$ with $f(x) = 0$ for $-1 \leq x \leq 0$ and $f(x) = x$ for $0 \leq x \leq 1$ and

$$Bf(x) = 2^{-\omega} \sum_{j=0}^{\omega} \binom{\omega}{j}(1+x)^j(1-x)^{\omega-j} f(-1+2j/\omega).$$

To simplify calculations we take $\omega = 2\nu$ even. Rather unexpectedly, the first kernel

function simplifies to the even function
$$\delta_0(x) = (2\nu)!\nu/(4^\nu(\nu)!^2)(1-x^2)^{\nu-1},$$
having an infinite peak of order $O(\sqrt{\nu})$ at the origin, and covering an area of 1. This function is the infinite version of **Landau's Kernel** for singular integrals. The second kernel function is found to be
$$\delta_1(x) = (2\nu)!(\nu-1)/(4^\nu(\nu)!^2)(1-x^2)^{\nu-2}.$$
Starting with the third kernel function
$$\delta_2(x) = (2\nu)!(\nu-1)/(4^\nu(\nu)!^2 3\nu)(1-x^2)^{\nu-3}(3\nu - 4 - \nu x^2)$$
an extra polynomial appears besides the power of $1-x^2$. This factor would definitely modify the general appearance of the kernel functions if ω were finite. For infinite ω the effect of the extra factor is negligible, and all the kernel functions are infinitesimal perturbations of δ_0, even in the monad of zero, where values are infinite. Thus we are entitled to say that this kernel for the delta-distribution has values up to infinitesimals.

Appendix A

The proof of Theorem A relies on a lemma, in which we use differences of higher order, defined as
$$\Delta_h^p g(x_0) = g(x_0 + ph) - \binom{p}{1} g(x_0 + (p-1)h) + \cdots + (-1)^p g(x_0).$$

If x_0 is displayed as a fraction m/n we set $h = 1/n$ and omit h from the notation. The mean value theorem for higher order differences says that $\Delta_h^p g(x_0) = h^p g^{(p)}(\xi)$ for some $x_0 < \xi < x_0 + ph$ if g is continuous on $[0,1]$ and $g^{(p)}$ exists in $]0,1[$.

Lemma. *For continuous f, natural p, k and hypernatural j with $0 \leq j \leq \omega - k - p$ we have*
$$\frac{\omega!}{(\omega-p)!} \Delta^{k+p}(I^p f)(\frac{j}{\omega}) \approx \Delta^k f(\frac{j}{\omega-p})$$

Proof. The mean value theorem implies that
$$\omega^p \Delta^p (I^p f)(j/\omega) = f(\xi) \text{ with } j/\omega \leq \xi \leq (j+p)/\omega.$$
By assumption, $p + j \leq \omega$, and so $0 \leq j/\omega \leq j/(\omega-p) \leq (j+p)/\omega \leq 1$. It follows that $f(\xi) \approx f(j/(\omega-p))$. Hence
$$\omega^p \Delta^p (I^p f)(j/\omega) \approx f(j/(\omega-p)).$$
Multiplying by $(1 - 1/\omega)(1 - 2/\omega)\ldots(1 - (p-1)/\omega) \approx 1$ we find
$$\frac{\omega!}{(\omega-p)!} \Delta^p (I^p f)(\frac{j}{\omega}) \approx f(\frac{j}{\omega-p}).$$
This is the formula we want for $k = 0$.

For $k = 1$ we have

$$\frac{\omega!}{(\omega-p)!}\Delta^{p+1}(I^p f)(\frac{j}{\omega}) = \frac{\omega!}{(\omega-p)!}\Delta^p(I^p f)(\frac{j+1}{\omega}) - \frac{\omega!}{(\omega-p)!}\Delta^p(I^p f)(\frac{j}{\omega})$$
$$\approx f((j+1)/(\omega-p)) - f(j/(\omega-p))$$
$$= \Delta f(j/(\omega-p)).$$

The general formula follows after a finite induction.

Proof of Theorem A. For the first kernel function $D^k Bf$ we have

$$\frac{D^k Bf(x)}{\omega^k} = (1 - \frac{1}{\omega})(1 - \frac{2}{\omega})\ldots(1 - \frac{k-1}{\omega}) \sum_{j=0}^{\omega-k} \binom{\omega-k}{j} x^j (1-x)^{\omega-k-j} \Delta^k f(\frac{j}{\omega})$$

by a well known derivation formula for Bernstein polynomials. Combining the continuity of f, which gives $\Delta f(j/\omega) \approx 0$, and the equation

$$\Delta^{m+1} f(\frac{j}{\omega}) = \Delta^m f(\frac{(j+1)}{\omega}) - \Delta^m f(\frac{j}{\omega})$$

we see that all the $\Delta^k f(j/\omega)$'s are infinitesimal. Let $\varepsilon \approx 0$ be the maximum of the $|\Delta^k f(j/\omega)|$'s. Then

$$|\sum_{j=0}^{\omega-k} \binom{\omega-k}{k} x^j(1-x)^{\omega-k-j}\Delta^k f(fracj\omega)| \le \varepsilon \sum_{j=0}^{\omega-k}\binom{\omega-k}{j}x^j(1-x)^{\omega-k-j} = \varepsilon,$$

hence $|D^k Bf(x)/\omega^k| \approx \varepsilon \approx 0$.

For the p-th kernel function we have to prove that $D^{k+p}BI^p f/\omega^k \approx 0$. The derivation formula for Bernstein polynomials gives us

$$\frac{D^{k+p}BI^p f(x)}{\omega^k} = \frac{\frac{\omega!}{(\omega-k-p)!}}{\omega^k} \sum_{j=0}^{\omega-k-p} \binom{\omega-k-p}{j} x^j(1-x)^{\omega-k-p-j} \Delta^{k+p}(I^p f)(\frac{j}{\omega})$$

and, if we apply it to $B_{\omega-p}f$ (note the exceptional order) we get

$$\frac{D^k B_{\omega-p} f(x)}{\omega^k} = \frac{\frac{\omega!}{(\omega-k-p)!}}{\omega^k} \sum_{j=0}^{\omega-k-p} \binom{\omega-k-p}{j} x^j(1-x)^{\omega-k-p-j} \Delta^k f(\frac{j}{\omega-p}).$$

Hence, taking

$$\eta_j = \frac{\omega!}{(\omega-p)!}\Delta^{k+p}(I^p f)(\frac{j}{\omega}) - \Delta^k f(\frac{j}{\omega-p}),$$

$$\frac{D^{k+p}BI^p f(x) - D^k B_{\omega-p}f(x)}{\omega^k} = \prod_{j=0}^{k-1}(\omega - \frac{p+j}{\omega}) \sum_{j=0}^{\omega-k-p} \binom{\omega-k-p}{j} x^j(1-x)^{\omega-k-p-j} \eta_j$$

the summation in absolute value being less than $\sum_{j=0}^{\omega-k-p} \binom{\omega-k-p}{j} x^j(1-x)^{\omega-k-p-j}\eta = \eta$, with $\eta = \max_{0 \le j \le \omega-k-p} |\eta_j|$ The Lemma tells us that η is the maximum of a hyperfinite number of infinitesimals, which implies that $\eta \approx 0$. We conclude that

$$\frac{|D^{k+p}BI^p f(x) - D^k B_{\omega-p}f(x)|}{\omega^k} \le \prod_{j=0}^{k-1}(\omega - \frac{p+j}{\omega})\eta \approx 0,$$

and thus we obtain

$$\frac{D^{k+p}BI^p f(x)}{\omega^k} \approx \frac{D^k B_{\omega-p} f(x)}{\omega^k} = \frac{(1-\frac{p}{\omega})^k D^k B_{\omega-p} f(x)}{(\omega-p)^k} \approx \frac{D^k B_{\omega-p} f(x)}{(\omega-p)^k}.$$

This expression is infinitesimal, for the same reason as $D^k Bf/\omega^k$ was.

Appendix B

To prove Theorem B we need a standard lemma concerning the numbers

$$E_m^n = (m+1)^n - \binom{n+1}{1} m^n + \binom{n+1}{2}(m-1)^n + \cdots + (-1)^m \binom{n+1}{m} 1^n,$$

known in combinatorial theory as **Eulerian numbers** (not to be confused with Euler numbers). They are related to Stirling numbers of the second kind (the first ones being $S_n^n = 1$, $S_{n-1}^n = \binom{n}{2}$, $S_{n-2}^n = \binom{n}{3}\frac{3n-5}{4}$) through the equation $\sum_{k=0}^{n-1} E_k^n \binom{k}{n-m} = m! S_m^n$. Substituting first $n = p+1 = m$, then $n = p+1, m = p$ and finally $n = p+1, m = p-1$ gives the values

$$b_0 = \sum_{j=0}^{p} E_j^{p+1} = (p+1)!,$$

$$b_1 = \sum_{j=0}^{p} E_j^{p+1} j = (p+1)! p/2$$

$$b_2 = \sum_{j=0}^{p} E_j^{p+1} j(j-1) = (p+1)!(p-1)(3p-2)/12.$$

Lemma. If, for $p \in \mathbb{N}$, we set $A_p(x) = \frac{1}{(p+1)!} \sum_{j=0}^{p} E_j^{p+1} (1+x)^j (1-x)^{p-j}$, then $A_p(x) = a_p^{(0)} + a_p^{(1)} x^2 + a_p^{(2)} x^4 + \cdots + a_p^{([p/2])} x^{2[p/2]}$, with $a_p^{(0)} = 1$ and $a_p^{(1)} = -(p-1)/3$.

Proof. The symmetry property of Eulerian numbers $E_m^n = E_{n-m-1}^n$ allows one to combine the j-th and $(p-j)$-th terms of $A_p(x)$ into even functions. If p is even, this procedure leaves an isolated middle term which is even by itself. In any case, $A_p(x)$ is even. Using the values of b_0, b_1 and b_2 we find

$$a_p^{(0)} = A_p(0) = \sum_{j=0}^{p} E_j^{p+1}/(p+1)! = 1$$

$$a_p^{(1)} = A_p''(0)/2 = \sum_{j=0}^{p} E_j^{p+1}(j(j-1) - 2j(p-j) + (p-j)(p-j-1)/(2(p+1)!)$$
$$= (4b_2 - 4(p-1)b_1 + p(p-1)b_0)/(2(p+1)!)$$
$$= -(p-1)/3.$$

In the next proof, we set $F_p(x)$ equal to 0 for $-1 \leq x \leq 0$ and to $x^{p+1}/(p+1)!$ for $0 \leq x \leq 1$.

127

Theorem B.
(1) *The kernel* $\{\delta_0, \delta_1, \ldots\}$ *of the distribution* $\delta : \varphi \mapsto 0, \forall \varphi \in \mathcal{D}_W$ *is given explicitly by* $(p = 0, 1, \ldots)$

$$\delta_p(x) = \frac{(2\nu)!\nu}{4^\nu(\nu!)^2}(1-x^2)^{\nu-p-1}C_p(x)$$

$$C_p(x) = \frac{(\nu!)^2}{\nu^{p+2}(p+1)!}\sum_{j=0}^p E_j^{p+1}\frac{(1+x)^j}{(\nu-1-j)!}\frac{(1-x)^{p-j}}{(\nu-1-(p-j))!} = \sum_{j=0}^{[p/2]} c_p^{(j)} x^{2j},$$

with all the $c_p^{(j)}$'s finite, and in particular $c_p^{(0)} \approx 1$ and $c_p^{(1)} \approx -(p-1)/3$.
(2) $\delta_0 \approx \delta_1 \approx \ldots$ in $*W$.
(3) *Every* δ_p *has at the origin a value which is infinitesimally less than the infinite number* $\delta_0(0) = \frac{\nu(2\nu)!}{4^\nu(\nu!)^2} = O(\sqrt{\nu})$, *and falls symmetrically and monotonously to infinitesimal values outside the monad of zero.*

Proof. (1) Using the differentiation formula for Bernstein polynomials, we find

$$\delta_p(x) = D^{p+2}BF_p(x)$$

$$= \frac{\omega!}{(\omega-p-2)!}\frac{1}{2^\omega}\sum_{j=0}^{\omega-p-2}\binom{\omega-p-2}{j}(1+x)^j(1-x)^{\omega-p-2-j}\Delta^{p+2}F_p\left(-1+\tfrac{j}{\nu}\right).$$

For $j < \nu-p-1$, the definition of Δ^{p+2} uses only zero values of F_p and for $j > \nu$ only the values $F_p(x) = x^{p+1}/(p+1)!$ enter, while $\Delta^m(x^n) = 0$ for $m > n$. What remains can be written as

$$\delta_p(x) = \frac{\omega!}{2^\omega}(1-x^2)^{\nu-p-1}\sum_{j=0}^p \frac{(1+x)^j}{(\nu-1-j)!}\frac{(1-x)^{p-j}}{(\nu-1-(p-j))!}\Delta^{p+2}F_p\left(\tfrac{j-p-1}{\nu}\right),$$

the difference reducing to the definition of $E_j^{p+1}/((p+1)!\nu^{p+1})$. Only the development of C_p in powers of x remains to be checked.

The symmetry in Eulerian numbers implies that $C_p(x)$ is even, and comparing degrees shows that the highest power possibly occurring is p. The coefficient $c_p^{(j)}$ is a sum of $p+1$ contributions, obtained successively from $(1-x)^p$, $(1+x)(1-x)^{p-1}$, $\ldots$, $(1+x)^p$. Each of these contributions is the same as for the coefficient $a_p^{(j)}$ in the standard polynomial $A_p(x)$, except for a finite number of factors $\left(1 - \tfrac{i}{\nu}\right)$ appearing in places where the corresponding contribution for $A_p(x)$ has the trivial factor 1. So there exist finite hyperreals $b_1^{(j)}, \ldots, b_p^{(j)}$ for which

$$c_p^{(j)} = a_p^{(j)} + b_1^{(j)}/\nu + b_2^{(j)}/\nu^2 + \cdots + b_p^{(j)}/\nu^p \approx a_p^{(j)}.$$

The values of $a_p^{(0)} = 1$ and $a_p^{(1)} = -(p-1)/3$ give the estimates for $c_p^{(0)}$ and $c_p^{(1)}$.
(2) Let us consider

$$\delta_p(x) - \delta_0(x) = (2\nu)!\,\nu(1-x^2)^{\nu-p-1}\left(C_p(x) - (1-x^2)^p\right)/(4^\nu(\nu!)^2).$$

Using the decomposition of the coefficients $c_p^{(j)}$ we infer that
$$\nu\left(C_p(x) - (1-x^2)^p\right) = \nu\left(A_p(x) - (1-x^2)^p\right) + R(x) = \nu\left((2p+1)x^2/3 + \ldots\right) + R(x),$$
where $R(x)$ is a polynomial of degree $2[p/2]$ with finite coefficients. Combining this with Wallis' estimate $2^{2n}(n!)^2/(2n)! = \sqrt{\pi(2n+\theta)/2}$ ($0 < \theta < 1$) we get
$$\delta_p(x) - \delta_0(x) \approx \sqrt{2}(1-x^2)^{\nu-p-1}\left(\nu(2p+1)x^2/3 + \ldots\right)/\sqrt{\pi(2\nu+\theta)} \text{ for } |x| \leq 1.$$
Inspecting the extreme values of $X^k(1-X)^\nu\sqrt{\nu}$ on $0 \leq X \leq 1$ one finds that this function is infinitesimal for finite $k \geq 1$, infinite ν and infinitesimal $X \gg 0$. Hence
$$(1-x^2)^{\nu-p-1}\nu x^2/\sqrt{2\nu+\theta} \approx (1-x^2)^{\nu-p-1}\nu x^4/\sqrt{2\nu+\theta} \approx \ldots$$
$$\approx (1-x^2)^{\nu-p-1}\nu x^{2[p/2]}/\sqrt{2\nu+\theta} \approx 0$$
for $x \approx 0$. It follows that $\delta_p(x) - \delta_0(x) \approx 0$ for $x \approx 0$. Outside the monad of zero every $\delta_p(x)$ is infinitesimal, as $C_p(x)$ is finite and $\nu(1-x^2)^{\nu-p-1}/\sqrt{2\nu+\theta} \approx 0$.

(3) Evidently, the factors $1-x^2$ fall off symmetrically and monotonously from a peak value at the origin. As for $C_p(x)$, its coefficients are infinitely close to those of the standard polynomial $A_p(x)$, so that both polynomial s stay infinitely close to each other for finite x. Since $A_p(x)$ is symmetric and falls off from a maximum at zero (its coefficient in x^2, calculated in the Lemma, being negative), $C_p(x)$ also falls off from a peak at zero, on a standard interval that extends well outside the monad of zero. (Falling can only stop at the nearest nonzero root of C_p', which is infinitely close to the nearest nonzero root of A_p', a standard number). Multiplying, we find that δ_p is symmetric and monotonously decreasing in the monad of zero.

Finally, $c_p^{(0)} < a_p^{(0)} = 1$ implies that $\delta_p(0) = \delta_0(0)C_p(0) < \delta_0(0)$. This concludes the proof.

References

[1] A.E. Hurd and P.A. Loeb *An introduction to nonstandard real analysis*, Academic Press, New York, 1985
[2] I.P. Natanson *Constructive Function Theory*, vol. I, F. Ungar, New York, 1964
[3] A.H. Zemanian *Distribution Theory and Transform Analysis*, Dover, New York, 1987

Department of Pure Mathematics and Computer Algebra
University of Gent
Galglaan 2
B-9000 Gent
Belgium

E-mail ci@cage.rug.ac.be

MICHAEL OBERGUGGENBERGER

Contributions of nonstandard analysis to partial differential equations

The aim of this paper is to present a number of typical applications of the nonstandard theory of generalized functions to partial differential equations. We are interested in nonlinear partial differential equations with singular data, like discontinuous coefficients and/or measures or even distributions as initial data. Such problems require a set-up allowing for nonlinear operations on distributions. Factor algebras of the nonstandard algebra $^*\mathcal{E}(\Omega)$ of internal smooth functions on open subsets Ω of $\mathbb{R}^n$ provide a suitable framework. In addition, we use Nonstandard Analysis here as a source of new objects to model generalized data and solutions.

The exposition covers part of a general program which might be called *Nonlinear Theory of Generalized Functions*. This program has been developed to some extent directly in the framework of Nonstandard Analysis, to a larger extent by standard methods, but implicitly using a substantial amount of ideas from Nonstandard Analysis, like internal functions, extended number systems, and transfer. The infinite powers and asymptotic properties employed in the standard sequential models translate into ultrapowers and estimates involving infinitesimals in the nonstandard setting. In this presentation, taking the nonstandard point of view, we intend to show some of the powerful principles of Nonstandard Analysis at work, yielding elegant arguments and structural insight not available in the standard approach.

Some remarks on the history of the subject, more details concerning the relation between standard and nonstandard models, and relevant literature can be found in the Appendix.

The plan of exposition is as follows: The first section serves to review some generalized functions theory, in particular, the construction of the nonstandard algebra $^\rho\mathcal{E}(\Omega)$ to be employed. Section 2 elaborates on generalized solutions to semilinear wave equations and delta-waves, while Section 3 will be dedicated to quasilinear hyperbolic systems and shock-waves. The final section contains an overview of further applications. Various other types of algebras of generalized functions are collected in the Appendix. The reader is assumed to be familiar with Nonstandard Analysis to the extent as covered, for example, in the introduction to the subject by Lindstrøm [52].

1. Generalized functions

To fix notation, let Ω be an open subset of $\mathbb{R}^n$. The algebra of infinitely differentiable functions on Ω is denoted by $\mathcal{E}(\Omega)$, the subalgebra of compactly supported elements by $\mathcal{D}(\Omega)$. A *distribution* w is a linear, continuous map

$$w : \mathcal{D}(\Omega) \to \mathbb{C}.$$

The value of w at an element $\varphi \in \mathcal{D}(\Omega)$ will be denoted by $\langle w, \varphi \rangle$. For example, the Dirac measure δ is the map
$$\langle \delta, \varphi \rangle = \varphi(0)$$
while any locally integrable function f defines a distribution by
$$\langle f, \varphi \rangle = \int_\Omega f(x)\varphi(x)\,dx.$$
The space of distributions is denoted by $\mathcal{D}'(\Omega)$. The theory of distributions, as developed by Schwartz [97], has become the basic framework in the field of partial differential equations as well as many other branches of analysis and will be our starting point. Thus we will not present another method for introducing distributions. Rather, our aim is to construct (nonstandard) algebras of generalized functions which contain the space of distributions as a subspace, thereby enabling nonlinear operations with distributions. An important aspect will be the study of the imbedding properties of the space of distributions into algebras.

We shall need the concept of *regularization*. Let $w \in \mathcal{D}'(\Omega)$ have compact support and let $\psi \in \mathcal{E}(\Omega)$. The *convolution* $w \star \psi$ is defined as the function
$$w \star \psi(x) = \langle w(y), \psi(x-y) \rangle, \quad x \in \Omega$$
and belongs to $\mathcal{E}(\Omega)$. If $\psi \in \mathcal{D}(\mathbb{R}^n)$ with $\int \psi(x)\,dx = 1$, and $\psi_\varepsilon(x) = \varepsilon^{-n}\psi(\varepsilon^{-1}x)$, then $w \star \psi_\varepsilon$ converges to w weakly, that is,
$$\langle w \star \psi_\varepsilon, \varphi \rangle = \langle w, \varphi \star \check\psi_\varepsilon \rangle \to \langle w, \varphi \rangle$$
as $\varepsilon \to 0$, where $\check\psi_\varepsilon(x) = \psi_\varepsilon(-x)$. This way the distribution w is approximated by a sequence of regularizations. In case the distribution w does not have compact support, the same procedure still works after inserting cut-off functions converging to one uniformly on compact subsets of Ω. These observations motivate our development of nonstandard representations for distributions.

We work in a polysaturated model over the complex numbers $\mathbb{C}$. A *nonstandard mollifier* is an internal smooth function $\Theta \in {}^*\mathcal{E}(\mathbb{R}^n)$ with support contained in ${}^*K$ for some standard compact set K, and such that $\int \Theta(x)\,dx = 1$. Given a positive infinitesimal $\rho \sim 0$, we define the *nonstandard delta function* Θ_ρ by
$$\Theta_\rho(x) = \rho^{-n}\Theta(\rho^{-1}x).$$
Clearly,
$$\int \Theta_\rho(x){}^*\varphi(x)\,dx \sim {}^*\varphi(0)$$
for all (standard) test functions $\varphi \in \mathcal{D}(\mathbb{R}^n)$. Finally, let χ be an element of ${}^*\mathcal{D}(\Omega)$ which is identically equal to one on the nearstandard points of ${}^*\Omega$. Given a (standard) distribution $w \in \mathcal{D}'(\Omega)$ we define the element $\iota(w)$ of ${}^*\mathcal{E}(\Omega)$ by
$$\iota(w) = ({}^*w\,\chi) \star \Theta_\rho \mid {}^*\Omega.$$
For any $\varphi \in \mathcal{D}(\Omega)$, ${}^*\varphi \star \check\Theta_\rho$ belongs to the monad of ${}^*\varphi$ with respect to the topology of $\mathcal{D}(\Omega)$. Observing that χ is identically equal to one on the support of ${}^*\varphi \star \check\Theta_\rho$ and

invoking the continuity of w it follows that

$$\int \iota(w)(x) \, {}^*\varphi(x) \, dx = \langle {}^*w, {}^*\varphi \star \check{\Theta}_\rho \rangle \sim \langle w, \varphi \rangle$$

for all $\varphi \in \mathcal{D}(\Omega)$. Whence the assignment $w \to \iota(w)$ yields a $\mathbb{C}$-linear imbedding of $\mathcal{D}'(\Omega)$ into ${}^*\mathcal{E}(\Omega)$. We also have the standard copy imbedding $\sigma : f \to {}^*f$ of $\mathcal{E}(\Omega)$ into ${}^*\mathcal{E}(\Omega)$. As it stands, the mapping ι does not have good imbedding properties: neither does it commute with derivatives nor does it extend the standard copy map σ. For this reason we go over to a factor algebra of ${}^*\mathcal{E}(\Omega)$ as follows. We fix a positive infinitesimal number ρ. The differential subalgebra $\mathcal{E}_M(\Omega)$ is defined as the set of all $u \in {}^*\mathcal{E}(\Omega)$ such that: for all (standard) compact sets $K \subset \Omega$ and for all (standard) $\alpha \in \mathbb{N}_0^n$ there is a (standard) $p \in \mathbb{N}$ with

$$\sup_{x \in {}^*K} |\partial^\alpha u(x)| \leq \rho^{-p}.$$

The ideal $\mathcal{N}(\Omega)$ comprises all $u \in {}^*\mathcal{E}(\Omega)$ such that: for all (standard) compact sets $K \subset \Omega$, for all (standard) $\alpha \in \mathbb{N}_0^n$ and for all (standard) $q \in \mathbb{N}$ it holds that

$$\sup_{x \in {}^*K} |\partial^\alpha u(x)| \leq \rho^q.$$

We let

$$^\rho\mathcal{E}(\Omega) = \mathcal{E}_M(\Omega)/\mathcal{N}(\Omega).$$

This is a nonstandard version of one of the constructions developed by Colombeau [23, 24]. It was introduced in [65] and in Todorov [100] and considered also by Hoskins and Sousa Pinto [44].

In a similar fashion, we can define the algebra $^\rho\mathcal{E}(\bar\Omega)$ on the closure of an open set Ω; in the definition, we allow compact sets $K \subset \bar\Omega$. The zero-dimensional algebra $^\rho\mathcal{E}(\mathbb{R}^0)$ coincides with the field $^\rho\mathbb{C}$ of *asymptotic numbers* introduced by Robinson [83].

To imbed the distributions $\mathcal{D}'(\Omega)$ into $^\rho\mathcal{E}(\Omega)$, we take a nonstandard mollifier $\Theta \in {}^*\mathcal{D}(\mathbb{R}^n)$ with the additional property that all its moments vanish:

(1) $$\int_{{}^*\mathbb{R}^n} \Theta(x) \, x^\alpha \, dx = 0, \text{ for all } \alpha \in \mathbb{N}_0^n, \, |\alpha| \geq 1.$$

The fact that such a mollifier exists is proved in a joint paper with Todorov [74] and constitutes a typical application of the saturation principle; note that no standard mollifier could possibly enjoy property (1) for more than finitely many exponents α. Using the corresponding delta function Θ_ρ, we define an imbedding

$$\iota : \mathcal{D}'(\Omega) \to {}^\rho\mathcal{E}(\Omega)$$
$$w \to \text{class of } ({}^*w \chi \star \Theta_\rho \mid {}^*\Omega).$$

The standard copy map σ is extended similarly to

$$\sigma : \mathcal{E}(\Omega) \to {}^\rho\mathcal{E}(\Omega).$$

Due to the localization built into the factor algebra, we now have that

$$\iota(\partial^\alpha w) = \partial^\alpha \iota(w).$$

for all $w \in \mathcal{D}'(\Omega), \alpha \in \mathbb{N}_0^n$, that is, the derivatives on ${}^\rho\mathcal{E}(\Omega)$ extend the distributional ones. Further,

(2) $$\iota \mid \mathcal{E}(\Omega) = \sigma.$$

The argument giving the latter assertion is most easily seen in the case of $f \in \mathcal{D}(\Omega)$, for which

$$({}^*f \star \Theta_\rho - {}^*f)(x) = \int ({}^*f(x - \rho y) - {}^*f(x)) \Theta(y)\,dy.$$

Taylor expansion up to order q and using property (1) of Θ shows that this expression is of order ρ^{q+1} on each ${}^*K$. Applying the same argument to the derivatives, we see that ${}^*f \star \Theta_\rho - {}^*f \in \mathcal{N}(\Omega)$. For general $f \in \mathcal{E}(\Omega)$ an additional argument involving the cut-off χ is needed.

At any rate, from (2) we infer that ι constitutes a homomorphism of algebras on $\mathcal{E}(\Omega)$, meaning that the pointwise product of smooth functions is preserved by the imbedding of the distributions. We arrive at the following list of properties of the algebra ${}^\rho\mathcal{E}(\Omega)$:

(a) It is an associative, commutative differential algebra with unit $\mathbf{1}$ and derivatives $\partial_{x_1}, \ldots \partial_{x_n}$.
(b) The space of distributions $\mathcal{D}'(\Omega)$ is $\mathbb{C}$-linearly imbedded in it.
(c) The derivatives $\partial_{x_1}, \ldots \partial_{x_n}$ coincide with the usual ones on $\mathcal{D}'(\Omega)$.
(d) The multiplication map coincides with the pointwise product on $\mathcal{E}(\Omega)$, imbedded as a subspace of $\mathcal{D}'(\Omega)$.

We note that this list of properties is optimal in the sense that given (a) - (c), the consistency result (d) cannot be improved on, in view of the impossibility result of Schwartz [96]. In particular, the classical product of continuous functions is necessarily changed in every algebra satisfying (a) - (c), a fact we just have to live with.

Two additional properties of the algebra ${}^\rho\mathcal{E}(\Omega)$ needed in the sequel are: It is invariant under superposition with smooth functions of polynomial growth, and restrictions to lower dimensional subspaces are well-defined. Thus we can make sense of nonlinear terms in differential equations as well as initial- and boundary-value problems.

Finally, we need the concept termed *association*, which is nothing but weak distributional infinitesimality. We say that two elements u, v of ${}^\rho\mathcal{E}(\Omega)$ are *associated*, if

$$\int (u(x) - v(x))\, {}^*\varphi(x)\,dx \sim 0$$

for all (standard) test functions $\varphi \in \mathcal{D}(\Omega)$. This will be denoted by

$$u \approx v.$$

The intuition going with the concept of association is that it brings the information incorporated in internal smooth functions down to the level of distribution theory - things are identified, if their actions on standard test functions differ only infinitesimally. As a first application, we may state a consistency result with classical products: If f, g are continuous functions, then

$$\iota(f)\iota(g) \approx \iota(fg),$$

that is, their product in $^\rho\mathcal{E}(\Omega)$ is associated with their classical product.

A discussion of further algebras containing the distributions is in the Appendix. Let us note, however, that the algebra $^\rho\mathcal{E}(\Omega)$ is currently the only known one enjoying all of the properties (a) - (d) as well as having a field as its ring of constants.

2. Semilinear wave equations

As a first application, we look at semilinear wave equations with distributional initial data. The emphasis is on conveying the ideas, so proofs will be rather sketchy. References to the literature are given at the end of the section.

We consider the initial value problem

(3) $$\partial_t^2 u - \Delta u = F(u) + h \quad \text{on } \mathbb{R}^n \times [0,\infty),$$
$$u \mid \{t=0\} = a, \quad \partial_t u \mid \{t=0\} = b \quad \text{on } \mathbb{R}^n$$

where $\Delta = \partial_{x_1}^2 + \ldots + \partial_{x_n}^2$ is the Laplace operator, F is smooth, polynomially bounded, and globally Lipschitz with $F(0) = 0$. For simplicity of exposition, we consider only the cases $n = 1, 2, 3$. Classically, with smooth data, problem (3) has a unique smooth solution u determined as the fixed point of the integral equation

(4) $$u(t) = \frac{d}{dt} E(t) \star a + E(t) \star b + \int_0^t E(t-s) \star (F(u(s)) + h(s)) \, ds$$

in terms of the fundamental solution $E \in \mathcal{E}([0,\infty) : \mathcal{D}'(\mathbb{R}^n))$, which is given by

$$E(t,x) = \begin{cases} \frac{1}{2} H(t - |x|), & n = 1 \\ \frac{1}{2\pi\sqrt{t^2 - |x|^2}} H(t - |x|), & n = 2 \\ \frac{1}{4\pi|x|} \delta(t - |x|), & n = 3 \end{cases}$$

where H denotes the Heaviside function. The following two observations will be important tools for the next result:

(i) $E \in L^\infty([0,T] : \mathcal{M}^1(\mathbb{R}^n))$ for $T > 0$,

that is, the fundamental solution is a measure with finite total mass for each $t \geq 0$, and we have the estimate

(ii) $$\left\| \frac{d}{dt} E(t) \star a \right\| \leq C \|(a, \nabla a)\|$$

for some constant $C > 0$, where $\|\cdot\|$ denotes the supremum norm.

Theorem 1. *Let $a, b \in {}^\rho\mathcal{E}(\mathbb{R}^n)$, $h \in {}^\rho\mathcal{E}(\mathbb{R}^n \times [0,\infty))$, and let F be as described above. Then problem (3) has a unique solution u in $^\rho\mathcal{E}(\mathbb{R}^n \times [0,\infty))$.*

Proof: To prove existence, let $\underline{a}, \underline{b}, \underline{h} \in \mathcal{E}_M \subset {}^*\mathcal{E}$ be representatives of the data. By transfer from the classical existence result, problem (3) has a unique solution $\underline{u} \in {}^*\mathcal{E}$ with these data. We are going to show that $\underline{u}$ belongs to $\mathcal{E}_M$; then its class u in $^\rho\mathcal{E}$ will define a solution. For this we shall derive estimates on $\underline{u}$ in terms of the data, using the corresponding integral equation (4). Let us assume for simplicity that $\underline{a}(x), \underline{b}(x), \underline{h}(x,t)$

vanish for $|x| \geq r$ for some standard $r > 0$. By transfer and the properties (i) and (ii) indicated above, we have the following estimates in terms of the supremum norm

$$||\underline{u}(.,t)|| \leq C||(\underline{a}, \nabla \underline{a}, \underline{b})|| + C||\underline{h}|| + \int_0^t CD||\underline{u}(.,s)|| \, ds$$

where D is the Lipschitz constant of F and C is a constant coming from properties (i) and (ii). The representatives of a, b, h are bounded by ρ^{-p_0} for some $p_0 \geq 0$. Thus, applying Gronwall's inequality, we get

$$\sup_{0 \leq t \leq T} ||\underline{u}(.,t)|| \leq \rho^{-p}$$

for some $p \geq 0$, so $\underline{u}$ satisfies the zero-th order $\mathcal{E}_M$- estimate. The derivatives are estimated inductively. Uniqueness is proven in the same way, this time involving a right hand term belonging to the ideal $\mathcal{N}$. For general $\underline{a}, \underline{b}, \underline{h}$ not satisfying the above support property, a partition of unity argument can be employed. □

Note that the proof prominently employs transfer (plus factorization): the basis for the existence of generalized solutions is the existence of classical solutions. Summing up, we have existence and uniqueness of solutions to a nonlinear equation where the data are allowed to be generalized functions, in particular, distributions. Such a result has no counterpart in the classical setting, where in general nonlinear operations with distributions are impossible. However, in certain cases classical or weak solution do exist. In these situations the question arises what is the relation of the generalized to the classical solutions. For example, if $a \in \mathcal{C}^1(\mathbb{R}^n)$, $b \in \mathcal{C}^0(\mathbb{R}^n)$ and $h \in \mathcal{C}^0(\mathbb{R}^n \times [0, \infty))$, then problem (3) has a unique solution $v \in \mathcal{C}^0(\mathbb{R}^n \times [0, \infty))$.

Proposition 1. *(a) If a, b, h are classical $\mathcal{C}^\infty$-functions, then $u = \iota(v)$ in $^\rho\mathcal{E}(\mathbb{R}^n \times [0, \infty))$, that is, generalized and classical solutions coincide.*

(b) If $a \in \mathcal{C}^1(\mathbb{R}^n)$, $b \in \mathcal{C}^0(\mathbb{R}^n)$, $h \in \mathcal{C}^0(\mathbb{R}^n \times [0, \infty))$, then $u \approx \iota(v)$, that is, generalized and classical solutions are associated.

Proof: (a) Since $\iota \mid \mathcal{E} = \sigma$, we may take $\sigma(v) = \,^*v$ as a representative for the generalized solution u.

(b) A representative of $\iota(a)$ is the internal function $^*a \star \Theta_\rho$, which is infinitely close to $^*a$ in the $\mathcal{C}^1$-topology; similarly for b and h. The assertion follows from the fact that the classical solution operator is continuous in the respective topologies. □

We should like to emphasize that the consistency result (a) for smooth solutions says slightly more than transfer, as it involves the imbedding ι and not just the standard copy imbedding σ. Note also that a strengthening of (b) to equality is in general impossible in algebras containing the distributions, for reasons similar to those underlying the Schwartz impossibility result, see [67, Appendix] for details.

In some cases associated distributions exist even for more singular data. We assume subsequently that $h \equiv 0$. A few basic results on Loeb integration will enter here, for which we refer to Lindstrøm [52, Section II.2].

Proposition 2. Let $a, b \in \mathcal{D}'(\mathbb{R}^n)$ have discrete support. Let $u \in {}^\rho\mathcal{E}(\mathbb{R}^n \times [0, \infty))$ be the generalized solution to problem (3). Assume further that F is bounded. Then $u \approx \iota(v) + \iota(w)$ where $v \in \mathcal{E}([0, \infty) : \mathcal{D}'(\mathbb{R}^n))$ solves

(5)
$$\partial_t^2 v - \Delta v = 0,$$
$$v \mid \{t = 0\} = a, \ \partial_t v \mid \{t = 0\} = b$$

and $w \in \mathcal{C}([0, \infty) : L^1(\mathbb{R}^n))$ solves

(6)
$$\partial_t^2 w - \Delta w = F(w + \mathbf{1}_\Lambda v),$$
$$w \mid \{t = 0\} = 0, \ \partial_t w \mid \{t = 0\} = 0$$

where Λ is the complement of the singular support of v (that is $v \mid \Lambda \in \mathcal{E}(\Lambda)$) and $\mathbf{1}_\Lambda v$ denotes the function which equals v on Λ and zero otherwise.

Proof: Let $\underline{u}, \underline{v} \in {}^*\mathcal{E}$ be the solutions of (3), respectively (5) with data $\underline{a} = {}^*a \star \Theta_\rho$, $\underline{b} = {}^*b \star \Theta_\rho$. By linear propagation of singularities, the singular support of the (standard) distributional solution v of (5) consists of the finitely many light-cone surfaces emanating from the support of a, b. By classical continuous dependence, we may assert that $\underline{v}(x)$ is infinitely close to ${}^*v(x)$ if x belongs to some set ${}^*K, K \subset \Lambda$ compact, that is, for every nearstandard $x \in {}^*\Lambda$. We have

$$(\underline{u} - \underline{v} - {}^*w)(., t) = \int_0^t {}^*E(t-s) \star ({}^*F(\underline{u}) - {}^*F({}^*w + \underline{v}))(., s)\, ds$$
$$+ \int_0^t {}^*E(t-s) \star ({}^*F({}^*w + \underline{v}) - {}^*F({}^*w + {}^*\mathbf{1}_\Lambda v))(., s)\, ds.$$

Using the $\mathcal{M}^1$-property of the fundamental solution E, we get at fixed $t > 0$

$$\int_{-\infty}^\infty |\underline{u} - \underline{v} - {}^*w|\, dx \leq CD \int_0^t \int_{-\infty}^\infty |\underline{u} - \underline{v} - {}^*w|\, dxds$$
$$+ C \int_0^t \int_{-\infty}^\infty |{}^*F({}^*w + \underline{v}) - {}^*F({}^*w + {}^*\mathbf{1}_\Lambda v)|\, dxds.$$

To see that the last integral is infinitesimal, we work on a fixed strip $\mathbb{R}^n \times [0, T]$, on which all integrands have compact support. By hypothesis, ${}^*F$ is limited. Thus the standard part of the last integral equals

$$\int_0^t \int_{-\infty}^\infty {}^\circ|{}^*F({}^*w + \underline{v}) - {}^*F({}^*w + {}^*\mathbf{1}_\Lambda v)|\, L(dxds)$$

for $0 \leq t \leq T$ where $L(dxds)$ denotes Loeb measure. But as noted above, the integrand vanishes except on the monad of the singular support of v, which has Lebesgue measure zero. Thus the integral is infinitesimal. Finally, applying Gronwall's inequality yields that

$$\sup_{0 \leq t \leq T} \int_{-\infty}^\infty |\underline{u} - \underline{v} - {}^*w|\, dx$$

is infinitesimal. Now $\underline{u}$ is a representative of the generalized solution u to (3), and $\underline{v}$ is weakly infinitely close to ${}^*v$, by classical linear continuous dependence of distributional solutions on the data. It follows that $u - \iota(v) - \iota(w)$ is associated with zero, as desired.

More detailed arguments can be found in [69, §15 - §17]. Colombeau type solutions in the one-dimensional case are constructed in [64] and Biagioni [15]; the three-dimensional nonlinear Klein-Gordon equation is treated in Colombeau [24]; for linear hyperbolic first order systems with generalized functions as coefficients in any space dimension see [48]. For the computation of associated distributions as in Proposition 2, commonly called *delta waves*, we refer to [36, 63, 75, 76], Gramchev [38], Rauch and Reed [82].

3. Quasilinear hyperbolic systems

This section serves to present further methods for constructing generalized solutions to nonlinear partial differential equations. These methods are not based on the existence of classical solutions and the transfer principle, but produce and adjust nonstandard objects in direct ways. The ideas can be nicely demonstrated by means of systems of the form

$$\partial_t u + G(u)\partial_x u = 0, \quad x \in \mathbb{R}, t > 0$$
$$u(x,0) = a(x), \quad x \in \mathbb{R}$$

where $u = (u_1, \ldots, u_n)$ and G is a smooth, $(n \times n)$-matrix valued function such that $G(u)$ has real and distinct eigenvalues for all $u \in \mathbb{R}^n$ (strict hyperbolicity). An important special class is constituted by systems of conservation laws, which can be written in gradient form as

$$\partial_t u + \partial_x F(u) = 0, \quad x \in \mathbb{R}, t > 0$$
$$u(x,0) = a(x), \quad x \in \mathbb{R}.$$

In a conservation law, the total quantity $\int u(x,t)\,dx$ is constant in time. Two typical examples are the Euler system of isentropic gas dynamics

$$\rho_t + (\rho v)_x = 0$$
$$(\rho v)_t + (\rho v^2)_x + p(\rho)_x = 0$$

where ρ is the density, v the velocity and p the pressure function of a one-dimensional fluid in spatial coordinates (as a general reference, see Whitham [105]). The second example arises in elasto-plasticity:

$$\rho_t + (\rho v)_x = 0$$
$$(\rho v)_t + (\rho v^2)_x = \sigma_x$$
$$\sigma_t + v\,\sigma_x = v_x$$

where σ is the stress and ρ, v are as above (for a derivation, see [70]). The term $v\,\sigma_x$ in the last line cannot be written as a gradient, hence the system is non-conservative.

The reasons for the need of generalized solutions are twofold. First, almost all solutions of interest to hyperbolic systems develop discontinuities in finite time, called shock waves, even if the data are smooth. In the conservative case, this can be handled by the classical weak solution concept: First apply the nonlinear term F to the discontinuous function u, then take distributional derivatives. But in the nonconservative case, unavoidable terms involving products of jump functions with Dirac measures arise, which have no classical meaning. Second, even in conservative systems, classical weak solutions may fail to exist, while regularization methods lead to delta-functions superimposing the shock waves. These effects require a framework of generalized functions.

The first method of constructing generalized solutions applies to the so-called Riemann problem, where the initial data are piecewise constant. Classically, solutions are sought in the shape of piecewise constant travelling waves. Thus we put

$$u(x,t) = u_l + (u_r - u_l) Y(x - ct)$$

where Y is an element of $^*\mathcal{E}(\mathbb{R})$, associated with the Heaviside function, the discontinuity is along the line $x = ct$ with c to be determined, while u_l, u_r are the constant values of the initial data to the left, respectively right of $x = 0$. In a sense, Y may be viewed as a preassigned shape function describing the infinitesimal behavior at the shock front. Inserting this $u(x,t)$ into the equation

$$\partial_t u + G(u) \partial_x u = 0$$

or into

$$\partial_t u + G(u) \partial_x u \approx 0$$

allows the calculation of the so-called Rankine-Hugoniot conditions relating the quantities u_l, u_r and the speed c, even in the nonconservative case, and it allows to model delta-functions placed on top of the shock curves.

The procedure may be demonstrated with the aid of a simple example, the nonconservative system

(7)
$$\begin{aligned} u_t + u\,u_x &= \sigma_x \\ \sigma_t + u\,\sigma_x &\approx u_x \,. \end{aligned}$$

We seek solutions of the form

(8)
$$\begin{aligned} u(x,t) &= u_l + (u_r - u_l) Y(x - ct) \\ \sigma(x,t) &= \sigma_l + (\sigma_r - \sigma_l) Z(x - ct) \end{aligned}$$

where both Y and Z are associated with the Heaviside function (actually, we require that $Y^2 \approx Y, Y^3 \approx Y$ in addition). To be sure, Y and Z need not be the same; in fact, it will turn out that Y determines Z functionally. Inserting (8) in the first line of (7) gives

$$(-c + u_l)(u_r - u_l) Y' + (u_r - u_l)^2 Y Y' = (\sigma_r - \sigma_l) Z'.$$

Using that $Y' \approx \delta, Z' \approx \delta$, and $YY' = \left(\tfrac{1}{2} Y^2\right)' \approx \tfrac{1}{2} Y' \approx \tfrac{1}{2}\delta$ we arrive at the first Rankine-Hugoniot condition

$$(-c + u_l)(u_r - u_l) = -\frac{1}{2}(u_r - u_l)^2 + (\sigma_r - \sigma_l).$$

On the other hand, the first line in (7) defines Z as a function of Y, and we compute
$$(\sigma_r - \sigma_l) Y Z' = (-c + u_l)(u_r - u_l) Y Y' + (u_r - u_l)^2 Y^2 Y'.$$
Using this in the second line of (7) together with $Y^2 Y' \approx \frac{1}{3}\delta$ we finally obtain the second Rankine-Hugoniot condition
$$(\sigma_r - \sigma_l)^2 = (u_r - u_l)^2 \left(1 - \frac{1}{12}(u_r - u_l)^2\right).$$
These two conditions relate $u_l, u_r, \sigma_l, \sigma_r$, and c. This is used to solve the Riemann problem (with given $u_l, u_r, \sigma_l, \sigma_r$) with two shocks along the two straight lines $x = c_1 t$, $x = c_2 t$ with intermediate values u_i, σ_i. The four unknowns c_1, c_2, u_i, σ_i are determined from the four Rankine-Hugoniot conditions along the shocks. Solutions of this type are employed in numerical schemes of Godunov type.

We point out that it was essential to write system (7) with equality in the first line and association in the second. Had we put equality twice, shock wave solutions would not exist (because we could algebraically transform the equation and this way obtain contradicting Rankine-Hugoniot conditions, a well-known effect already present in the classical theory of conservation laws). Had we put association twice, Z would not be determined by Y, and uniqueness of Rankine-Hugoniot conditions would be lost. A sample of references to this first method of constructing solutions to the Riemann problem, together with numerical schemes, are Aragona and Villarreal [8], Biagioni [16], Colombeau and coworkers [25, 26, 33, 34], see also [70]. Concerning the occurrence of delta-functions on shock curves we refer to Berger, Colombeau and Moussaoui [13, 14], Joseph [46], Keyfitz and Kranzer [47], Tan and Zhang [99], see also [69, Example 20.9].

The second method of constructing generalized solutions to conservation laws is a nonstandard version of the viscosity method. We complement the system with a second order term making it parabolic:
$$\begin{aligned}\partial_t u + \partial_x F(u) &= \mu \partial_x^2 u \\ u(x, 0) &= a(x)\end{aligned} \qquad (9)$$
where $\mu \in {}^*\mathbb{R}, \mu > 0$. The point is that we can take μ infinitely small, this way incorporating the concept of *vanishing viscosity* in our notion of solution.

For the time being, we consider (9) as a scalar conservation law ($n = 1$). In the classical setting, adjoining the second order term $\mu \partial_x^2 u$ brings about that unique bounded smooth solutions exist for bounded smooth initial data. To exploit this fact we work in a modified algebra ${}^\rho \mathcal{E}_g(\mathbb{R}^n \times [0, \infty))$, where the bounds in terms of ρ^{-p} and ρ^q hold globally on all strips ${}^*\mathbb{R} \times [0, T]$, and not just on compacts.

Theorem 2. *Let $a \in {}^\rho \mathcal{E}_g(\mathbb{R})$ and assume that either $\sup_{x \in {}^*\mathbb{R}} |a(x)|$ is limited or F is globally Lipschitz. Then (9) has a solution $u \in {}^\rho \mathcal{E}_g(\mathbb{R} \times [0, \infty))$.*

Proof: It follows as in Section 2 by rewriting (9) as an integral equation, employing transfer on representatives, and using either the maximum principle or the Lipschitz property of F to deduce the zero-order bound in terms of ρ^{-p}. □

There are also intermediate cases where the growth of F at infinity is related to the order p appearing in the bound ρ^{-p} of the initial data. Uniqueness can be obtained for limited initial data as well as general ones subject to some mild additional conditions. Systems with globally Lipschitz nonlinearity are treated in the same fashion. In general, we need invariant regions as a substitute for the maximum principle in the case of systems.

As in Section 2, the method produces generalized solutions with generalized functions as initial data, which have no counterpart in the classical setting. The following consistency results are available in cases where classical solutions do exist:

(1) $\mu \in \mathbb{R}, \mu > 0, a \in \mathcal{E}(\mathbb{R})$. In this case, the generalized solution is equal with the classical smooth solution.
(2) $\mu \in \mathbb{R}, \mu > 0, a \in L^\infty(\mathbb{R})$. In this case, the generalized solution is associated with the classical solution.
(3) $\mu \in {}^*\mathbb{R}, \mu > 0, \mu \sim 0, a \in L^\infty(\mathbb{R})$. In this case, the generalized solution is associated with that classical weak solution to the conservation law

$$\partial_t v + \partial_x F(v) = 0$$

with the same data, which satisfies the so-called entropy condition (meaning for example that shocks are stable).

This latter case is noteworthy in view of the notorious failure of uniqueness of classical weak solutions. Uniqueness can be brought about, in the classical setting, only by imposing additional constraints; this is the role of the entropy condition (see Smoller [98]). In the nonstandard framework, shock waves can be modeled by adding an infinitely small viscosity term, and then automatically the entropy condition obtains. We refer to joint work with Biagioni [17] and Wang [77], see also [69, 70].

The third method of constructing generalized solutions is by means of *regularized derivatives*. We rewrite the quasilinear system in the form

(10) $$\begin{aligned} \partial_t u + G(u)\tilde{\partial}_x u &= 0 \\ u(x,0) &= a(x) \end{aligned}$$

where the regularized derivative $\tilde{\partial}_x$ is defined by

$$\tilde{\partial}_x u = \text{class of } \underline{u} \star \partial_x \Theta_{h(\rho)} \text{ in } {}^\rho\mathcal{E}_g(\mathbb{R} \times [0,\infty)).$$

Here the convolution is effected in the x-variable only, $\underline{u}$ is a representative of u, Θ_ρ is a nonstandard delta-function as in Section 1, and $h(\rho)$ is an infinitesimal obtained from ρ by a suitable scaling function h. On the level of representatives, (10) is an ordinary integro-differential equation:

$$\partial_t \underline{u}(x,t) + {}^*G(\underline{u}) \int_{-\infty}^\infty \underline{u}(x-y) \partial_x \Theta_{h(\rho)}(y)\, dy \in \mathcal{N}(\mathbb{R} \times [0,\infty)),$$
$$\underline{u}(x,0) - \underline{a}(x) \in \mathcal{N}(\mathbb{R}),$$

the solution of which can be obtained by classical methods and transfer. Depending on the properties of G, the infinitesimal $h(\rho)$ can be adjusted so as to have unique solutions in the factor algebra ${}^\rho\mathcal{E}_g(\mathbb{R} \times [0,\infty))$.

Consistency results with classical solutions are available, among others, in the case of linear and nonlinear symmetric hyperbolic systems as well as scalar conservation laws

(11) $$\partial_t u + \tilde{\partial}_x F(u) = 0.$$

In the latter case, the generalized solution is again associated with the classical weak solution satisfying the entropy condition, provided the support of the mollifier Θ is chosen on the appropriate side of zero, depending on the direction of wave propagation in (11). References are Colombeau and Heibig [28], Heibig and Moussaoui [40] as well as the joint work [29, 30, 31]. The concept of regularized derivatives was used independently by Rosinger [86] to develop unconditionally stable numerical schemes.

4. Further applications

Here is a brief survey of further applications that have been undertaken so far.

(1) *Generalized solutions.* Various other classes of equations have been treated, for example:

- hyperbolic equations with discontinuous coefficients, arising e.g. in transmission problems in discontinuous media [48, 66];
- semilinear parabolic equations with singular data, as for example (see Colombeau and Langlais [32], Langlais [49]

$$\partial_t u - \Delta u = f(u), \ u \mid \{t = 0\} = \delta \, ;$$

- dispersive equations, as the Korteweg de-Vries equation [18]

$$u_t + u\, u_x + u_{xxx} = 0 \, ;$$

- equations from kinetic gas theory of the form

$$u_t + A\, u_x = Q(u, u) \, ,$$

for which we refer to the general results of Arkeryd [9], see also [36, 68].
- Schrödinger operators $-\Delta + V$ with multiples or powers of delta functions as potential V: Albeverio, Fenstad, Høegh-Krohn [1], Albeverio, Fenstad, Høegh-Krohn, Lindstrøm [2] and references therein, Nelson [61], Rosinger [84];
- special conservative and nonconservative systems from elasticity and hydrodynamics, see the references given in Section 3, especially Colombeau [26];
- analytic partial differential equations: existence of global solutions in the algebra of *nowhere dense type*, see Rosinger [87, 88, 89] and the joint work [73];
- linear partial differential equations with constant, possibly generalized coefficients, fundamental solutions, local solvability; see work of Nedeljkov, Pilipović, and Scarpalezos [58, 60, 81], Todorov [102, 103]. For results on the $\bar{\partial}$-operator and generalized analytic functions we refer to Aragona [6], the references therein, as well as [7], Colombeau and Galé [27].

(2) *Regularity theory and pseudodifferential operators.* In the classical situation, we have elliptic regularity saying for example:

If $u \in \mathcal{D}'(\Omega)$ and $\Delta u \in \mathcal{E}(\Omega)$, then $u \in \mathcal{E}(\Omega)$.

Trivially, this is not true if we replace $\mathcal{D}'(\Omega)$ by $^*\mathcal{E}(\Omega)$ or $^\rho\mathcal{E}(\Omega)$ (take nonstandard constants as solutions to $\Delta u = 0$). However, in the Colombeau setting there is a substitute for $\mathcal{E}(\Omega)$, a subalgebra $\mathcal{G}^\infty(\Omega)$ of the standard Colombeau algebra $\mathcal{G}(\Omega)$ with $\mathcal{G}^\infty \cap \mathcal{D}' = \mathcal{E}$, such that:

If $u \in \mathcal{G}(\Omega)$ and $\Delta u \in \mathcal{G}^\infty(\Omega)$, then $u \in \mathcal{G}^\infty(\Omega)$.

This can be generalized to arbitrary elliptic equations, propagation of singularities in hyperbolic equations, and a symbolic calculus of pseudodifferential operators is possible, see [69], joint work with Gramchev [39], Nedeljkov and Pilipović [59], Pilipović [80], Scarpalezos [94].

(3) *Calculus of variations.* For Lagrangian functionals we can assert: If

$$\int L(u, \partial u, \ldots)\, dx \in {}^\rho\mathbb{R}$$

is minimal, then $u \in {}^\rho\mathcal{E}(\Omega)$ satisfies the Euler-Lagrange equations. Thus generalized solutions of partial differential equations may arise as minimizing points of Lagrangian functionals, as in the classical case. Two examples:

$$\iint \left(\frac{1}{2}|u_t|^2 - \frac{1}{2}|\nabla u| - V(u)\right) dx dt, \quad u \in {}^\rho\mathcal{E}(\mathbb{R}^{n+1})$$

is the Lagrange functional of the Klein-Gordon equation

$$\partial_t^2 u - \Delta u + V'(u) = 0 \text{ in } {}^\rho\mathcal{E}(\mathbb{R}^{n+1}).$$

The second example concerns singular potentials in ordinary differential equations:

$$\int_{t_0}^{t_1} \left(\frac{1}{2}|\dot{x}|^2 + \delta(x)\right) dt$$

is the Lagrangian for

$$\ddot{x}(t) - \delta'(x(t)) = 0 \text{ in } {}^\rho\mathcal{E}(\mathbb{R}).$$

There is joint work in progress with Hermann [42]. A geometric theory in the framework of Nonstandard Analysis has been developed recently by Tuckey [104]. For optimization by nonstandard methods, see Rubio [92].

(4) *Lie groups and generalized functions.* The idea, due to Rosinger, is to study nonlinear transformation groups on algebras of generalized functions (Lie theory). Classical Lie groups act by transforming the graphs of sufficiently differentiable functions. For distributions, such operations make no sense in general (distributions do not have graphs). But generalized functions do have generalized graphs, and thus a generalized Lie theory is possible again. Investigations in the Colombeau framework as well as the algebras of *nowhere dense type* have been started in [73] and Rosinger and Walus [91]. The exploitation of nonstandard ideas in this respect appears as a promising task, remaining to be done.

(5) *Numerical methods.* Many of the theoretical ideas have resulted in new numerical schemes. We mention the generalized Riemann solvers of Section 2; the different interpretations of differential equations with equality or association; and regularized derivatives. The latter lead to a class of schemes inhabiting the region between finite difference methods and spectral methods. We refer to Colombeau [26], the references therein, and Rosinger [86].

(6) *Generalized stochastic analysis.* Similar to algebras of generalized functions, one can construct algebras of generalized stochastic processes. The idea is that generalized processes like *white noise* $\dot{W}$, the distributional derivative of Brownian motion, have representatives as ∗-smooth processes. This way one can make sense of general nonlinear stochastic ordinary differential equations

$$\dot{X}_t = F(X_t, \dot{W}_t)$$

as well as stochastic partial differential equations which do not have classical processes as solutions, like

$$u_{tt} - \Delta u = f(u) + \dot{W};$$
$$u_t + u\,u_x = \mu\,u_{xx} + \dot{W}.$$

In the latter two examples, $\dot{W}$ denotes space-time white noise, obtained by taking (n+1) derivatives of the Brownian sheet on $\mathbb{R}^{n+1}$. This approach takes a position quite opposite to the hyperfinite setting (which constitutes a central line of development in Nonstandard Analysis: the term *radically elementary probability theory* was coined by Nelson [62]). The generalized functions approach we refer to here works with a continuous time line, but generalized stochastic processes are represented as ∗-smooth objects, the main ingredient being regularization (not discretization). This could be called *radically smooth stochastic analysis*, and allows applications of ideas from the nonlinear theory of generalized functions. We refer to [72], Albeverio, Haba and Russo [3], Lozanov-Crvenković and Pilipović [54], Russo [93], Holden, Lindstrøm, Øksendal, Ubøe, Zhang [43].

This paper has focussed on the study of generalized solutions to nonlinear partial differential equations with singular data. Of course, Nonstandard Analysis has found many other important and beautiful applications to classical questions arising in this field. We think of the work of Loeb in potential theory [53], see also Bliedtner and Loeb [19] for a recent presentation in a general framework; the work of Arkeryd on the Boltzmann equation and related topics [9, 10, 11, 12]; the work of Capiński and Cutland [20, 21, 22] on the Navier-Stokes equation. Certainly the perspective taken here covers only part of the broad spectrum, but hopefully has exhibited some of the new thinking about the nature of generalized functions and generalized solutions to partial differential equations.

Appendix

The purpose of this appendix is to describe some of the important standard algebras containing the distributions as well as their nonstandard counterparts. To achieve a unified presentation we take the infinite power

$$\mathcal{F}(\Omega) = (\mathcal{E}(\Omega))^{(0,\infty)}$$

as underlying structure. Its elements are sequences $(u_\varepsilon)_{\varepsilon>0}$ of smooth functions. With the operations defined componentwise, $\mathcal{F}(\Omega)$ is a differential algebra. The sequential models of generalized functions will be factor algebras of the form

$$\mathcal{A}(\Omega)/\mathcal{I}(\Omega),$$

where $\mathcal{A}(\Omega)$ is a subalgebra of $\mathcal{F}(\Omega)$ and $\mathcal{I}(\Omega)$ is a differential ideal in $\mathcal{A}(\Omega)$.

Credit for the first construction in this spirit is due to Schmieden and Laugwitz [95], Laugwitz [50, 51], based on the ring $\mathbb{R}^{(0,\infty)}$ as extended number system. The notions of infinitesimally, internal functions, as well as representations of distributions by internal smooth functions are available, and it is surprising to see how much of elementary Nonstandard Analysis can be done in this setting. For example, Robinson's sequential lemma holds (see [69, Exercise 23.5] and the detailed investigations of Palmgren [78, 79]). Generalized functions in the sense of Laugwitz and Schmieden are obtained by taking

$$\mathcal{A}(\Omega) = \mathcal{F}(\Omega)$$
$$\mathcal{I}(\Omega) = \{(u_\varepsilon)_{\varepsilon>0} \in \mathcal{A}(\Omega) : \exists \eta > 0 \text{ such that } u_\varepsilon \equiv 0 \text{ for } 0 < \varepsilon < \eta\}.$$

As noted before, localization is a useful ingredient when imbedding properties of $\mathcal{D}'(\Omega)$ are under consideration. A corresponding construction was considered by Egorov [37], namely

$$\mathcal{A}(\Omega) = \mathcal{F}(\Omega)$$
$$\mathcal{I}(\Omega) = \{(u_\varepsilon)_{\varepsilon>0} \in \mathcal{A}(\Omega) : \forall K \subset \Omega \text{ compact } \exists \eta > 0 \text{ such that } u_\varepsilon \mid K \equiv 0 \text{ for } 0 < \varepsilon < \eta\}.$$

Rosinger [84, 85, 87, 88] developed the general sequential theory of algebras containing the space of distributions. Specifically, he considered ideals defined by vanishing properties (zero-set-filters) of the elements $(u_\varepsilon)_{\varepsilon>0}$ and succeeded in characterizing those ideals $\mathcal{I}(\Omega)$ which admit imbeddings of the distributions into $\mathcal{F}(\Omega)/\mathcal{I}(\Omega)$, see [87, 88, 90]. The emphasis on studying imbedding properties of $\mathcal{D}'(\Omega)$ as well as addressing the question of consistency with classically definable nonlinear operations is due to Colombeau and Rosinger. In fact, Colombeau introduced his construction for the purpose of achieving properties (a) - (d) of Section 1.

In its elementary version, the algebra of Colombeau can be described by

$$\mathcal{A}(\Omega) = \{(u_\varepsilon)_{\varepsilon>0} \in \mathcal{F}(\Omega) : \forall K \subset \Omega \text{ compact } \forall \alpha \in \mathbb{N}_0^n$$
$$\exists p > 0 \, \exists \eta > 0 \text{ such that}$$
$$\sup\nolimits_{x \in K} |\partial^\alpha u_\varepsilon(x)| \leq \varepsilon^{-p} \text{ for } 0 < \varepsilon < \eta\}.$$

$$\mathcal{I}(\Omega) = \{(u_\varepsilon)_{\varepsilon>0} \in \mathcal{A}(\Omega) : \forall K \subset \Omega \text{ compact } \forall \alpha \in \mathbb{N}_0^n$$
$$\forall q > 0 \, \exists \eta > 0 \text{ such that}$$
$$\sup\nolimits_{x \in K} |\partial^\alpha u_\varepsilon(x)| \leq \varepsilon^q \text{ for } 0 < \varepsilon < \eta\}.$$

Various versions of the latter algebra are obtained by changing the parameter set $(0, \infty)$, see Colombeau [23, 24], Colombeau and Meril [35], or by replacing the defining seminorms by other ones, like L^p-norms (joint work with Biagioni [18]). Ultimately, this leads to a topological vector space setting for constructions of this type, see Antonevich and Radyno [4, 5]. The ideal $\mathcal{I}(\Omega)$ above was also considered by Maslov and Tsupin [56, 57] for the purpose of describing negligible sequences in asymptotic expansions.

Transition to nonstandard versions is achieved by introducing a free ultrafilter $\mathcal{U}$ on $(0, \infty)$ containing all terminal sets $(0, \eta), \eta > 0$ and replacing the defining estimate in $\mathcal{A}(\Omega)$:

$$\sup\nolimits_{x \in K} |\partial^\alpha u_\varepsilon(x)| \leq \varepsilon^{-p} \text{ for } 0 < \varepsilon < \eta$$

by

$$\{\varepsilon > 0 : \sup\nolimits_{x \in K} |\partial^\alpha u_\varepsilon(x)| \leq \varepsilon^{-p}\} \in \mathcal{U},$$

and similarly for $\mathcal{I}(\Omega)$. Thereby, the resulting algebra $\mathcal{A}(\Omega)/\mathcal{I}(\Omega)$ is realized as a factor algebra in the elementary ultrapower $^*\mathcal{E}(\Omega) = \mathcal{E}(\Omega)^{(0,\infty)}/\sim_\mathcal{U}$. Letting u be the class of $(u_\varepsilon)_{\varepsilon>0}$ in $^*\mathcal{E}(\Omega)$ and ρ the infinitesimal number defined by the sequence $(\varepsilon)_{\varepsilon>0}$ we arrive at the previously given definition of $^\rho\mathcal{E}(\Omega)$. At this stage, we can dispense with the specific ultrapower construction of $^\rho\mathcal{E}(\Omega)$ and work in polysaturated models to have the full range of nonstandard reasoning at our disposal. As remarked in Section 1, the number field $^\rho\mathbb{R}$ was introduced earlier by Robinson [83]; the same kind of factorization in the context of normed spaces was investigated by Luxemburg [Lux].

Here are nonstandard versions of the other algebras mentioned above. The algebra of Laugwitz and Schmieden becomes nothing but $^*\mathcal{E}(\Omega)$ itself. The algebra of Egorov turns into $^*\mathcal{E}(\Omega)/\mathcal{J}(\Omega)$, where

$$\mathcal{J}(\Omega) = \{u \in {}^*\mathcal{E}(\Omega) : u \mid {}^*K \equiv 0 \text{ for all compact } K \subset \Omega\}$$
$$= \{u \in {}^*\mathcal{E}(\Omega) : u(x) = 0 \text{ for all nearstandard } x \in {}^*\Omega\}.$$

In this version it was employed by Todorov [103]. The algebra of *nowhere dense type* referred to in Section 4, introduced by Rosinger [87, 88], is of importance in the study of shock waves and analytic partial differential equations. Its nonstandard analogue is obtained as $^*\mathcal{E}(\Omega)/\mathcal{J}(\Omega)$ with the *nowhere dense ideal*

$$\mathcal{J}(\Omega) = \{u \in {}^*\mathcal{E}(\Omega) : \exists \Gamma \subset \Omega \text{ closed, nowhere dense, such that} \\ u(x) = 0 \text{ for all } x \in {}^*\Omega \setminus {}^*\Gamma \text{ with } d(x,{}^*\Gamma) > \rho\},$$

where d denotes the distance function and ρ is some fixed positive infinitesimal. A discussion and comparison of the respective properties of various algebras can be found in [71], see also Todorov [101].

Here is a list of monographs available on the nonlinear theory of generalized functions (standard viewpoint): Biagioni [16], Colombeau [23, 24, 26], Hermann [41], Pilipović [80], Rosinger [84, 85, 87, 88]. The monograph [69] exhibits many of the applications outlined above and contains indications on nonstandard methods. A more thorough presentation of nonstandard algebras of generalized functions is in Hoskins and Sousa Pinto [45]. Applications of Nonstandard Analysis to ordinary/partial differential equations as well as linear/nonlinear operations on distributions appear to be more scattered in the nonstandard literature. A wealth of material can be found in Albeverio, Fenstad, Høegh-Krohn, Lindstrøm [2] and Rubio [92] as well as the extensive bibliographies there.

References

[1] S. ALBEVERIO, J.E. FENSTAD, R. HØEGH-KROHN: *Singular perturbations and nonstandard analysis.* Trans. Am. Math. Soc. **252**(1979), 275 - 295.

[2] S. ALBEVERIO, J.E. FENSTAD, R. HØEGH-KROHN, T. LINDSTRØM: *Nonstandard Methods in Stochastic Analysis and Mathematical Physics.* Acad. Press, Orlando 1986.

[3] S. ALBEVERIO, Z. HABA, F. RUSSO: *Trivial solutions for a non-linear two-space dimensional wave equation perturbed by space-time white noise.* Prépublication 93-17, Université de Provence, Marseille 1993.

[4] A.B. ANTONEVICH, YA.V. RADYNO: *On a general method for constructing algebras of generalized functions.* Soviet Math. Dokl. **43**(1991), 680 - 684.

[5] A.B. ANTONEVICH, YA.V. RADYNO: *Sur la méthode générale de l'injection d'espaces topologiques vectorielles dans les algèbres.* Preprint 1994.

[6] J. ARAGONA: *Some results for the $\bar{\partial}$ operator on generalized differential forms.* J. Math. Anal. Appl. **180**(1993), 458 - 468.

[7] J. ARAGONA, J.F. COLOMBEAU: *On the $\bar{\partial}$-Neumann problem for generalized functions.* J. Math. Anal. Appl. **110**(1985), 179 - 199.

[8] J. ARAGONA, F. VILLARREAL: *Colombeau's theory and shock waves in a problem of hydrodynamics.* J. d'Analyse Math. **61**(1993), 113 - 144.

[9] L. ARKERYD: *Some examples of NSA methods in kinetic theory.* In: L. ARKERYD, C. CERCIGNANI, P.L. LIONS, D.A. MARKOVICH, M. PULVIRENTI, S.R.S. VARADHAN (EDS.), *Nonequilibrium Problems in Many-Particle Systems.* Lecture Notes Math. Vol. 1551, Springer-Verlag, Berlin 1993.

[10] L. ARKERYD: *The nonlinear Boltzmann equation far from equilibrium.* In: N. CUTLAND (ED.), *Nonstandard Analysis and its Applications.* Cambridge Univ. Press, Cambridge 1988, 321 - 340.

[11] L. ARKERYD: *Loeb solutions of the Boltzmann equation.* Arch. Rat. Mech. Anal. **86**(1984), 85 - 97.

[12] L. ARKERYD: *On the generation of measure-valued solutions on $I\!R^n$ from Loeb solutions on $^*I\!R^n$ for the Boltzmann equation.* Transport Theory Statist. Phys. **18**(1989), 133 - 140.

[13] F. BERGER, J.F. COLOMBEAU, M. MOUSSAOUI: *Solutions mesures de Dirac de systèmes de lois de conservation et applications numériques.* C. R. Acad. Sci. Paris Sér. I **316**(1993), 989 - 994.

[14] F. BERGER, J.F. COLOMBEAU, M. MOUSSAOUI: *Delta shocks and numerical applications.* Preprint 1994.

[15] H.A. BIAGIONI: *The Cauchy problem for semilinear hyperbolic systems with generalized functions as initial conditions.* Results Math. **14**(1988), 231 - 241.

[16] H.A. BIAGIONI: *A Nonlinear Theory of Generalized Functions.* Lecture Notes Math. Vol. 1421, Springer-Verlag, Berlin 1990.

[17] H.A. BIAGIONI, M.OBERGUGGENBERGER: *Generalized solutions to Burgers' equation.* J. Diff. Eqs. **97**(1992), 263 - 287.

[18] H.A. BIAGIONI, M. OBERGUGGENBERGER: *Generalized solutions to the Korteweg - de Vries and the regularized long-wave equations.* SIAM J. Math. Anal. **23**(1992), 923 - 940.

[19] J. BLIEDTNER, P. LOEB: *A reduction technique for limit theorems in analysis and probability.* Ark. mat. **30**(1992), 25 - 43.

[20] M. CAPIŃSKI, N. CUTLAND: *Statistical solutions of PDEs by nonstandard densities.* Monatshefte Math. **111**(1991), 99 - 117.

[21] M. CAPIŃSKI, N. CUTLAND: *A simple proof of existence of weak and statistical solutions of Navier-Stokes equations.* Proc. Royal Soc. London **436**(1992), 1 - 11.

[22] M. CAPIŃSKI, N. CUTLAND: *Nonstandard Methods for Stochastic Fluid Mechanics.* World Scientific Publ., Singapore, to appear.

[23] J.F. COLOMBEAU: *New Generalized Functions and Multiplication of Distributions.* North-Holland Math. Studies Vol. 84, North-Holland, Amsterdam 1984.

[24] J.F. COLOMBEAU: *Elementary Introduction to New Generalized Functions.* North-Holland Math. Studies Vol. 113, North-Holland, Amsterdam 1985.

[25] J.F. COLOMBEAU: *The elastoplastic shock problem as an example of the resolution of ambiguities in the multiplication of distributions.* J. Math. Phys. **30**(1989), 2273 - 2279.

[26] J.F. COLOMBEAU: *Multiplication of Distributions. A tool in mathematics, numerical engineering and theoretical physics.* Lecture Notes Math. Vol. 1532, Springer-Verlag, Berlin 1992.

[27] J.F. COLOMBEAU, J.E. GALÉ: *Holomorphic generalized functions.* J. Math. Anal. Appl. **103**(1984), 117 - 133.

[28] J.F. COLOMBEAU, A. HEIBIG: *Generalized solutions to Cauchy problems.* Monatshefte Math. **117**(1994), 33 - 49.

[29] J.F. COLOMBEAU, A. HEIBIG, M. OBERGUGGENBERGER: *Generalized solutions to partial differential equations of evolution type.* Preprint 1993.

[30] J.F. COLOMBEAU, A. HEIBIG, M. OBERGUGGENBERGER: *Le problème de Cauchy dans un espace de fonctions généralisées I*, C. R. Acad. Sci. Paris Ser. I **317**(1993), 851 - 855.

[31] J.F. COLOMBEAU, A. HEIBIG, M. OBERGUGGENBERGER: *Le problème de Cauchy dans un espace de fonctions généralisées II*, C. R. Acad. Sci. Paris Ser. I, to appear.

[32] J.F. COLOMBEAU, M. LANGLAIS: *Existence and uniqueness of solutions of nonlinear parabolic equations with Cauchy data distributions.* J. Math. Anal. Appl. **145**(1990), 186 - 196.

[33] J.F. COLOMBEAU, A.Y. LE ROUX: *Multiplications of distributions in elasticity and hydrodynamics.* J. Math. Phys. **29**(1988), 315 - 319.

[34] J.F. COLOMBEAU, A.Y. LE ROUX, A. NOUSSAIR, B. PERROT: *Microscopic profiles of shock waves and ambiguities in multiplications of distributions.* SIAM J. Num. Anal. **26**(1989), 871 - 883.

[35] J.F. COLOMBEAU, A. MERIL: *Generalized functions and multiplication of distributions on C^∞ manifolds.* J. Math. Anal. Appl.**186**(1994), 357 - 364.

[36] J.F. COLOMBEAU, M. OBERGUGGENBERGER: *On a hyperbolic system with a compatible quadratic term: generalized solutions, delta waves, and multiplication of distributions.* Comm. Part. Diff. Eqs. **15**(1990), 905 - 938.

[37] YU. V. EGOROV: *A contribution to the theory of generalized functions.* Russian Math. Surveys **45:5**(1990), 1 - 49.

[38] T. GRAMCHEV: *Semilinear hyperbolic systems and equations with singular initial data.* Monatshefte Math. **112**(1991), 99 - 113.

[39] T. GRAMCHEV, M. OBERGUGGENBERGER: *Regularity theory and pseudodifferential operators in algebras of generalized functions.* Preprint 1995.

[40] A. HEIBIG, M. MOUSSAOUI: *Generalized and classical solutions of nonlinear parabolic equations.* Nonlinear Anal., to appear.

[41] R. HERMANN: *C-O-R Generalized Functions, Current Algebras and Control.* Interdisciplinary Math. Vol. 30, Math Sci Press, Brookline, to appear.

[42] R. HERMANN, M. OBERGUGGENBERGER: *Generalized functions, ordinary differential equations, and calculus of variations.* Preprint 1995.

[43] H. HOLDEN, T. LINDSTRØM, B. ØKSENDAL, J. UBØE, T.-S. ZHANG: *The Burgers equation with a noisy force and the stochastic heat equation.* Comm. Part. Diff. Eqs. **19**(1994), 119 - 141.

[44] R.F. HOSKINS, J. SOUSA PINTO: *Nonstandard treatments of new generalized functions.* In: R.S. PATHAK (ED.), *Generalized Functions and Their Applications.* Plenum Press, New York 1993.

[45] R.F. HOSKINS, J. SOUSA PINTO: *Distributions, Ultradistributions and Other Generalised Functions.* Ellis Horwood, New York 1994.

[46] K.T. JOSEPH: *A Riemann problem whose viscosity solutions contain δ-measures.* Preprint 1992.

[47] B.L. KEYFITZ, H.C. KRANZER: *A viscosity approximation to a system of conservation laws with no classical Riemann solution.* In: C. CARASSO, P. CHARRIER, B. HANOUZET, J.-L. JOLY (EDS.), *Nonlinear Hyperbolic Problems.* Lecture Notes Math. Vol. 1402, Springer-Verlag, Berlin 1989.

[48] F. LAFON, M. OBERGUGGENBERGER: *Generalized solutions to symmetric hyperbolic systems with discontinuous coefficients: the multidimensional case.* J. Math. Anal. Appl. **160**(1991), 93 - 106.

[49] M. LANGLAIS: *Generalized functions solutions of monotone and semilinear parabolic equations.* Monatshefte Math. **110**(1990), 117 - 136.

[50] D. LAUGWITZ: *Eine Einführung der δ-Funktionen.* Sitzungsb. Bayer. Akad. Wiss., Math.-nat. Kl. 1959, 41 - 59.

[51] D. LAUGWITZ: *Anwendung unendlich kleiner Zahlen. I. Zur Theorie der Distributionen.* J. reine angew. Math. **207**(1961), 53 - 60.

[52] T. LINDSTRØM: *An invitation to nonstandard analysis.* In: N. CUTLAND (ED.), *Nonstandard Analysis and its Applications.* Cambridge Univ. Press, Cambridge 1988, 1 - 105.

[53] P. LOEB: *Applications of nonstandard analysis to ideal boundaries.* Israel J. Math. **25**(1976), 154 - 187.

[54] Z. LOZANOV CRVENKOVIĆ, S. PILIPOVIĆ: *Gaussian Colombeau generalized random processes.* Preprint 1994.

[55] W.A.J. LUXEMBURG: *On a class of valuation fields introduced by A. Robinson.* Israel J. Math. **25**(1976), 189 - 201.

[56] V. P. MASLOV, V. A. TSUPIN: *δ-shaped Sobolev generalized solutions of quasilinear equations.* Russion Math. Surveys **34:1**(1979), 231 - 232.

[57] V. P. MASLOV, V. A. TSUPIN: *Necessary conditions for the existence of infinitesimally narrow solitons in gas dynamics.* Sov. Phys. Dokl. **24**(5)(1979), 354 - 356.

[58] M. NEDELJKOV, S. PILIPOVIĆ: *Linear partial differential equations in Colombeau's space.* Preprint 1994.

[59] M. NEDELJKOV, S. PILIPOVIĆ: *Hypoelliptic pseudo-differential operators in Colombeau's space.* Preprint 1994.

[60] M. NEDELJKOV, S. PILIPOVIĆ, D. SCARPALÉZOS: *Division problem and partial differential equations with constant coefficients in Colombeau's space of new generalized functions.* Preprint 1994.

[61] E. NELSON: *Internal set theory: a new approach to nonstandard analysis.* Bull. Am. Math. Soc. **83**(1977), 1165 - 1198.

[62] E. NELSON: *Radically Elementary Probability Theory.* Annals Math. Studies Vol. 117, Princeton University Press, Princeton 1987.

[63] M. OBERGUGGENBERGER: *Weak limits of solutions to semilinear hyperbolic systems.* Math. Ann. **274**(1986), 599 - 607.

[64] M. OBERGUGGENBERGER: *Generalized solutions to semilinear hyperbolic systems.* Monatshefte Math. **103**(1987), 133 - 144.

[65] M. OBERGUGGENBERGER: *Products of distributions: nonstandard methods.* Zeitschrift Anal. Anw. **7**(1988), 347 - 365. Corrections to this article: Zeitschr. Anal. Anw. **10**(1991), 263 - 264.

[66] M. OBERGUGGENBERGER: *Hyperbolic systems with discontinuous coefficients: Generalized solutions and a transmission problem in acoustics.* J. Math. Anal. Appl. **142**(1989), 452 - 467.

[67] M. OBERGUGGENBERGER: *Semilinear wave equations with rough initial data.* In: P. ANTOSIK, A. KAMIŃSKI (EDS.), *Generalized Functions and Convergence.* World Scientific Publ., Singapore 1990, 181 - 203.

[68] M. OBERGUGGENBERGER: *The Carleman system with positive measures as initial data.* Transport Theory Statist. Phys. **20**(1991), 177 - 197.

[69] M. OBERGUGGENBERGER: *Multiplication of Distributions and Applications to Partial Differential Equations.* Pitman Research Notes Math. Vol. 259, Longman, Harlow 1992.

[70] M. OBERGUGGENBERGER: *Case study of a nonlinear, nonconservative, non-strictly hyperbolic system.* Nonlinear Anal. **19**(1992), 53–79.

[71] M. OBERGUGGENBERGER: *Nonlinear theories of generalized functions.* In: S. ALBEVERIO, W.A.J. LUXEMBURG, M.P.H. WOLFF (EDS.), *Advances in Analysis, Probability, and Mathematical Physics - Contributions from Nonstandard Analysis.* Kluwer, Dordrecht, to appear.

[72] M. OBERGUGGENBERGER: *Generalized functions and stochastic processes.* In: Proceedings of the Seminar on Stochastic Analysis, Random Fields and Applications, Ascona 1993. Birkhäuser-Verlag, to appear.

[73] M. OBERGUGGENBERGER, E. E. ROSINGER: *Solution of Continuous Nonlinear PDEs through Order Completion.* North-Holland Math. Studies Vol. 181, North-Holland, Amsterdam 1994.

[74] M. OBERGUGGENBERGER, T. TODOROV: *Schwartz distributions in an algebra of generalized functions.* Preprint 1994.

[75] M. OBERGUGGENBERGER, Y.-G. WANG: *Delta-waves for semilinear hyperbolic Cauchy problems.* Math. Nachr. **166**(1994), 317 - 327.

[76] M. OBERGUGGENBERGER, Y.-G. WANG: *Reflection of delta-waves for nonlinear wave equations in one space variable.* Nonlinear Anal. **22**(1994), 983 - 992.

[77] M. OBERGUGGENBERGER, Y.-G. WANG: *Generalized solutions to conservation laws.* Zeitschr. Anal. Anw. **13**(1994), 7 - 18.

[78] E. PALMGREN: *A constructive approach to nonstandard analysis.* Ann. Pure Appl. Logic, to appear.

[79] E. PALMGREN: *A constructive approach to nonstandard analysis II.* Preprint 1994.

[80] S. PILIPOVIĆ: *Colombeau's Generalized Functions and Pseudo-differential Operators.* Lectures Math. Sciences Vol. 4, Univ. Tokyo 1994.

[81] S. PILIPOVIĆ, D. SCARPALÉZOS: *Differential operators with generalized constant coefficients.* Preprint 1994.

[82] J. RAUCH AND M. REED: *Nonlinear superposition and absorption of delta waves in one space dimension.* J. Funct. Anal. **73**(1987), 152 - 178.

[83] A. ROBINSON: *Function theory on some nonarchimedean fields.* Am. Math. Monthly **80**(6), Part II: Papers in the Foundations of Mathematics (1973), 87 - 109.

[84] E.E. ROSINGER: *Distributions and Nonlinear Partial Differential Equations.* Lecture Notes Math. Vol. 684, Springer-Verlag, Berlin 1978.

[85] E.E. ROSINGER: *Nonlinear Partial Differential Equations. Sequential and Weak Solutions.* North-Holland Math. Studies Vol. 44, North-Holland, Amsterdam 1980.

[86] E.E. ROSINGER: *Nonlinear Equivalence, Reduction of PDEs to ODEs and Fast Convergent Numerical Methods.* Pitman Research Notes Math. Vol. 77, Pitman, Boston 1982.

[87] E. E. ROSINGER: *Generalized Solutions of Nonlinear Partial Differential Equations.* North-Holland Math. Studies Vol. 146, North-Holland, Amsterdam 1987.

[88] E.E. ROSINGER: *Non-linear Partial Differential Equations. An Algebraic View of Generalized Solutions.* North-Holland Math. Studies Vol. 164, North-Holland, Amsterdam 1990.

[89] E.E. ROSINGER: *Global version of the Cauchy-Kovalevskaia theorem for nonlinear PDEs.* Acta Appl. Math. **21**(1990), 331 - 343.

[90] E.E. ROSINGER: *Characterization for the solvability of nonlinear partial differential equations.* Trans. Am. Math. Soc. **330**(1992), 203 - 225.

[91] E.E. ROSINGER, Y.E. WALUS: *Group invariance of generalized solutions obtained through the algebraic method.* Nonlinearity **7**(1994), 837 - 859.

[92] J.E. RUBIO: *Optimization and Nonstandard Analysis.* M. Dekker, New York 1994.

[93] F. RUSSO: *Colombeau generalized functions and stochastic analysis.* Preprint 1994.

[94] D. SCARPALÉZOS: *Colombeau's generalized functions: topological structures, microlocal properties. A simplified point of view.* Preprint 1993.

[95] C. SCHMIEDEN, D. LAUGWITZ: *Eine Erweiterung der Infinitesimalrechnung.* Math. Zeitschr. **69**(1958), 1 - 39.

[96] L. SCHWARTZ: *Sur l'impossibilité de la multiplication des distributions.* C. R. Acad. Sci. Paris **239**(1954), 847 - 848.

[97] L. SCHWARTZ: *Théorie des distributions.* Nouvelle ed., Hermann, Paris 1966.

[98] J.A. SMOLLER: *Shock Waves and Reaction-Diffusion Equations.* Grundlehren math. Wiss. Vol. 258, Springer-Verlag, New York 1983.

[99] D. TAN, T. ZHANG: *Two-dimensional Riemann problem for a system of conservation laws.* J. Diff. Eqs. **111**(1994), 203 - 254.

[100] T. TODOROV: *Colombeau's generalized functions and nonstandard analysis.* In: B. STANKOVIĆ, E. PAP, S. PILIPOVIĆ, V.S. VLADIMIROV (EDS.), *Generalized Functions, Convergence Structures, and Their Applications.* Plenum Press, New York 1988, 327 - 339.

[101] T. TODOROV: *Nonstandard asymptotic analysis and nonlinear theory of generalized functions.* Preprint 1993.

[102] T. TODOROV: *An existence result for a class of partial differential equations with smooth coefficients.* In: S. ALBEVERIO, W.A.J. LUXEMBURG, M.P.H. WOLFF (EDS.), *Advances in Analysis, Probability, and Mathematical Physics - Contributions from Nonstandard Analysis.* Kluwer, Dordrecht, to appear.

[103] T. TODOROV: *Existence of solutions for linear PDEs with C^∞-coefficients in a sheaf of algebras of generalized functions.* Trans. Am. Math. Soc., to appear.

[104] C. TUCKEY: *Nonstandard Methods in the Calculus of Variations.* Pitman Research Notes Math. Vol. 297, Longman Harlow 1993.

[105] G.B. WHITHAM: *Linear and Nonlinear Waves.* Wiley-Interscience, New York 1974.

Institut für Mathematik und Geometrie
Universität Innsbruck
A - 6020 Innsbruck Austria

E-mail michael@mat0.uibk.ac.at

NIGEL CUTLAND
Loeb measure theory

1. Introduction

Loeb measures, discovered by Peter Loeb in 1975 [32], are very rich yet tractable measure spaces, which play a central and powerful rôle in many applications of nonstandard analysis – in measure and probability theory, stochastic analysis, functional analysis, mathematical physics, economics and mathematical finance theory. At the same time their richness and simple construction makes them a new and fascinating class of structures worthy of study in their own right.

In this article we give an introduction to the basics of Loeb measure and integration theory, designed to make the extensive literature that uses these notions accessible. In particular we will illustrate the way in which Loeb measures can be used to construct or represent standard measures (such as Lebesgue measure or Wiener measure) in a very simple way – often by means of counting measures. This is in many cases the key to their power and usefulness. In the final section we give a sample of applications to differential equations.

We work with a nonstandard universe that is $\aleph_1$-saturated, and where necessary assume extra saturation. To expand on this slightly, we suppose given a superstructure

$$V(S) = \bigcup_{n \in \mathbb{N}} V_n(S)$$

where S is a non-empty set (usually $S = \mathbb{R}$ is sufficient), $V_0(S) = S$, and $V_{n+1}(S) = V_n(S) \cup \mathcal{P}(V_n(S))$. (Here $\mathcal{P}$ denotes the power set operation.)

A nonstandard universe corresponding to this is then given by an extension ${}^*S \supseteq S$ and a mapping ${}^* : V(S) \to V({}^*S)$ that satisfies the *transfer principle* for bounded quantifier statements of the language of set theory. The nonstandard universe is the structure

$$^*V(S) = \bigcup_{n \in \mathbb{N}} {}^*V_n(S),$$

which is a substructure of $V({}^*S)$. The sets in ${}^*V(S)$ are called *internal*, and those sets in $V({}^*S) \setminus {}^*V(S)$ are called *external*.

If κ is a cardinal number, then the κ-*saturation principle* states that if I is an index set with cardinality $|I| < \kappa$ and $(A_i)_{i \in I}$ is a family of internal subsets of an internal set A having the finite intersection property, then $\bigcap_{i \in I} A_i \neq \emptyset$. An alternative and very useful formulation of $\aleph_1$-saturation is that given any sequence $(A_n)_{n \in \mathbb{N}}$ of internal subsets of an internal set A, there is an *internal* sequence $(A_n)_{n \in {}^*\mathbb{N}}$ of subsets of A that extends the original sequence.

Saturation is a kind of compactness property, and plays an important rôle in nonstandard methodology. Basic Loeb measure theory requires only $\aleph_1$-saturation. For applications involving topological spaces, we need κ-saturation, where the topological space in question has a subbase of open sets of cardinality less than κ. This is needed so that the basic nonstandard characterizations of topological notions can be used; we assume familiarity with these. In particular we will use the standard part mapping st : $^*X \to X$ for a Hausdorff space (X, τ); this is the generalization of the standard part mapping from the finite elements of $^*\mathbb{R}$ to $\mathbb{R}$. For further information, consult [9] or [AFHL] for example.

2. Loeb Measures

A *Loeb measure* is a measure constructed from a nonstandard measure by the following construction of P. Loeb [32]. We confine our attention in this paper mainly to finite (or bounded) Loeb measures.

Suppose that an internal set Ω and an internal algebra $\mathcal{A}$ of subsets of Ω are given, and suppose further that M is a finite internal finitely additive measure on $\mathcal{A}$. This means that M is an internal mapping

$$M : \mathcal{A} \to {}^*[0, \infty)$$

with $M(A \cup B) = M(A) + M(B)$ for disjoint $A, B \in \mathcal{A}$, and that $M(\Omega)$ finite. Thus $M(A)$ is finite for each $A \in \mathcal{A}$, so we may define the mapping

$$°M : \mathcal{A} \to [0, \infty)$$

by $°M(A) = °(M(A))$. Clearly $°M$ is finitely additive, so that $(\Omega, \mathcal{A}, °M)$ is a *standard finitely additive measure space*.

However, in general this is *not* a measure space, because $\mathcal{A}$ is not σ-additive unless $\mathcal{A}$ is finite. To see this, if $\mathcal{A}$ is infinite there is a countable collection of pairwise disjoint nonempty sets $(A_n)_{n \in \mathbb{N}}$ all belonging to $\mathcal{A}$. Then $\aleph_1$-saturation ensures that the set $A = \bigcup_{n \in \mathbb{N}} A_n$ does not belong to $\mathcal{A}$ (in fact it is not even internal). For otherwise, putting $B_n = A \setminus \bigcup_{m \leq n} A_n$, the family $(B_n)_{n \in \mathbb{N}}$ is a family of internal sets with the finite intersection property, yet $\bigcap_{n \in \mathbb{N}} B_n = \emptyset$, which contradicts $\aleph_1$-saturation.

Nevertheless, if $(A_n)_{n \in \mathbb{N}}$ is a family of sets from $\mathcal{A}$, then the set $\bigcup_{n \in \mathbb{N}} A_n$ is "almost" in $\mathcal{A}$ – it differs from a set in $\mathcal{A}$ by a "null" set (a notion to be defined shortly); see the Key Lemma (Lemma 2.2) and its corollary below. This is what lies at the heart of the following fundamental result proved by Loeb.

Theorem 2.1. *There is a unique σ-additive extension of $°M$ to the σ-algebra $\sigma(\mathcal{A})$ generated by $\mathcal{A}$. The completion of this measure is the* Loeb measure, *denoted M_L and the completion of $\sigma(\mathcal{A})$ is the* Loeb σ-algebra, *denoted $L(\mathcal{A})$.*

Proof For a quick proof we can apply Caratheodory's extension theorem. It is only necessary to check σ-additivity of $°M$ on $\mathcal{A}$. Suppose that $(A_n)_{n \in \mathbb{N}}$, is a sequence of pairwise disjoint sets from $\mathcal{A}$ such that

$$\bigcup_{n \in \mathbb{N}} A_n \in \mathcal{A}.$$

By $\aleph_1$-saturation, arguing as above, there is $m \in \mathbb{N}$ such that

$$\bigcup_{n \in \mathbb{N}} A_n = \bigcup_{n=1}^{m} A_n.$$

So $A_k = \emptyset$ for $k > m$, and

$$^\circ M\left(\bigcup_{n \in \mathbb{N}} A_n\right) = {}^\circ M\left(\bigcup_{n=1}^{m} A_n\right) = \sum_{n=1}^{m} {}^\circ M(A_n) = \sum_{n \in \mathbb{N}} {}^\circ M(A_n).$$

using finite additivity. Caratheodory's theorem (see [40] for example) now gives the result. □

It is quite straightforward and rather more illuminating to prove Loeb's theorem from "first principles" and here is one way to proceed – based around the idea of a Loeb null set. First we have:

Lemma 2.2 (Key Lemma). *Let $(A_n)_{n \in \mathbb{N}}$ be a family of sets, with each A_n in $\mathcal{A}$. Then there is a set $A \in \mathcal{A}$ such that*

(i) $\bigcup_{n \in \mathbb{N}} A_n \subseteq A$;
(ii) $^\circ M(A) \leq \sum_{n \in \mathbb{N}} {}^\circ M(A_n)$;
(iii) if the sets $(A_n)_{n \in \mathbb{N}}$ are disjoint, then

$$^\circ M(A) = \sum_{n \in \mathbb{N}} {}^\circ M(A_n).$$

Proof Let $\alpha = \sum_{n \in \mathbb{N}} {}^\circ M(A_n)$; if α is infinite, there is nothing to prove (simply take $A = \Omega$). Otherwise, let $B_n = \bigcup_{m \leq n} A_m$. For each finite n, by finite additivity,

$$M(B_n) \leq \sum_{m \leq n} M(A_m) \leq \sum_{m \leq n} {}^\circ M(A_n) + \frac{1}{n} \leq \alpha + \frac{1}{n}.$$

If we now take an internal sequence $(A_n)_{n \in {}^*\mathbb{N}}$ of sets in $\mathcal{A}$ extending the sequence $(A_n)_{n \in \mathbb{N}}$, then for some infinite N we must have

$$M(B_N) \leq \alpha + \frac{1}{N}$$

where $B_N = \bigcup_{n \leq N} A_n$. (By finite additivity $B_N \in \mathcal{A}$.) Let $A = B_N$. Then $^\circ M(A) \leq \alpha$, so (i) and (ii) hold. For (iii), if the sets $(A_n)_{n \in \mathbb{N}}$ are disjoint then for each finite n

$$\sum_{m \leq n} {}^\circ M(A_m) = {}^\circ \sum_{m \leq n} M(A_m) = {}^\circ M\left(\bigcup_{m \leq n} A_m\right) \leq {}^\circ M(A)$$

and so $\alpha \leq {}^\circ M(A)$. This gives (iii). □

In view of part (iii) of this lemma, it is natural to think of the difference $A \setminus \bigcup_{n \in \mathbb{N}} A_n$ as a *null set*; we can formalize this with the following definition.

Definition 2.3. *Let $B \subseteq \Omega$ (not necessarily internal). We say that B is a Loeb null set if for each real $\varepsilon > 0$ there is a set $A \in \mathcal{A}$ with $B \subseteq A$ and $M(A) < \varepsilon$.*

Remarks 1. A subset of a Loeb null set is Loeb null.
2. If B is Loeb null and $B \in \mathcal{A}$ then $M(B) \approx 0$.

We can now express the Key Lemma in the following way.

Corollary 2.4. *Let $(A_n)_{n \in \mathbb{N}}$ be a family of sets, with each A_n in $\mathcal{A}$. Then there is a set $A \in \mathcal{A}$ such that*

(i) $\bigcup_{n \in \mathbb{N}} A_n \subseteq A$ and $A \setminus \bigcup_{n \in \mathbb{N}} A_n$ is Loeb null;
(ii) if the sets $(A_n)_{n \in \mathbb{N}}$ are disjoint, then

$$^\circ M(A) = \sum_{n \in \mathbb{N}} {}^\circ M(A_n).$$

Proof Assume first that the sets $(A_n)_{n \in \mathbb{N}}$ are disjoint. Let A be the set given by the Key Lemma. To see that the set $B = A \setminus \bigcup_{n \in \mathbb{N}} A_n$ is a Loeb null set, take real $\varepsilon > 0$ and find $n \in \mathbb{N}$ with $\sum_{m>n} {}^\circ M(A_m) < \varepsilon$. Let $D = A \setminus \bigcup_{m \leq n} A_m \in \mathcal{A}$. Then $B \subseteq D$ and $M(D) = M(A) - \sum_{m \leq n} M(A_m) \approx \alpha - \sum_{m \leq n} {}^\circ M(A_m) = \sum_{m>n} {}^\circ M(A_m) < \varepsilon$.

If the sets $(A_n)_{n \in \mathbb{N}}$ are not disjoint, then simply work with the sets A'_n defined by $A'_n = A_n \setminus \bigcup_{m<n} A_m \in \mathcal{A}$, and note that $\bigcup_{n \in \mathbb{N}} A'_n = \bigcup_{n \in \mathbb{N}} A_n$. □

Now observe that

Lemma 2.5. *A countable union of Loeb null sets is Loeb null.*

Proof Suppose that $(B_n)_{n \in \mathbb{N}}$ is a sequence of Loeb null sets and $\varepsilon > 0$ is real. Take sets A_n with $B_n \subseteq A_n \in \mathcal{A}$ and $M(A_n) < \varepsilon/2^n$. Let $A \in \mathcal{A}$ be the set given by the Key Lemma; i.e. $A \supseteq \bigcup_{n \in \mathbb{N}} A_n \supseteq \bigcup_{n \in \mathbb{N}} B_n$ and

$$^\circ M(A) \leq \sum_{n \in \mathbb{N}} {}^\circ M(A_n) \leq \sum_{n \in \mathbb{N}} \varepsilon 2^{-n} = \varepsilon.$$

Hence $\bigcup_{n \in \mathbb{N}} B_n$ is Loeb null. □

We are ready to make the following definition.

Definition 2.6. *Let $B \subseteq \Omega$. We say that B is Loeb measurable if there is a set $A \in \mathcal{A}$ such that $A \Delta B$ $(= (A \setminus B) \cup (B \setminus A))$ is Loeb null. Denote the collection of all Loeb measurable sets by $L(\mathcal{A})$.*

Remarks. 1. $\mathcal{A} \subseteq L(\mathcal{A})$ and all Loeb null sets are Loeb measurable.

2. The relation "$A \Delta B$ is null" (for *any* subsets of Ω) is transitive, since for any sets A, B, C we have $A \Delta C \subseteq (A \Delta B) \cup (B \Delta C)$.

3. If $B \in L(\mathcal{A})$ and $B \Delta C$ is Loeb null then $C \in L(\mathcal{A})$; simply take $A \in \mathcal{A}$ with $A \Delta B$ Loeb null, and then by remark 2 above $A \Delta C$ is Loeb null.

4. If $B \in L(\mathcal{A})$ and $A_1, A_2 \in \mathcal{A}$ with $B \Delta A_i$ both Loeb null, then $A_1 \Delta A_2$ is Loeb null and so $M(A_1) \approx M(A_2)$.

In view of Remark 4 above, we may make the following definition.

Definition 2.7. *Let $B \in L(\mathcal{A})$. Define a function $M_L : L(\mathcal{A}) \to \mathbb{R}$ by*
$$M_L(B) = {}^\circ M(A)$$
for any $A \in \mathcal{A}$ such that $B \Delta A$ is Loeb null. In anticipation of the theorem below, we say that $M_L(B)$ is the Loeb measure of B.

Clearly for $A \in \mathcal{A}$, $M_L(A) = {}^\circ M(A)$. Note also that

Proposition 2.8. *For any $B \subseteq \Omega$, B is Loeb null if and only if $B \in L(\mathcal{A})$ and $M_L(B) = 0$.*

Proof If B is Loeb null, then take $A = \emptyset$ in the definition of Loeb measurable, so that $B \in L(\mathcal{A})$ and $M_L(B) = {}^\circ M(\emptyset) = 0$. Conversely, if $B \in L(\mathcal{A})$ and $M_L(B) = 0$, take $A \in \mathcal{A}$ with $A \Delta B$ Loeb null. Then $M(A) \approx 0$ and A is Loeb null; hence so is B. □

The corollary to the Key Lemma (Corollary 2.4) tells that

Lemma 2.9. *Let $A_n \in \mathcal{A}$ for $n \in \mathbb{N}$. Then $\bigcup_{n \in \mathbb{N}} A_n \in L(\mathcal{A})$ and if the sets A_n are pairwise disjoint, then*
$$M_L(\bigcup_{n \in \mathbb{N}} A_n) = \sum_{n \in \mathbb{N}} M_L(A_n).$$

Proof Immediate from Corollary 2.4. □

Now we can prove:

Theorem 2.10. *$L(\mathcal{A})$ is a σ-algebra, and M_L is a complete (σ-additive) measure on $L(\mathcal{A})$.*

Proof To see that $L(\mathcal{A})$ is a σ-algebra it is sufficient to show that $L(\mathcal{A})$ is closed under complements and countable unions.

If $B \in L(\mathcal{A})$ and $A \Delta B$ is Loeb null, with $A \in \mathcal{A}$, then $A^c \Delta B^c = A \Delta B$ and $A^c \in \mathcal{A}$, so $B^c \in L(\mathcal{A})$.

For countable unions, suppose that (B_n) is a sequence of Loeb measurable sets, and $A_n \in \mathcal{A}$ with $A_n \Delta B_n$ Loeb null. Then $\bigcup_{n \in \mathbb{N}} A_n$ is Loeb measurable (by Lemma 2.9); and
$$(\bigcup_{n \in \mathbb{N}} A_n) \Delta (\bigcup_{n \in \mathbb{N}} B_n) \subseteq \bigcup_{n \in \mathbb{N}} (A_n \Delta B_n)$$
which is Loeb null by Lemma 2.5. So by the third remark above, $\bigcup_{n \in \mathbb{N}} B_n$ is Loeb measurable.

To see that M_L is a complete measure on $L(\mathcal{A})$, first check that M_L is σ-additive. Take sequences $B_n \in L(\mathcal{A})$ and $A_n \in \mathcal{A}$ as above, with B_n pairwise disjoint. By trimming if necessary we may assume that the sets A_n are also pairwise disjoint. Since $(\bigcup_{n \in \mathbb{N}} A_n) \Delta (\bigcup_{n \in \mathbb{N}} B_n)$ is Loeb null, we have
$$M_L(\bigcup_{n \in \mathbb{N}} B_n) = M_L(\bigcup_{n \in \mathbb{N}} A_n) = \sum_{n \in \mathbb{N}} M_L(A_n) = \sum_{n \in \mathbb{N}} M_L(B_n)$$
where we have used Lemma 2.9.

It is clear that $L(\mathcal{A})$ is complete, since a subset of a Loeb null set is Loeb null. □

The measure space $\boldsymbol{\Omega} = (\Omega, L(\mathcal{A}), M_L)$ is called the *Loeb space* given by $(\Omega, \mathcal{A}, M)$. Of course $L(\mathcal{A})$ depends on both $\mathcal{A}$ and M, so strictly we should write $L(\mathcal{A}, M)$, but usually it is clear which measure is intended. If $M(\Omega) = 1$ then $\boldsymbol{\Omega}$ is a *Loeb probability space* and M_L is the *Loeb probability measure* given by M.

The following give alternative characterizations of Loeb measurable sets, and are often taken as the fundamental definition (see [AFHL], [14] or [Li] for example). First define some more terms.

Definition 2.11. *Let $B \subseteq \Omega$.*
  1. B is M-approximable if for every real $\varepsilon > 0$ there are sets $A, C \in \mathcal{A}$ with $A \subseteq B \subseteq C$ and $M(C \setminus A) < \varepsilon$.
  2. The inner and outer Loeb measure of B, $\underline{M}(B)$ and $\overline{M}(B)$ are given by
$$\underline{M}(B) = \sup\{^\circ M(A) : A \subseteq B, A \in \mathcal{A}\}$$
$$\overline{M}(B) = \inf\{^\circ M(A) : A \supseteq B, A \in \mathcal{A}\}$$

Then we have

Theorem 2.12. *The following are equivalent:*
  (i) B is Loeb measurable.
  (ii) B is M-approximable.
  (iii) $\overline{M}(B) = \underline{M}(B)$.

Proof If B is Loeb measurable, take $A \in \mathcal{A}$ with $A \triangle B$ null and let $A \triangle B \subseteq D \in \mathcal{A}$ with $M(D) < \varepsilon$. Then $A \setminus D \subseteq B \subseteq A \cup D$ and $M((A \cup D) \setminus (A \setminus D)) = M(D) < \varepsilon$, so B is M-approximable.

If B is M-approximable, then taking $A, C \in \mathcal{A}$ as in the definition of M-approximable we see that $\overline{M}(B) - \underline{M}(B) < \varepsilon$ for any real $\varepsilon > 0$, and so $\overline{M}(B) = \underline{M}(B)$.

Finally, if $\overline{M}(B) = \underline{M}(B) = \alpha$, say, then take sets $A_n \subseteq B \subseteq C_n$ with $A_n, C_n \in \mathcal{A}$ and $M(A_n) > \alpha - n^{-1}$ and $M(C_n) < \alpha + n^{-1}$. Let $A = \bigcup_{n \in \mathbb{N}} A_n$ and $C = \bigcap_{n \in \mathbb{N}} C_n$. Then A and C are Loeb measurable and $M_L(A) \geq \alpha \geq M_L(C)$. On the other hand
$$A \subseteq B \subseteq C$$
which shows that B is Loeb measurable also (by the completeness of M_L). □

Here are some examples of Loeb measures.

Examples
  (1) Let Ω be a hyperfinite set and $\mathcal{A}$ the family of all internal subsets of Ω. The *normalized counting measure* on Ω is defined by
$$M(A) = \frac{\#(A)}{\#(\Omega)}$$
where $\#(A)$ denotes the internal cardinality of $A \in \mathcal{A}$.

The Loeb measure M_L in this case is called the *Loeb counting measure* on Ω, or the *Loeb counting probability* (since $M(\Omega) = 1$).

(2) Let $\Omega = {}^*[0,1]$ and $\mathcal{A} = {}^*\mathcal{M}$, where $\mathcal{M}$ denotes the family of Lebesgue measurable subsets of $[0,1]$. Let λ the Lebesgue measure on $[0,1]$. The set function ${}^*\lambda$ is finitely additive. The corresponding Loeb measure ${}^*\lambda_L$ is called the *uniform Loeb measure* on ${}^*[0,1]$.

(3) Generalizing (2), we can take any standard measure space $(X, \mathcal{F}, \mu)$ and let $\Omega = {}^*X$, $\mathcal{A} = {}^*\mathcal{F}$ and $M = {}^*\mu$. This gives the Loeb space

$$({}^*X, L({}^*\mathcal{F}), {}^*\mu_L).$$

Unbounded Loeb measures. If the internal measure M is unbounded, then Loeb [32] showed that ${}^\circ M$ (taking the value ∞ when necessary) can be extended to the σ-algebra $\sigma(\mathcal{A})$ by defining $M_L(B) = \overline{M}(B)$, and Henson [24] showed that this extension is unique. The class of Loeb measurable sets is defined in this case by $B \in L(\mathcal{A})$ if $B \cap A$ is Loeb measurable in the sense above for each $A \in \mathcal{A}$ with $M(A)$ finite. The reader is referred to [Li, 41] for more information about unbounded Loeb measures.

Loeb measures have been used extensively for *representation* of measures (by internal measures that are easier to handle) and for *construction* of measures for a wide variety of purposes. We continue with a sample of results of the first kind.

3. Representation of Measures

First recall some definitions. Let (X, τ) be a Hausdorff topological space, and write $\mathcal{B} = \mathcal{B}(\tau)$ for the Borel sets given by τ. Let μ be a finite Borel measure on X.

Definition 3.1.

(1) The measure μ is regular *if for each* $B \in \mathcal{B}$

$$\mu(B) = \sup\{\mu(F) : F \subseteq B, \ F \text{ is closed}\}$$
$$= \inf\{\mu(U) : B \subseteq U, \ U \text{ is open }\}$$

(2) The measure μ is Radon *if for each* $B \in \mathcal{B}$

$$\mu(B) = \sup\{\mu(K) : K \subseteq B, \ K \text{ is compact}\}$$
$$= \inf\{\mu(U) : B \subseteq U, \ U \text{ is open }\}$$

Remark In view of the finiteness of μ, the second condition in each of these definitions is redundant.

One of the most fundamental representation results is the following, due to Anderson [7]

Theorem 3.2. Let $(X, \mathcal{B}, \mu)$ be a Radon measure on a Hausdorff space X, and let $\mathcal{C}$ be the μ-completion of $\mathcal{C}$. Then
  (a) $C \in \mathcal{C}$ if and only if $\operatorname{st}^{-1}(C) \in L(^*\mathcal{B})$;
  (b) if $C \in \mathcal{C}$ then
$$\mu(C) = {}^*\mu_L(\operatorname{st}^{-1}(C))$$
(where st denotes the standard part mapping $\operatorname{st} : {}^*X \to X$.

Hence ${}^*\mu_L({}^*X \setminus \operatorname{ns}({}^*X)) = 0$ (and we say that μ_L is nearstandardly concentrated; $\operatorname{ns}({}^*X)$ denotes the nearstandard elements of ${}^*X$).

Proof Take $C \in \mathcal{C}$ and compact sets K_n and open sets U_n with $K_n \subseteq C \subseteq U_n$ and $\mu(U_n \setminus K_n) < n^{-1}$. Then
$${}^*K_n \subseteq \operatorname{st}^{-1}(K_n) \subseteq \operatorname{st}^{-1}(C) \subseteq \operatorname{st}^{-1}(U_n) \subseteq {}^*U_n$$
by the characterization of compact and open sets, and so $\operatorname{st}^{-1}(C)$ is ${}^*\mu$-approximable (since ${}^*\mu({}^*U_n \setminus {}^*K_n) = \mu(U_n \setminus K_n) < n^{-1}$). So $\operatorname{st}^{-1}(C) \in L(^*\mathcal{B})$ and it is clear that ${}^*\mu_L(\operatorname{st}^{-1}(C)) = \mu(C)$. In particular ${}^*\mu_L(\operatorname{ns}({}^*X)) = \mu(X)$.

Conversely, suppose that $B = \operatorname{st}^{-1}(C) \in L(^*\mathcal{B})$, and set ${}^*\mu_L(B) = \alpha$. Take $A \in {}^*\mathcal{B}$ with $A \subseteq B$ and ${}^*\mu_L(A) > \alpha - n^{-1}$. Then $F = \operatorname{st}(A)$ is closed and $F \subseteq C$. Now $A \subseteq \operatorname{st}^{-1}(F)$, and using the first part,
$$\mu(F) = {}^*\mu_L(\operatorname{st}^{-1}(F)) \geq {}^*\mu_L(A) > \alpha - n^{-1}.$$
Now consider the complement $C^c = X \setminus C$; we have $\operatorname{st}^{-1}(C^c) = \operatorname{ns}({}^*X) \setminus \operatorname{st}^{-1}(C) \in L(^*\mathcal{B})$ since $\operatorname{ns}({}^*X) \in L(^*\mathcal{B})$ (by the first part). Moreover
$${}^*\mu_L(\operatorname{st}^{-1}(C^c)) = {}^*\mu_L(\operatorname{ns}({}^*X)) - {}^*\mu_L(\operatorname{st}^{-1}(C)) = \mu(X) - \alpha.$$
As for the set C, we can find a closed set $F' \subseteq C^c$ with $\mu(F') > \mu(X) - \alpha - n^{-1}$. This is sufficient to show that $C \in \mathcal{C}$ (and $\mu(C) = \alpha$). $\square$

The following is a useful corollary to this proof.

Corollary 3.3. Let $C \in \mathcal{C}$. Then ${}^*C \triangle \operatorname{st}^{-1}(C)$ is Loeb null.

Proof The set concerned is a subset of each ${}^*U_n \setminus {}^*K_n$, where K_n and U_n are the compact and open sets approximating C in the proof of Theorem 3.2. $\square$

Remark A number of variations of Theorem 3.2 can be formulated. For example, examination of the proof shows that it is only necessary to assume that μ is regular and that $\operatorname{ns}({}^*X) \in L(^*\mathcal{B})$. It is easy to check that when μ is Radon, $\operatorname{ns}({}^*X) \in L(^*\mathcal{B})$. It is also clear from the proof that it is sufficient to work with any internal subalgebra $\mathcal{A} \subseteq {}^*\mathcal{B}$ with the property that either ${}^*U \in \mathcal{A}$ or $\operatorname{st}^{-1}(U) \in \mathcal{A}$ for each open set U. See [7] or [AFHL] for more details.

The above representation result can be refined in a number of ways to allow a measure μ on X to be represented by a Loeb measure on a "simpler" subspace of X. Here are some examples, which will all follow from a general result (Theorem 3.9) below.

Theorem 3.4. *Let $N \in {}^*\mathbb{N}$ be infinite and let $\Delta t = N^{-1}$. Let $\mathbf{T}$ be the hyperfinite time line*
$$\mathbf{T} = \{k\Delta t : 0 \leq k < N\}$$
and let ν be the counting measure on $\mathbf{T}$ with algebra $\mathcal{A} = {}^*\mathcal{P}(\mathbf{T})$. Then for $A \subseteq [0,1]$
  (a) *A is Lebesgue measurable if and only if $\mathrm{st}^{-1}(A) \cap \mathbf{T} \in L(\mathcal{A})$.*
  (b) *if A is Lebesgue measurable then $\lambda(A) = \nu_L(\mathrm{st}^{-1}(A) \cap \mathbf{T})$.*

Remark This representation theorem for Lebesgue measure can be regarded as a *construction* of Lebesgue measure – see the remark following Theorem 4.1 below.

The next more general "hyperfinite representation theorem", is due to Anderson [7].

Theorem 3.5. *Let (X,τ) be a Hausdorff space and μ a Radon measure on X. Let $\mathcal{C}$ be the μ-completion of $\mathcal{B}$. There is a hyperfinite subset $Z \subseteq {}^*X$ and an internal measure M on Y (carried on the internal algebra $\mathcal{Z} = {}^*\mathcal{P}(Z)$) such that*
  (a) *$C \in \mathcal{C}$ if and only if $\mathrm{st}^{-1}(C) \cap Z \in L(\mathcal{Z})$;*
  (b) *if $C \in \mathcal{C}$ then*
$$\mu(C) = M_L(\mathrm{st}^{-1}(C) \cap Z)$$

In the next example, H is a separable Hilbert space, with o.n. basis $(e_n)_{n \in \mathbb{N}}$. Let H_n be the subspace of dimension n spanned by $e_1, \ldots, e_n$.

Theorem 3.6. *Let μ be a Borel measure on H and for each n let $\mu^n = \mu \circ \mathrm{Pr}_n^{-1}$ where Pr_n is the projection on H_n. Then, again putting $\mathcal{C} =$ the μ-completion of $\mathcal{B}$:*
  (a) *$C \in \mathcal{C}$ if and only if $\mathrm{st}^{-1}(C) \cap H_N \in L(\mathcal{B}_N)$;*
  (b) *if $C \in \mathcal{C}$ then*
$$\mu(C) = \mu_L^N(\mathrm{st}^{-1}(C) \cap H_N)$$

In [9] this result was used to represent a Borel measure on H by a nonstandard density against *Lebesgue measure on H_N. By smoothing μ_N, it is possible to obtain an internal measure ν on H_N that is absolutely continuous with respect to the internal *Lebesgue measure on H_N, and such that
$$\mu(B) = \nu_L(\mathrm{st}^{-1}(B) \cap H_N)$$
for all $B \in \mathcal{B}$. The measure ν then has an internal density Φ so that for all *Borel $D \subseteq H_N$,
$$\nu(D) = \int_D \Phi(X) dX$$
where the integral is the nonstandard Lebesgue integral on $H_N \cong {}^*\mathbb{R}^N$. This can be used to solve equations for time-evolving measures by re-casting them as equations for the time evolution of their nonstandard densities. See [13] for details.

We now describe a general setting for the above representation results. Let (X, τ) be a Hausdorff space and μ a finite Radon measure on X, and denote by $\mathcal{C}$ the μ-completion of $\mathcal{B}$ (as in Theorem 3.5). Write ns for ns($^*X$). Suppose that a *measurable mapping $\pi : {}^*X \to {}^*X$ is given with the property that

(1) $\qquad z \in \text{ns} \;\Rightarrow\; \pi(z) \in \text{ns} \;\text{ and }\; {}^\circ\pi(z) = {}^\circ z.$

Let $Z = \pi(^*X)$. Finally define $\text{st}_Z = \text{st}|Z$ (the restriction to Z), so that $\text{st}_Z^{-1}(C) = \text{st}^{-1}(C) \cap Z$ for any $C \subseteq X$. It is easy to check that the property (1) implies that $\text{st}^{-1}(C) = \pi^{-1}(\text{st}_Z^{-1}(C)) \cap \text{ns}$.

The examples of this that we have in mind are:

(a) (for Theorem 3.4) $X = [0,1]$ with $\mu = \lambda =$ Lebesgue measure, $\pi =$ the function taking $x \in {}^*X$ to the nearest point in $\mathbf{T}$ to the left of x; i.e. $\pi(x) = \mathbf{t} \in \mathbf{T}$ where $\mathbf{t} \leq x < \mathbf{t} + \Delta t$; so $Z = \mathbf{T}$;

(b) $X = H$, as in Theorem 3.6, and $\pi = \Pr_N$, so $Z = H_N$ in this case.

Returning to the general setting, define an internal algebra $\mathcal{Z}$ on Z by

$$Y \in \mathcal{Z} \;\Leftrightarrow\; \pi^{-1}(Y) \in {}^*\mathcal{B}$$

for $Y \subseteq Z$, and a measure M on $\mathcal{Z}$ by

$$M(Y) = {}^*\mu(\pi^{-1}(Y)).$$

The question we now address is under what circumstances we have μ represented (via st_Z^{-1}) by the "simpler" measure M_L on the "simpler" set Z? i.e. when do we have

(2) $\qquad C \in \mathcal{C} \;\Leftrightarrow\; \text{st}_Z^{-1}(C) \in L(\mathcal{Z}) = L(\mathcal{Z}, M)$

and

$$\mu(C) = M_L(\text{st}_Z^{-1}(C)) \qquad \text{for all } C \in \mathcal{C}.$$

A related question is whether we have

(3) $\qquad Y \in L(\mathcal{Z}) \;\Leftrightarrow\; \pi^{-1}(Y) \in L(^*\mathcal{B})$

and

$$M_L(Y) = {}^*\mu_L(\pi^{-1}(Y)) \qquad \text{for all } Y \in L(\mathcal{Z}).$$

It is easy to check that (3) implies (2), using Theorem 3.2 and the next two results.

First notice that

Proposition 3.7.
If $Y \in L(\mathcal{Z})$ then $\pi^{-1}(Y) \in L(^*\mathcal{B})$, and $M_L(Y) = {}^*\mu_L(\pi^{-1}(Y))$.

Proof Take $Y_1, Y_2 \in \mathcal{Z}$ with $Y_1 \subseteq Y \subseteq Y_2$ and $M(Y_2 \setminus Y_1) < \varepsilon$; then we have $\pi^{-1}(Y_1) \subseteq \pi^{-1}(Y) \subseteq \pi^{-1}(Y_2)$ and $^*\mu(\pi^{-1}(Y_2) \setminus \pi^{-1}(Y_1)) < \varepsilon$. The result now follows. $\square$

Corollary 3.8. *If $\text{st}_Z^{-1}(C) \in L(\mathcal{Z})$ then $C \in \mathcal{C}$ and $M_L(\text{st}_Z^{-1}(C)) = \mu(C)$.*

Proof From the previous proposition, $\pi^{-1}(\mathrm{st}_Z^{-1}(C)) \in L(^*\mathcal{B})$, and $M_L(\mathrm{st}_Z^{-1}(C)) = {}^*\mu_L(\pi^{-1}(\mathrm{st}_Z^{-1}(C)))$.

Now $\mathrm{st}^{-1}(C) = \pi^{-1}(\mathrm{st}_Z^{-1}(C)) \cap \mathrm{ns}$, so from Theorem 3.2, $\mathrm{st}^{-1}(C) \in L(^*\mathcal{B})$ and $C \in \mathcal{C}$. Moreover $\mu(C) = {}^*\mu_L(\mathrm{st}^{-1}(C)) = {}^*\mu_L(\pi^{-1}(\mathrm{st}_Z^{-1}(C))) = M_L(\mathrm{st}_Z^{-1}(C))$. (We have used the fact that ns is a full set in $^*X$.) $\square$

We now have:

Theorem 3.9. (a) If $\pi(^*K) \in \mathcal{Z}$ for all compact $K \subseteq X$, then (2) holds, and so $\mu(C) = M_L(\mathrm{st}_Z^{-1}(C))$ for all $C \in \mathcal{C}$.
 (b) If $\pi(K) \in \mathcal{Z}$ for all * compact $K \subseteq {}^*X$, then (3) holds.

Proof (a) From Corollary 3.8 we only need to show that if $C \in \mathcal{C}$ then $\mathrm{st}_Z^{-1}(C) \in L(\mathcal{Z})$. Let K_1, K_2 be a compact sets with $K_1 \subseteq C \subseteq K_2^c$ and $\mu(K_1 \cup K_2) > \mu(X) - \varepsilon$, where ε is a positive real. Then

$$\pi(^*K_1) \subseteq \mathrm{st}_Z^{-1}(C) \subseteq Z \setminus \pi(^*K_2)$$

(using property (1) and the compactness of K_1, K_2). Now $\pi(^*K_i) \in \mathcal{Z}$ by hypothesis, and, putting $K = K_1 \cup K_2$,

$$\begin{aligned} M(\pi(^*K_1) \cup \pi(^*K_2)) = M(\pi(^*K)) &= {}^*\mu(\pi^{-1}(\pi(^*K))) \\ &\geq {}^*\mu(^*K) \\ &= \mu(K) \\ &> \mu(X) - \varepsilon \\ &= M(Z) - \varepsilon. \end{aligned}$$

So $\mathrm{st}_Z^{-1}(C)$ is M-approximable, and hence belongs to $L(\mathcal{Z})$ as required.

 (b) By Proposition 3.7, we only have to show that if $\pi^{-1}(Y) \in L(^*\mathcal{B})$, then $Y \in L(\mathcal{Z})$, for $Y \subseteq Z$.

For such Y, given real positive ε, there are sets $A_1, A_2 \in {}^*\mathcal{B}$ with $A_1 \subseteq \pi^{-1}(Y) \subseteq A_2$ and $^*\mu(A_2 \setminus A_1) < \varepsilon$. Since $^*\mu$ is * Radon, we may assume that $A_1 = K_1$ and $A_2 = {}^*X \setminus K_2$, where K_1, K_2 are * compact; i.e.

$$K_1 \subseteq \pi^{-1}(Y) \subseteq K_2^c.$$

Then

$$\pi(K_1) \subseteq Y \subseteq Z \setminus \pi(K_2)$$

with $\pi(K_1)$ and $Z \setminus \pi(K_2)$ both in $\mathcal{Z}$ by hypothesis. We have (putting $K = K_1 \cup K_2$)

$$\begin{aligned} M(\pi(K_1) \cup \pi(K_2)) = M(\pi(K)) &= {}^*\mu(\pi^{-1}(\pi(K))) \\ &\geq {}^*\mu(K) \\ &\geq {}^*\mu(^*X) - \varepsilon \\ &= M(Z) - \varepsilon \end{aligned}$$

and so Y is M-approximable; i.e. $Y \in L(\mathcal{Z})$. $\square$

Note that the hypothesis in (b) of this theorem is fulfilled in the following two particular instances:

(i) if Z is hyperfinite (since then $Z = {}^*\mathcal{P}(Z)$ and $\pi(A) \in Z$ for all internal $A \subseteq {}^*X$);

(ii) if π is * continuous, since then $\pi(K)$ is *compact for *compact $K \subseteq {}^*X$, and so $\pi^{-1}(\pi(K)) \in {}^*\mathcal{B}$ – which ensures that $\pi(K) \in Z$.

These two cases give Theorems 3.4 and 3.6 above.

For Theorem 3.5, simply take a hyperfinite algebra $\mathcal{B}_0 \subseteq {}^*\mathcal{B}$ such that ${}^*B \in \mathcal{B}_0$ for all sets $B \in \mathcal{B}$. Let Z be a hyperfinite set containing exactly one point from each atom of $\mathcal{B}_0$, and define $\pi(x)$ to be the point in Z that is in the atom $A(x)$ of $\mathcal{B}_0$ that contains x. Clearly π is *-measurable. If x is nearstandard, then $A(x) \subseteq {}^*U$ for each open neighbourhood U of x, and so $A(x) \subseteq \text{monad}(x)$; thus ${}^\circ\pi(x) = {}^\circ x$. So the property (1) holds; thus Theorem 3.9 applies, and Theorem 3.5 follows.

There are many other representation results in the literature that extend those we have mentioned. For example, Ross [39] showed that if $(X, \mathcal{B}, \mu)$ is a *compact* finite measure space, then there is a surjection $\varphi : {}^*X \to X$ such that μ is represented by the Loeb measure ${}^*\mu_L$ via φ; i.e. $\mu(B) = {}^*\mu_L \circ \varphi^{-1}(B)$ for $B \in \mathcal{B}$. (A compact measure is a non-topological analogue of a Radon space – one that is inner regular with respect to a sub-family of $\mathcal{B}$ that behaves in certain respects like the compact sets in a topological space. In general compact measures need not be topological.) Ross goes on to show that if the space is *supercompact* then the mapping φ can be used to define a topology on X that makes the measure space Radon.

Landers & Rogge [LR] used the *outer* measure $\overline{{}^*\mu}$ defined by ${}^*\mu$ to represent a measure μ on a Hausdorff space that is regular and τ-*smooth* (not necessarily Radon). In the paper [4] Aldaz found a way to generalize the representation results of Anderson, Ross and Landers & Rogge.

4. Construction of Measures

The procedure for using Loeb measures to construct measures on topological spaces is the reverse of that discussed in the previous section. Here we *begin* with an internal measure on ${}^*X$ (perhaps concentrated on some "simpler" subspace) and then use the standard part to "push" the corresponding Loeb measure down to X. Here is a typical example of a general "pushing down" theorem:

Theorem 4.1. *Let X be a Hausdorff space and suppose that M is an internal finitely additive measure on an internal algebra $\mathcal{A}$ of subsets of ${}^*X$, such that $\text{st}^{-1}(F) \in L(\mathcal{A})$ for each closed $F \subseteq X$. Define a measure μ on sets C with $\text{st}^{-1}(C) \in L(\mathcal{A})$ by*

$$\mu(C) = M_L(\text{st}^{-1}(C))$$

Then μ is the completion of a regular Borel measure on X.

Proof Clearly if B is Borel, then $\text{st}^{-1}(B) \in L(\mathcal{A})$ and so μ is defined for a σ-algebra extending $\mathcal{B}$. Suppose now that C is any set with $\text{st}^{-1}(C) \in L(\mathcal{A})$. Then, exactly as in the last part of the proof of Theorem 3.2 we can find closed F with $F \subseteq C$ and $\mu(F) > \mu(C) - \varepsilon$ for any given real $\varepsilon > 0$. This is sufficient to establish the result. $\square$

Remark Note that the measure μ constructed in this theorem need not have the same total mass as $°M$ – for this we would need the extra condition that $M_L(\mathrm{ns}(^*X)) = °M(^*X)$ – which can fail if some of the mass of M is carried by the non-nearstandard points of $^*X$.

A well-known example of Theorem 4.1 is when $M = \nu$, the measure concentrated on $\mathbf{T} = \{k\Delta t : 0 \le k < N\} \subseteq {}^*[0,1]$ with $\nu(t) = N^{-1} = \Delta t$ for each $t \in \mathbf{T}$ (i.e. ν is the counting measure on $\mathbf{T}$). In this case, $M_L(\mathrm{st}^{-1}(B)) = M_L(\mathrm{st}_\mathbf{T}^{-1}(B))$. It is easy to check that the measure μ is then Lebesgue measure λ on $[0,1]$ – i.e. that $\mu(I) = \mathrm{length}(I)$ for any interval $I \subseteq [0,1]$. Note that this is a *construction* of Lebesgue measure, as compared with the result Theorem 3.4, which is a *representation* result because it presupposes the existence of Lebesgue measure.

The following is essentially Anderson's famous construction of Brownian motion.

Example 4.2. (Anderson [6]) *Let $\Omega \subseteq {}^*C[0,1]$ be the set of polygonal paths starting at 0 with step sizes $\pm\sqrt{\Delta t}$ for time increments Δt. Let M be the probability measure concentrated on Ω that gives weight 2^{-N} to each path $\omega \in \Omega$. Then the measure W defined on $C[0,1]$ by*

$$W(B) = M_L(\mathrm{st}^{-1}(B))$$

is Wiener measure.

For a proof of this see any of the references [AFHL, 6, 14, Li]. The crucial (and hardest) part of the proof is to show that M_L-almost all paths $X \in \Omega$ are S-continuous (equivalently, nearstandard in the uniform topology on $^*C[0,1]$).

A variation on this construction using the nonstandard Gaussian Γ measure on $^*\mathbb{R}^\mathbf{T}$ was given in [17]. The internal probability measure Γ is defined by

$$\Gamma(A) = (2\pi\Delta t)^{-\frac{N}{2}} \int_A \exp\left(-\tfrac{1}{2\Delta t} \sum_{t\in\mathbf{T}} x_t^2\right) dx$$

for $A \subseteq {}^*\mathbb{R}^\mathbf{T}$, where dx denotes *Lebesgue measure. Now think of the variables x_t as the increments of a path $\Sigma(x)$ in $^*C[0,1]$, given by

$$\Sigma(x)(t) = \sum_{s\in\mathbf{T}, s<t} x_s$$

for $t \in \mathbf{T}$ and fill in linearly. Then, via the standard part map we have Wiener measure:

Theorem 4.3. [17] *Define a mapping $\pi : {}^*\mathbb{R}^\mathbf{T} \to C[0,1]$ by $\pi(x) = \mathrm{st} \circ \Sigma(x)$. Then*

$$W(B) = \Gamma_L(\pi^{-1}(B))$$

is Wiener measure.

Another variation (see [20]) constructs Wiener measure from the uniform probability on an infinite dimensional sphere. Denote by M the uniform probability measure on $S^{N-1}(1) \subseteq {}^*\mathbb{R}^\mathbf{T}$ (i.e. the surface of the sphere of unit radius in $^*\mathbb{R}^\mathbf{T}$). Then

Theorem 4.4.
$$W(B) = M_L(\pi^{-1}(B))$$

is Wiener measure.

Suppose now that β is a (suitably scaled) *Brownian motion on $S^{N-1}(1)$, with $\beta(0)$ uniformly distributed. Then under the mapping $\pi : S^{N-1}(1) \to C[0,1]$ this gives the infinite dimensional Ornstein-Uhlenbeck process (([21]); cf. the paper [31], which constructs this process from Anderson's Brownian motion).

The book [AFHL] gives many examples of measures obtained by pushing down, including pleasant proofs of several classical results concerning the existence of measures.

With weaker conditions on the internal measure M, a measure can be obtained by pushing down, but it may not be defined on all Borel sets. In the papers [22, 37], for example, the Riesz representation theorem was proved by pushing a Loeb measure down to give a Baire measure on a compact space.

Given an internal measure M on an internal algebra $\mathcal{A}$ on *X, where X is Hausdorff, the measure obtained by pushing M_L down via the standard part map st : *$X \to X$ depends crucially on the collection of sets B for which $\text{st}^{-1}(B)$ is Loeb measurable (with respect to M_L). Put another way, this is the question of the measurability of the mapping st. A closely related issue is (for the case $B = X$), whether $\text{st}^{-1}(X)\ (= \text{ns}(^*X))$ is Loeb measurable?

The papers [33, 34, LR, 3, 36] contain a wealth of information about this; see also [25].

In [LR], following Loeb [33], Landers and Rogge make the following definitions.

Definition 4.5.

(1) *Given an algebra $\mathcal{A}$ on a set X, a set $Y \subseteq {}^*X$ is universally Loeb measurable if $Y \in L_M(^*\mathcal{A})$ for every Loeb algebra given by a finite internal finitely additive measure M on *$\mathcal{A}$. Write $L_u(^*\mathcal{A})$ for the universally measurable sets (relative to *$\mathcal{A}$).*

(2) *The standard part map* st : *$X \to X$ *is universally Loeb measurable if* $\text{st}^{-1}(B)$ *is universally Loeb measurable for all Borel sets B.*

A sample of the results proved in [LR] is:

Theorem 4.6. *Let (X, τ) be a Hausdorff space with Borel algebra $\mathcal{B}$. Then*
 (a) *If X is either locally compact, or σ-compact or complete metric, then*
$$\text{ns}(^*X) \in L_u(^*\mathcal{B}).$$
 (b) *If X is regular, then for every $B \in \mathcal{B}$*
$$\text{st}^{-1}(B) \in L_u(^*\mathcal{B}) \cap \text{ns}(^*X).$$

(c) *If X is regular, then* st *is universally Loeb measurable if (and only if) the set* $\text{ns}(^*X)$ *is universally Loeb measurable.*

Results along these lines make it easier to see when pushing down M_L will give a Borel measure – this is the case for example when X is locally compact or σ-compact or complete metric. Landers & Rogge also showed that for regular spaces, if st is universally Loeb measurable then the Borel measure $M_L \circ \text{st}^{-1}$ obtained by pushing down is necessarily Radon. In [LR] they took the pushing down idea a stage further, by considering the set functions $\overline{M} \circ \text{st}^{-1}$ and $\underline{M} \circ \text{st}^{-1}$, and showed that these give a

τ-smooth measure and a Radon measure respectively on the Borel subsets of X, even when st is not universally Loeb measurable.

The theory of pushing down Loeb measures is quite well understood for regular Hausdorff spaces (see the papers cited above) but in [5] Aldaz & Loeb give a number of counterexamples showing that the results for regular spaces do not extend in general to arbitrary Hausdorff spaces.

Weak standard parts of measures.

Related to the above discussion are some early results of Loeb [33] (see also Anderson & Rashid [8]), concerning weak standard parts of measures. The following is a simple illustration of the ideas.

Theorem 4.7. *Let X be a compact Hausdorff space and let M be an internal *Baire measure on $^*X$ with $M(^*X)$ finite. Then M is weakly nearstandard and its standard part is μ given by*

$$\mu(B) = M_L(\mathrm{st}^{-1}(B))$$

for each Baire set $B \subseteq X$.

Proof From results in the previous section it is clear that this defines a Baire measure on X. To see that it is the weak standard part of M, we have to show that for each $f \in C(X)$

$$\int_{^*X} {^*f}(y)dM(y) \approx \int_X f(x)d\mu(x).$$

Now $^*f(y) \approx f(^\circ y)$ for all $y \in {^*X}$ since f is continuous, so $^*f$ is a *lifting* (see the next section) of $f(^\circ y)$. This gives

$$\int_{^*X} {^*f}(y)dM(y) \approx \int_{^*X} f(^\circ y)dM_L(y) = \int_X f(x)d\mu(x)$$

using the definition of μ for the last equality, and for the first $\approx$ we use the theory of Loeb integration from the next section. □

Using this idea Anderson [6] used his construction of Wiener measure to give an elementary proof of Donsker's invariance principle, which shows that Wiener measure is the weak limit of the measures induced by a symmetric random walk. In similar vein, and more simply, the above construction of Lebesgue measure (or the representation result Theorem 3.4) shows that Lebesgue measure is the weak limit of uniform counting measures.

5. Loeb Integration

Loeb integration theory is simply the theory of the integral in the classical sense with respect to Loeb measures. The special features of this theory stem from the source of the measures, and not from any unusual definition of integration itself.

Loeb Measurable Functions

If $(\Omega, L(\mathcal{A}), M_L)$ is a Loeb measure space, then a function $f : \Omega \to \overline{\mathbb{R}} = [-\infty, \infty]$ is *Loeb measurable* (with respect to M_L) if f is measurable in the conventional sense i.e. $f^{-1}(B) \in L(\mathcal{A})$ for open $B \subseteq \mathbb{R}$.

The corresponding notion for internal functions $F : \Omega \to {}^*\mathbb{R}$ is that for any *open set $A \subseteq {}^*\mathbb{R}$ we have $F^{-1}(A) \in \mathcal{A}$. The connection between these notions uses the important notion of a *lifting*.

Definition 5.1. An internal $\mathcal{A}$-measurable function $F : \Omega \to {}^*\mathbb{R}$ is a *lifting* of f if $f(\omega) = {}^\circ F(\omega)$ for M_L-almost all ω.

Then we have

Theorem 5.2. *The function $f : \Omega \to \overline{\mathbb{R}}$ is Loeb measurable if and only if it has a lifting F. If f is bounded above (or below, or both) then F may be chosen with the same bound.*

Proof First note that if $F : \Omega \to {}^*\mathbb{R}$ is internal and $\mathcal{A}$-measurable then ${}^\circ F : \Omega \to \overline{\mathbb{R}}$ is Loeb measurable since for $r \in \mathbb{R}$

$$\{\omega \in \Omega : {}^\circ F(\omega) \leq r\} = \bigcap_{n \in \mathbb{N}} \{\omega \in \Omega : F(\omega) \leq r + \tfrac{1}{n}\} \in \sigma(\mathcal{A}) \subseteq L(\mathcal{A}).$$

If F is a lifting of f, then $f = {}^\circ F$ a.s. and so f is also Loeb measurable.

Conversely, assume that f is Loeb measurable. Let $(q_n)_{n \in \mathbb{N}}$ be an enumeration of all rationals. Put

$$B_n = \{\omega \in \Omega : f(\omega) \leq q_n\}.$$

For $n = 1, 2, \ldots$ choose internal sets $A_n \in \mathcal{A}$ such that $M_L(A_n \Delta B_n) = 0$ and $A_n \subseteq A_m$ whenever $q_n \leq q_m$. Extend the sequence $\{A_n\}$ to an internal sequence $\{A_n\}_{n \in {}^*\mathbb{N}}$ and by overflow find an $K \in {}^*\mathbb{N}\setminus\mathbb{N}$ such that for all $n, m \leq K$, if $q_n \leq q_m$ then $A_n \subseteq A_m$. The hyperfinite set $\{q_n\}_{n \leq K}$ can be ordered: $q_{i_1} < q_{i_2} < \ldots < q_{i_K}$ and we put

$$F(\omega) = \begin{cases} q_{i_j} & \text{for } \omega \in A_{i_j} \setminus A_{i_{j-1}} \\ q_{i_K} + 1 & \text{for } \omega \notin A_{i_K} \end{cases}$$

(let $A_{i_0} = \emptyset$). Outside the M_L-null set

$$\bigcup_{n \in \mathbb{N}} (A_n \Delta B_n)$$

we have $F(\omega) \leq q_n$ if and only if $f(\omega) \leq q_n$ for all $n \in \mathbb{N}$; hence ${}^\circ F(\omega) = f(\omega)$ a.s.

If $f \leq k$, then $F \wedge k$ is also a lifting of f with $F \wedge k \leq k$; and similarly for the other cases. $\square$

Remark It is easy to prove (as Anderson did in [7]) that this holds for Loeb-measurable functions with range in a Hausdorff space Y having a countable base of open sets (for example a separable metric space). In [38] Ross showed that Theorem 5.2 is true whenever the range of f is a metric space, using the fact that any such function takes its values almost surely in a separable subspace of the range. He also extended the idea of a lifting to more general range spaces Y, where it is natural to consider mappings φ more general than the standard part – for example if Y is the nonstandard hull of an internal metric space (Z,ρ) take φ to be the quotient mapping from Z to Y.

A very important and useful result regarding liftings is Anderson's 'Luzin' Theorem [7]:

Theorem 5.3. *Let $(X,\mathcal{C},\mu)$ be a (completed) Radon space and let $f : X \to \mathbb{R}$ be measurable. Then $^*f$ is a lifting of f with respect to $^*\mu_L$ i.e.*

$$^*f(x) \approx f(^\circ x)$$

for $(^*\mu)_L$-almost all $x \in {^*X}$.

Consequently, there is a set of full Loeb measure $Y \subseteq {^*X}$ such that for all $y_1, y_2 \in Y$

$$y_1 \approx y_2 \implies {^*f(y_1)} \approx {^*f(y_2)}.$$

Proof Let $(U_n)_{n \in \mathbb{N}}$ be an enumeration of all open intervals with rational endpoints. If $x \in \mathrm{ns}(^*X)$ and $^*f(x) \not\approx f(^\circ x)$ then there is some n with

$$f(^\circ x) \in U_n \quad \text{and} \quad {^*f(x)} \notin U_n.$$

Hence $x \in \mathrm{st}^{-1}(A_n) \setminus {^*A_n}$ where $A_n = f^{-1}(U_n)$. Now Corollary 3.3 shows that

$$^*\mu_L(\mathrm{st}^{-1}(A_n) \setminus {^*A_n}) = 0.$$

since $A_n \in \mathcal{C}$. So

$$\{x \in {^*X} : {^*f(x)} \not\approx f(^\circ x)\} \subseteq ({^*X} \setminus \mathrm{ns}(^*X)) \cup \bigcup_{n \in \mathbb{N}} A_n$$

and the set on the right is a countable union of null sets. □

Remark Anderson established this result for measurable functions into any Hausdorff space with countable base of open sets, and it is clear how to modify the proof.

Loeb Integration

If $F : \Omega \to {^*\mathbb{R}}$ is $\mathcal{A}$-measurable and *integrable (with respect to M), then we have the internal integral[1] $\int F dM$. The first connection with Loeb integration (i.e. integration in the standard classical sense with respect to M_L) is:

Proposition 5.4. *If F is a bounded internal measurable function then*

$$^\circ \int F dM = \int {^\circ F} dM_L.$$

---

[1] If M is only finitely additive, then this requires the theory of integration with respect to finitely additive measures. However, in most applications $\mathcal{A}$ is a $^*\sigma$-algebra and M is $^*\sigma$-additive on $\mathcal{A}$ so we can handle the integrals by transfer of the standard theory.

Proof Take any real $\varepsilon > 0$ and choose m such that $|F(\omega)| \leq m\varepsilon$. Let
$$A_k = \{\omega \in \Omega : k\varepsilon \leq F(\omega) < (k+1)\varepsilon\}, \quad -m \leq k \leq m,$$
and define
$$F_1(\omega) = k\varepsilon \quad \text{if} \quad \omega \in A_k,$$
$$F_2(\omega) = (k+1)\varepsilon \quad \text{if} \quad \omega \in A_k,$$
so that $F_1 \leq F \leq F_2$ and $0 \leq F_2 - F_1 \leq \varepsilon$. This implies
$$\int F_1 dM \leq \int F dM \leq \int F_2 dM$$
hence

(4) $$a = {}^\circ\!\!\int F_1 dM \leq {}^\circ\!\!\int F dM \leq {}^\circ\!\!\int F_2 dM = b,$$

say, and $b - a \leq \varepsilon M(\Omega)$. Similarly ${}^\circ F_1 \leq {}^\circ F \leq {}^\circ F_2$ a.s. and so

(5) $$a' = \int {}^\circ F_1 dM_L \leq \int {}^\circ F dM_L \leq \int {}^\circ F_2 dM_L = b',$$

say. Now
$${}^\circ\!\!\int F_2 dM = {}^\circ\!\left(\sum_{-m \leq k \leq m} \varepsilon(k+1)M(A_k)\right) = \sum_{-m \leq k \leq m} \varepsilon(k+1){}^\circ M(A_k)$$
$$= \sum_{-m \leq k \leq m} \varepsilon(k+1)M_L(A_k) = \int {}^\circ F_2 dM_L$$

and similarly for F_1. This gives $a = a'$, $b = b'$, so by (4) and (5)
$$\left|{}^\circ\!\!\int F dM - \int {}^\circ F dM_L\right| \leq b - a \leq \varepsilon M(\Omega)$$
which finishes the proof. $\square$

Corollary 5.5. *If F is a bounded lifting of a Loeb measurable f, then*
$$\int f dM_L = {}^\circ\!\!\int F dM.$$

We cannot in general expect equality of ${}^\circ \int F dM$ and $\int {}^\circ F dM_L$. In the example below F is large on a set of infinitesimal measure.

Example Consider $\Omega = {}^*[0,1]$ and define $F : {}^*[0,1] \to {}^*\mathbb{R}$ by
$$F(\tau) = \begin{cases} K & \text{for } \tau \leq \dfrac{1}{K} \\ 0 & \text{otherwise.} \end{cases}$$

So ${}^\circ F(\tau) = 0$ almost everywhere hence $\int {}^\circ F d\Lambda_L = 0$ but $\int F d\Lambda = 1$ (where $\Lambda = {}^*\lambda$ is *Lebesgue measure).

We always have

Proposition 5.6. *For any internal $\mathcal{A}$-measurable F with $F \geq 0$*

$$\int {}^\circ F dM_L \leq {}^\circ\!\!\int F dM,$$

where we allow the value ∞ on either side.

Proof By classical integration theory

$$\int {}^\circ F dM_L = \lim_{n \to \infty} \int ({}^\circ F \wedge n) dM_L$$
$$= \lim_{n \to \infty} {}^\circ\!\!\int (F \wedge n) dM \quad \text{by Proposition 5.4}$$
$$\leq {}^\circ\!\!\int F dM$$

as required. □

To obtain equality of ${}^\circ\!\int F dM$ and $\int {}^\circ F dM_L$ it is necessary to have some condition on F akin to standard integrability — roughly, so that F is not too big on small sets. The following is the appropriate condition.

Definition 5.7. *Let a function $F : \Omega \to {}^*\mathbb{R}$ be $\mathcal{A}$-measurable and internal and M an internal finite measure. Then F is S-integrable if*

(i) $\int_\Omega |F| dM$ *is finite*,

(ii) *if $A \in \mathcal{A}$ and $M(A) \approx 0$, then $\int_A |F| dM \approx 0$.*

Note If M is not finite an extra condition has to be added:

(iii) if $A \in \mathcal{A}$ and $F \approx 0$ on A, then $\int_A |F| dM \approx 0$.

This is always satisfied for a finite measure M: if $F \approx 0$ on A, then for any $\varepsilon > 0$, $\varepsilon \in \mathbb{R}$, $\int_A |F| dM < \varepsilon M(A) \approx 0$.

The function in the example above is not S-integrable because $A = [0, \frac{1}{K}]$ has $\Lambda(A) \approx 0$ but $\int_A F d\Lambda = 1$.

Note that F is S-integrable if and only if its positive and negative parts F^+ and F^- are S-integrable.

Theorem 5.8. *Let $F : \Omega \to {}^*\mathbb{R}$ be $\mathcal{A}$-measurable, with $F \geq 0$. Then the following conditions are equivalent:*

(i) F *is S-integrable*,

(ii) ${}^\circ F$ *is Loeb integrable and*

$$ {}^\circ\!\!\int F dM = \int {}^\circ F dM_L.$$

Proof (i) ⇒ (ii)   Suppose first that F is S-integrable and let

$$I = \int {}^\circ F dM_L$$

which is finite by Proposition 5.6. By Proposition 5.4
$$\int (F \wedge n) dM \leq \int (^\circ F \wedge n) dM_L + \tfrac{1}{n} \leq I + \tfrac{1}{n}$$
for finite n, so by overflow there is an infinite K with
$$\int (F \wedge K) dM \leq I + \frac{1}{K}.$$
So
$$\int F dM \leq \int_{\{F>K\}} F dM + \int (F \wedge K) dM \leq \int_{\{F>K\}} F dM + I + \frac{1}{K}.$$
Now $^\circ F(\omega) < \infty$ almost surely (since $^\circ F$ is integrable) so $M(\{F > K\}) \approx 0$. Since F is S-integrable, this gives $\int_{\{F>K\}} F dM \approx 0$. So
$$^\circ\!\!\int F dM \leq I = \int ^\circ F dM_L$$
which combines with Proposition 5.6 to give the result.

(ii) $\Rightarrow$ (i) If $M(A) \approx 0$, then
$$^\circ\!\!\int F dM = ^\circ\!\!\int_{\Omega \setminus A} F dM + ^\circ\!\!\int_A F dM \geq ^\circ\!\!\int_{\Omega \setminus A} F dM \geq \int_{\Omega \setminus A} {}^\circ F dM_L = \int ^\circ F dM_L$$
since $M_L(A) = 0$. So $\int_A F dM \approx 0$ since $^\circ \int F dM = \int ^\circ F dM_L$. □

The following is an equivalent formulation of S-integrability (the proof is left as an exercise).

Proposition 5.9. *An internal function F is S-integrable if and only if for all infinite K*
$$\int_{|F|>K} |F| dM \approx 0.$$

To complete the basic theory of Loeb integration we have:

Theorem 5.10. *Let $f : \Omega \to \overline{\mathbb{R}}$ be Loeb measurable. Then f is M_L-integrable if and only if then it has an S-integrable lifting $F : \Omega \to {}^*\mathbb{R}$.*

Proof If F is an S-integrable lifting of f then $f = {}^\circ F$ almost surely and by Theorem 5.8 we know that $^\circ F$ is integrable.

Suppose, conversely, that f is integrable. We may consider $f \geq 0$ and take a lifting $F \geq 0$ of f (Theorem 5.2). For each finite n, by Corollary 5.5
$$\int (F \wedge n) dM \leq \int (f \wedge n) dM_L + \frac{1}{n} \leq \int f dM_L + \frac{1}{n}$$
and so there is infinite K with
$$\int (F \wedge K) dM \leq \int f dM_L + \frac{1}{K}.$$
Clearly $F \wedge K$ is a lifting of f and
$$^\circ\!\!\int (F \wedge K) dM \leq \int f dM_L = \int ^\circ (F \wedge K) dM_L.$$

The reverse inequality is always true so we may apply Theorem 5.8 to deduce that $F \wedge K$ is S-integrable; it is clearly a lifting of f. □

Definition 5.11. *We say that $F : \Omega \to {}^*\mathbb{R}$ is $\mathbf{SL}^p$ ($p > 0$) if $|F|^p$ is S-integrable (so SL^1 means S-integrable).*

Here is a very useful test for S-integrability isolated by Lindstrøm [29] and frequently applied in the case $p = 2$.

Theorem 5.12. *Suppose $M(\Omega) < \infty$. If $F : \Omega \to {}^*\mathbb{R}$ is internal, $\mathcal{A}$-measurable, and*
$$\int_\Omega |F|^p dM < \infty$$
for some $p > 1$, $p \in \mathbb{R}$, then F is S-integrable.

Proof Clearly (since $M(\Omega)$ is finite) $\int_\Omega |F| dM$ is finite. Consider any $A \in \mathcal{A}$ with $M(A) \approx 0$. Then
$$\int_A |F| dM \leq \left(\int_\Omega \mathbb{I}_A dM\right)^{\frac{1}{q}} \left(\int_\Omega |F|^p dM\right)^{\frac{1}{p}}, \quad \frac{1}{p} + \frac{1}{q} = 1,$$
where $\mathbb{I}_A$ denotes the indicator function of the set A. Now $\int_\Omega |F|^p dM$ is finite and $\int_\Omega \mathbb{I}_A dM \approx 0$, hence the result. □

The following application of Theorem 5.10 gives a representation of the Lebesgue integral as a Riemann sum.

Theorem 5.13. *A function $f : [0,1] \to \mathbb{R}$ is Lebesgue integrable if and only if there is an S-integrable function $F : \mathbf{T} \to {}^*\mathbb{R}$ such that*
$$F(\mathbf{t}) \approx f(^\circ \mathbf{t})$$
for almost all $\mathbf{t} \in \mathbf{T}$. For any such F
$$\int f d\lambda \approx \sum_{\mathbf{t} \in \mathbf{T}} F(\mathbf{t}) \Delta t.$$

Proof Define $\hat{f} : \mathbf{T} \to \mathbb{R}$ by
$$\hat{f}(\mathbf{t}) = f(^\circ \mathbf{t}).$$
From Theorem 3.4, f is Lebesgue measurable if and only if $\hat{f}$ is Loeb measurable with respect to the Loeb counting measure ν_L, and the same is true for integrability. In the latter case,
$$\int f d\lambda = \int \hat{f} d\nu_L.$$
By Proposition 5.10, $\hat{f}$ is Loeb integrable if and only if it has an S-integrable lifting $F : \mathbf{T} \to {}^*\mathbb{R}$. For any such F we have $F(\mathbf{t}) \approx \hat{f}(\mathbf{t})$ for a.a. $\mathbf{t} \in \mathbf{T}$ and
$$\int \hat{f} d\nu_L \approx \int F d\nu = \sum_{\mathbf{t} \in \mathbf{T}} F(\mathbf{t}) \Delta t.$$
Thus F is as required. □

Remark The function F as in this theorem is called a *lifting* of f – but note that this is a slightly different usage of the term. (Some authors call this a two-legged lifting.)

Finally in this section, for many applications it is worth making the following elementary observation.

Proposition 5.14. *Suppose that T is finite and $F : {}^*[0,T] \to {}^*\mathbb{R}$ is S-integrable with respect to $\Lambda = {}^*\lambda$ (λ is Lebesgue measure). Then the function*

$$G(\tau) = \int_0^\tau F(\sigma)d\sigma$$

is S-continuous.

Proof It $\tau_1 \approx \tau_2$ and $\tau_1 \leq \tau_2$, say, then $\Lambda([\tau_1, \tau_2]) \approx 0$ and so

$$G(\tau_2) - G(\tau_1) = \int_{\tau_1}^{\tau_2} F(\sigma)d\sigma \approx 0$$

as required. □

6. Differential equations

One of the most fruitful applications of Loeb measures has been in the solution of differential equations, both deterministic and stochastic; often these are more correctly described as *integral* equations. There are a number of different Loeb space techniques available, and for any particular equation or purpose it is necessary to select the most useful or appropriate method. We will illustrate three of them below.

The first and perhaps most appealing technique is that of *hyperfinite difference equations*, pioneered by Keisler and used to great effect especially for stochastic differential equations – see [35]. Here is a typical and fairly general example of this approach in the deterministic case – a proof of Peano's fundamental existence theorem.

Theorem 6.1. *Suppose that $f : [0,1] \times \mathbb{R} \to \mathbb{R}$ is bounded, measurable, and continuous in the second variable, and let $x_0 \in \mathbb{R}$. Then there is a solution to the differential equation*

(6)
$$\begin{aligned} dx(t) &= f(t, x(t))dt \\ x(0) &= x_0 \end{aligned}$$

(Of course, what is meant is really the corresponding integral equation.)

Proof Without any loss of generality we may assume that $x_0 = 0$ (otherwise consider the equation for $x(t) - x_0$). Suppose that $|f| \leq K$. Define the measurable function $\hat{f} : [0,1] \to C([-K, K])$ by

$$\hat{f}(t)(z) = f(t, z).$$

for $|z| \leq K$. From this we obtain (using Theorem 3.4) a Loeb measurable function $\check{f} : \mathbf{T} \to C([-K, K])$ (where $\mathbf{T}$ is the hyperfinite time line, endowed with the counting measure ν as above) by

$$\check{f}(\mathbf{t}) = \hat{f}(°\mathbf{t}).$$

for t ∈ **T** (in this section we use sanserif symbols **t, s** for elements of **T** to distinguish them from those in $[0,1]$.) Taking the uniform topology on $C([-K, K])$ and the extension of Theorem 5.2 to separable metric spaces, we obtain a lifting $\hat{F} : \mathbf{T} \to {}^*C([-K, K])$ such that for almost all $\mathbf{t} \in \mathbf{T}$ (with respect to ν_L)

$$\hat{F}(\mathbf{t}) \approx \check{f}(\mathbf{t}) = \hat{f}(°\mathbf{t})$$

and $|\hat{F}| \leq K$. This means that for all such $\mathbf{t}$

$$\hat{F}(\mathbf{t})(X) \approx \check{f}(\mathbf{t})(°X) = \hat{f}(°\mathbf{t})(°X) = f(°\mathbf{t}, °X)$$

for all $|X| \leq K$. Now define $F : \mathbf{T} \times {}^*[-K, K] \to {}^*\mathbb{R}$ by

$$F(\mathbf{t}, X) = \hat{F}(\mathbf{t})(X).$$

Then $|F| \leq K$ and for almost all $\mathbf{t} \in \mathbf{T}$

(7) $$F(\mathbf{t}, X) \approx f(°\mathbf{t}, °X)$$

for all $|X| \leq K$. The *hyperfinite difference equation* corresponding to (6) is now

$$\Delta X(\mathbf{t}) = F(\mathbf{t}, X(\mathbf{t}))\Delta t,$$

where $\Delta X(\mathbf{t}) = X(\mathbf{t} + \Delta t) - X(\mathbf{t})$. This is an internal equation for an internal function $X : \mathbf{T} \to {}^*\mathbb{R}$, with solution $X(\mathbf{t})$ defined recursively by

$$\begin{cases} X(0) &= 0 \\ X(\mathbf{t} + \Delta t) &= X(\mathbf{t}) + F(\mathbf{t}, X(\mathbf{t}))\Delta t. \end{cases}$$

Clearly (as in Proposition 5.14) X is S-continuous, and $|X(\mathbf{t})| \leq K\mathbf{t} \leq K$ for all $\mathbf{t} \in \mathbf{T}$. So we may define a continuous function $x : [0, 1] \to \mathbb{R}$ by

$$x(t) = °X(\mathbf{t})$$

for any $\mathbf{t} \approx t$. Clearly $|x(t)| \leq K$. To see that $x(t)$ is a solution, observe that, by (7) and the definition of x, for almost all $\mathbf{t} \in \mathbf{T}$

$$F(\mathbf{t}, X(\mathbf{t})) \approx f(°\mathbf{t}, °X(\mathbf{t})) = f(°\mathbf{t}, x(°\mathbf{t})).$$

So, applying Theorem 5.13 to the function $g(t) = f(t, x(t))$ and its lifting $G(\mathbf{t}) = F(\mathbf{t}, X(\mathbf{t}))$ we have (putting $t = °\mathbf{t}$)

$$x(t) = °X(\mathbf{t}) = °\sum_{\mathbf{s} < \mathbf{t}} F(\mathbf{s}, X(\mathbf{s}))\Delta t$$

$$= \int_0^t f(s, x(s))ds$$

as required. □

A slightly different Loeb measure approach to differential equations is to work with an infinitesimal *delayed equation*, and we illustrate this with an alternative proof of the Peano theorem.

Alternative Proof of Theorem 6.1 Let $\Delta = \Delta t = N^{-1}$ as above, and define an internal function $X : {}^*[-\Delta, 1] \to {}^*\mathbb{R}$ by

$$X(\tau) = x_0 \qquad \text{for } -\Delta \leq \tau \leq 0$$
$$X(\tau) = x_0 + \int_0^\tau {}^*f(\sigma, X(\sigma - \Delta))d\sigma \qquad \text{for } 0 \leq \tau \leq 1$$

Note that $X(\tau)$ is defined recursively on $[k\Delta, (k+1)\Delta]$ for $k = 0, 1, 2, \ldots, N-1$.

Since f and hence ${}^*f$ is bounded, clearly X is S-continuous and we can define a standard function $x : [0, 1] \to \mathbb{R}$ by

$$x(t) = {}^\circ X(t) = {}^\circ X(\tau)$$

for any $\tau \approx t$. We claim that $x(t)$ is a solution to equation (6).

Let $\Lambda = {}^*\lambda = {}^*$ Lebesgue measure. Using the extension of Anderson's Luzin Theorem 5.3, mentioned above, and considering the function $\hat{f} : [0, 1] \to C(\mathbb{R})$ defined by $\hat{f}(t)(z) = f(t, z)$ we have that for almost all τ (with respect to Λ_L)

$${}^*f(\tau, y) \approx f({}^\circ\tau, {}^\circ y) \qquad \text{for all finite } y \in {}^*\mathbb{R}.$$

Hence, for almost all $\tau \in {}^*[0, 1]$

$${}^*f(\tau, X(\tau - \Delta)) \approx f({}^\circ\tau, {}^\circ X(\tau - \Delta)) = f({}^\circ\tau, x({}^\circ\tau))$$

since ${}^\circ(\tau - \Delta) = {}^\circ\tau$. Now this means that $G(\tau) = {}^*f(\tau, X(\tau - \Delta))$ is a bounded lifting of $g(\tau) = f({}^\circ\tau, x({}^\circ\tau))$ and so for any $t \in [0, 1]$

$$x(t) = {}^\circ X(t) = x_0 + {}^\circ\!\!\int_0^t G(\tau)d\tau = x_0 + \int_0^t g(\tau)d_L\tau$$

where $d_L\tau$ denotes integration with respect to Λ_L. Since $\lambda = \Lambda_L \circ \text{st}^{-1}$, we have

$$\int_0^t g(\tau)d_L\tau = \int_0^t f({}^\circ\tau, x({}^\circ\tau))d_L\tau = \int_0^t f(t, x(t))dt$$

which shows that $x(t)$ is a solution to equation (6). $\square$

The hyperfinite difference approach has been used to great effect in the solution of Itô stochastic differential equations (SDEs), based on Anderson's hyperfinite random walk construction of Brownian motion and the Itô integral [6]. This was pioneered in [35] and subsequently used by many authors, both in solving SDEs and in applications such as optimal control theory [16] and mathematical finance theory ([19] for example). The delay approach is also appropriate for certain SDEs – see [15]. Loeb space methods for SDEs have been extended to equations involving general stochastic integrals against martingales and semimartingales, beginning with Hoover & Perkins [26]

For partial differential equations (PDEs) – or, more generally, infinite dimensional differential equations, in addition to the above approaches there are new possibilities for constructing solutions using hyperfinite *dimensional* representation. The book [13] develops this idea in some detail for the Navier–Stokes equations, which are formulated as a differential equation in a certain separable Hilbert space H. Abstractly this equation has the form

(8) $$x(t) = x_0 + \int_0^t f(s, x(s))ds$$

(for $t \in [0,T]$, say) which is the same as equation (6) except that now $x(t) \in H$, and the integral here is the Bochner integral. The essence of the solution technique in [13] is to take an appropriate hyperfinite dimensional subspace $H_N \subseteq {}^*H$ and formulate the following differential equation in H_N:

$$X(\tau) = X(0) + \int_0^\tau F(\sigma, X(\sigma))d\sigma \tag{9}$$

where $X : {}^*[0,T] \to H_N$, $X(0) = \Pr_N x_0$ and $F = \Pr_N {}^* f$.

The strategy is to show that equation (9) has an (internal) solution X, by the transfer of standard finite dimensional ODE theory. Then, by some careful analysis, show that X has a "standard part" $x(t)$ in an appropriate sense, and that this solves (8). Of course, depending on the nature of the original equation, this can involve considerable effort. These ideas were extended to stochastic PDEs to solve some open existence problems for the stochastic Navier–Stokes equations – see [13].

Measure valued equations on an infinite dimensional space can also be treated successfully using hyperfinite dimensional representation, together with the idea of *nonstandard densities*. Such an equation for time evolving probability measures $(\mu_t)_{t \geq 0}$ on a Hilbert space H may take the form

$$\int_H \varphi(x)d\mu_t(x) = \int_H \varphi(x)d\mu_0(x) + \int_0^t \int_H (L\varphi)(x)d\mu_s(x)ds \tag{10}$$

where φ ranges over all *test functions* $\varphi : H \to \mathbb{R}$. One way to solve this is to take the corresponding equation for evolving internal measures $(M_\tau)_{\tau \geq 0}$ on H_N, and then formulate an equation for nonstandard densities $\rho(\tau, X)$ for M_τ against *Lebesgue measure on H_N. This turns out to be a *PDE, which can be solved by the transfer of standard results. From the solution $\rho(\tau, X)$ then *define* measures M_τ on H_N by

$$M_\tau(A) = \int_A \rho(\tau, X)dX$$

for $A \subseteq H_N$. Now, in appropriate circumstances, standard measures μ_t on H defined by

$$\mu_t(B) = (M_t)_L(\text{st}^{-1}(B))$$

give a solution to (10). See [13].

Loeb Differential Equations

The existence of the Loeb-Lebesgue measure $^*\lambda_L$ on $^*\mathbb{R}$ makes it possible (and natural) to formulate and solve *Loeb differential equations* for the "rich" time line $^*\mathbb{R}$. By this we mean *integral* equations of the following kind:

$$x(\tau) = x_0 + \int_0^\tau f(\sigma, x(\sigma))d_L\sigma$$

where $f : {}^*[0,1] \times \mathbb{R} \to \mathbb{R}$ is Loeb measurable in τ and continuous in the second variable. The solution $x(\tau)$ will be S-continuous *and* real valued, so it will really be a continuous function. Such equations occur in the study of optimal control theory, where it is natural to consider Loeb measurable controls. In particular, it can be shown [23] that a general optimal control problem will always have an optimal Loeb control, even when there is no optimal Lebesgue control.

References

[1] S.Albeverio, J.-E.Fenstad, R.Høegh-Krohn, & T.Lindstrøm, *Nonstandard Methods in Stochastic Analysis and Mathematical Physics*, Academic Press, New York 1986.
[2] S.Albeverio, W.A.J.Luxemburg and M.P.H.Wolff (eds.), *Advances in Analysis, Probability and Mathematical Physics – Contributions from Nonstandard Analysis*, Kluwer Academic Publishers, Dordrecht, Boston, London, 1995
[3] J.M.Aldaz, A characterisation of universal Loeb measurability for completely regular Hausdorff spaces, *Can. J. Math.* **44**(1992), 673-690.
[4] J.M.Aldaz, Representing abstract measures by Loeb measures: a generalization of the standard part map, *Proc. Amer. Math. Soc.*, to appear.
[5] J.M.Aldaz & P.A.Loeb, Counterexamples in nonstandard measure theory, *Bull. Can. Math. Soc.*, to appear.
[6] R.M.Anderson, A nonstandard representation for Brownian motion and Itô integration, *Israel Math. J.* **25**(1976), 15-46.
[7] R.M.Anderson, Star–finite representations of measure spaces, *Trans. Amer. Math. Soc.* **271**(1982), 667-687.
[8] R.M. Anderson & S.Rashid, A nonstandard characterisation of weak convergence, *Proc. Amer. Math. Soc.*, **69**(1978), 327-332.
[9] M.Capiński & N.J.Cutland, Statistical solutions of Navier–Stokes equations by nonstandard densities, *Mathematical Models and Methods in Applied Sciences* **1**:4 (1991), 447–460.
[10] M.Capiński & N.J.Cutland, Stochastic Navier–Stokes equations, *Acta Applicanda Mathematicae* **25** (1991), 59–85.
[11] M.Capiński & N.J.Cutland, A simple example of intrinsic turbulence, *Journal of Differential Equations* **98**:1 (1992), 19–33.
[12] M.Capiński & N.J.Cutland, A simple proof of existence of weak and statistical solutions of Navier–Stokes equations, *Proceedings of the Royal Society, London, Ser.A*, **436** (1992), 1–11.
[13] M.Capiński & N.J.Cutland, *Nonstandard Methods for Stochastic Fluid Mechanics*, World Scientific, Singapore, London, 1995.
[14] N.J.Cutland, Nonstandard measure theory and its applications, *J. Bull. London Math. Soc.* **15** (1983), 529-589.
[15] N.J.Cutland, Simplified existence for solutions to stochastic differential equations, *Stochastics* **14**(1985), 319-325.
[16] N.J.Cutland, Infinitesimal methods in control theory: deterministic and stochastic, *Acta Applicandae Mathematicae* **5**(1986), 105-135.
[17] N.J.Cutland, Infinitesimals in action, *J. London Math. Soc.* **35**(1987), 202-216.
[18] N.J.Cutland (ed.), *Nonstandard Analysis and its Applications*, Cambridge University Press, Cambridge 1988.
[19] N.J.Cutland, P.E. Kopp & W. Willinger, A nonstandard approach to option pricing, *Mathematical Finance* **1**(4)(1991), 1-38.
[20] N.J.Cutland & Siu-Ah Ng, The Wiener sphere and Wiener measure, *The Annals of Probability* **21** (1993), 1–13.
[21] N.J.Cutland, Brownian motion on the Wiener sphere is the infinite dimensional O-U process, in preparation.
[22] D.J.H.Garling, Another 'short' proof of the Riesz representation theorem, *Math. Proc. Camb. Phil. Soc.* **99**(1986), 261-262.
[23] D.R.Gordon, PhD thesis, University of Hull 1995.

[24] C.W.Henson, Unbounded Loeb measures, *Proc. Amer. Math. Soc.* **74**(1979), 143-150.
[25] C.W.Henson, Analytic sets, Baire sets and the standard part map, *Can. J. Math.* **31**(1979), 663-672.
[26] D.N.Hoover & E.Perkins, Nonstandard constructions of the stochastic integral and applications to stochastic differential equations I, II, *Trans. Amer. Math. Soc.* **275**, 1-58.
[27] A.E.Hurd & P.A.Loeb, *An Introduction to Nonstandard Real Analysis*, Academic Press, New York 1985.
[28] D.Landers & L.Rogge, Universal Loeb-measurability of sets and of the standard part map with applications, *Trans. Amer. Math. Soc.* **304**(1987), 229-243.
[29] T.L.Lindstrøm, Hyperfinite stochastic integration I, II, III, *Math. Scand.* **46** (1980), 265–333.
[30] T.L.Lindstrøm, An invitation to nonstandard analysis, in [18], 1-105.
[31] T.L.Lindstrøm, Anderson's Brownian motion and the infinite dimensional Ornstein–Uhlenbeck process, in [2].
[32] P.A.Loeb, Conversion from nonstandard to standard measure spaces and applications in probability theory, *Trans. Amer. Math. Soc.* **211** (1975), 113–122.
[33] P.A.Loeb, Weak limits of measures and the standard part map, *Proc. Amer. Math. Soc.* **77**(1979), 128-135.
[34] P.A.Loeb, A functional approach to nonstandard measure theory, *Contemporary Mathematics* **26**(1984), 251-261.
[35] H.J.Keisler, An infinitesimal approach to stochastic analysis, *Mem. Amer. Math. Soc.* **297**(1984).
[36] H.Render, Pusing down Loeb measures, *Math. Scand.* **72**(1993), 61-84.
[37] D.A.Ross, Yet another short proof of the Riesz representation theorem, *Math. Proc. Camb. Phil. Soc.* **105**(1989), 139-140.
[38] D.A.Ross, Lifting theorems in nonstandard measure theory, *Proc.Amer.Math.Soc.* **109**(1990), 809-822.
[39] D.A.Ross, Compact measures have Loeb preimages, *Proc. Amer. Math. Soc.* **115**(1992), 365-370.
[40] H.L.Royden, *Real Analysis*, Macmillan, New York 1968.
[41] K.D.Stroyan & J.M.Bayod, *Foundations of Infinitesimal Stochastic Analysis*, North-Holland, Amsterdam, New York, Oxford, 1986.

Dep. of Pure Mathematics and Statistics
University of Hull
HU6 7RX England

E-mail n.j.cutland@maths.hull.ac.uk

David A. Ross
Unions of Loeb nullsets: the context

1. Introduction

Over ten years ago I first posed the question, "Is Loeb measure compact?" This question was recently answered (in the negative) by another participant at this conference, Jesus Aldaz [1]. This seems as good a time as any to explain why the question is interesting in the first place.

Call a probability space $(X, \mathcal{B}, P)$ **ono** provided every point-finite, completely-measurable family $\mathcal{E}$ of nullsets has null union, $P(\bigcup \mathcal{E}) = 0$. ($\mathcal{E}$ is **point–finite** if for every $x \in X$, $\{E \in \mathcal{E} \mid x \in E\}$ is finite. $\mathcal{E}$ is **completely measurable** provided $\bigcup \mathcal{K} \in \mathcal{B}$ whenever $\mathcal{K} \subseteq \mathcal{E}$.)

While this term is new, the concept is old: Ulam knew that Lebesgue measure was ono provided measurable cardinals didn't exist; Fremlin [5] proved in 1981 that all Radon measures are ono (incidentally eliminating the 'no measurable cardinals' constraint from Ulam's result); an alternate proof of Fremlin's result by Prikry and Koumoullis [14] in 1983 applies in fact to all compact measures. It therefore followed that if Loeb measures are compact, then they are ono.

Since, as Aldaz has shown, Loeb measures need *not* be compact, the standard results do not apply to these spaces. In 1983, I gave a proof that Loeb spaces are ono [16] – under, however, extremely restrictive conditions on the nonstandard model. Recently, I have proved that in any reasonable nonstandard model, Loeb spaces are ono. The precise statement of this result is given below; the proof will appear elsewhere [20]. The goal of the current paper is to put this result in context – in particular, to describe some ramifications for both standard and nonstandard measure theory, and to point out some interesting open questions connected to the result.

Before proceeding it is useful to point out that the conditions in the definition of 'ono' are not as artificial as they at first might seem. The 'point-finite' condition on $\mathcal{E}$ can be understood as a mild generalization of 'disjoint', or of the topological condition 'locally finite'. A natural example of a completely measurable family is any collection of open sets in a topological measure space.

Here is an interesting illustrative example. Consider any ono probability space $(X, \mathcal{B}, P)$ in which singletons are nullsets (for example Lebesgue measure on $[0, 1]$, or normalized uniform Loeb measure on a hyperfinite set). Put $\mathcal{E} = \{\{x\} \mid x \in X\}$. $\mathcal{E}$ is a point-finite family of nullsets, and $P(\bigcup \mathcal{E}) = 1$; it follows that $\mathcal{E}$ is not completely measurable, which in turn means that there is a nonmeasurable subset of X.

After considering this example, the reader might conclude that point-finiteness and complete measurability are difficult conditions to meet simultaneously, and therefore that 'ono' is not a terribly interesting property for measure spaces. However, suppose that $\mathcal{E}$ is a point-finite collection of subsets of X in *any* complete measure space $(X, \mathcal{B}, P)$. It is easy to see that if $P(\bigcup \mathcal{E}) = 0$ then $\mathcal{E}$ is a completely-measurable family of P-nullsets. Thus, the conditions in the theorem are not only sufficient, they are in some sense necessary.

2. Precise formulation of results

The reader is assumed to be familiar with nonstandard analysis in general, and Loeb measures in particular, and is referred to [12] or [21] for background. All nonstandard models will be assumed $\aleph_1$-saturated. Further saturation requirements are discussed below.

For X a set, write $\mathcal{P}(X)$ for the power set of X, $\mathcal{P}_n(X)$ for $\{E \in \mathcal{P}(X) \mid \text{card}(E) = n\}$, and $\mathcal{P}_{\text{fin}}(X)$ for $\{E \in \mathcal{P}(X) \mid E \text{ finite}\} = \bigcup_{n=0}^{\infty} \mathcal{P}_n(X)$.

Say that $\mathcal{E} \subseteq \mathcal{P}(X)$ has the **finite intersection property** if $E_1 \cap \cdots \cap E_n \neq \emptyset$ whenever $\{E_1, \cdots, E_n\} \subseteq \mathcal{E}$, $n \in \mathbb{N}$. $\mathcal{E}$ is **compact** provided that for every subfamily $\mathcal{K} \subseteq \mathcal{E}$ with the finite intersection property, $\bigcap \mathcal{K} \neq \emptyset$.

Let X be an arbitrary set, and $\nu : \mathcal{P}(X) \to [0, \infty)$. A set $E \in \mathcal{P}(X)$ is an **atom** for ν if $\nu(E) \neq 0$ and $\{\nu(B), \nu(E - B)\} = \{0, \nu(E)\}$ for every $B \subseteq E$. If ν has no atoms it is **atomless**. If α is an infinite cardinal, ν is a finite measure on $(\alpha, \mathcal{P}(\alpha))$ for which (i) α is an atom, (ii) singletons are nullsets, and (iii) $\nu(\bigcup \mathcal{E}) = 0$ whenever $\mathcal{E}$ is a family of nullsets with card$(\mathcal{E}) < \alpha$, then α is a **measurable cardinal**.

Suppose α and κ are cardinals; say that α is **not measurably cofinal in** κ, NMC(α, κ), if either α is not a measurable cardinal or α is not cofinal in κ. Note that NMC(α, κ) holds for *every* α precisely when the cofinality of κ is not measurable.

Suppose that X is internal and $\mathcal{A} \subseteq {}^*\mathcal{P}(X)$. Call $\mathcal{A}$ κ-**terraced** provided

(i): $\kappa = \text{card}(\mathcal{A})$, and
(ii): For some nondecreasing sequence $\{\mathcal{A}_i\}_{i<\kappa}$ with each $\mathcal{A}_i \subseteq \mathcal{A}$ compact, $\mathcal{A} = \bigcup_{i<\kappa} \mathcal{A}_i$.

$(\Omega, \mathcal{A}, \mu)$ will always be an internal, *finitely–additive probability space. Denote by $(\Omega, \mathcal{A}_L, \mu_L)$ the (standard, complete) probability space generated from $(\Omega, \mathcal{A}, \mu)$ by the Loeb construction.

Here is the main result of [20]:

Theorem 1. *Let $(\Omega, \mathcal{A}, \mu)$ be an internal *finitely–additive probability space, and let $\mathcal{E} = \{A_i\}_{i<\alpha}$ be a point-finite, completely measurable family of μ_L-nullsets. Suppose that $\mathcal{A}$ is κ-terraced, and that NMC(α, κ) holds. Then $\mu_L(\bigcup_{i<\alpha} A_i) = 0$.*

The conditions on $\mathcal{A}$ in this theorem are nontrivial to verify. Fortunately, there are nonstandard models in which they *always* hold for *every* Loeb probability space. This follows if, for example, the nonstandard model satisfies the 'special model axiom' (see §3, below):

Theorem 2. *There are nonstandard models in which all Loeb spaces are ono.*

The reader is referred to [20] for proofs of these results.

Some application are discussed below, in §4 and 5.

3. Saturation properties

The proof of Theorem 2 has some relevance to the question of how much saturation nonstandard models need for doing 'interesting' mathematics.

In order to construct the Loeb measure itself, only $\aleph_1$ saturation is necessary, and most papers in nonstandard probability theory make this assumption alone.

For some applications in functional analysis, Henson [6] found it useful to introduce a stronger property, called the κ-**Isomorphism Property**, or IP_κ. This asserts that any two elementary-equivalent internally-presented $\mathcal{L}$-structures are isomorphic, where $\mathcal{L}$ is a first-order language with fewer than κ symbols. IP_κ has some interesting consequences; for example, any two (externally) infinite internal sets have the same external cardinality.

In a different context, Henson [7] constructed a horribly nonmeasurable set as a counterexample to a question about unbounded Loeb measures. The construction appeals directly to the details of a specific nonstandard model; the question naturally arose whether this example could be constructed in a model satisfying IP_κ for some κ.

In [18] I introduced an apparently stronger saturation property, which I called the **Special Model Axiom**, or SMA_κ. This asserts that any internally-presented $\mathcal{L}$-structure is a special model (in the sense of [4]), if $\mathcal{L}$ is a first-order language with fewer than κ symbols. With SMA_κ I could construct Henson's example, as well as many other interesting examples which had previously been produced under the assumption that the nonstandard model was *fully* saturated.

In an interesting and important pair of papers, Jin [9] and Jin and Shelah [10] have gone a long way toward showing that IP_κ is all the saturation one really needs for practical work. The first paper shows that SMA_κ is strictly stronger than IP_κ, and that IP_κ can be used for all the examples in [18], except possibly the Henson example. The second paper shows that IP_κ suffices for the Henson example as well; in addition, the authors find a nice reformulation of IP_κ which should make it easy to use on an everyday basis.

However, for the proof of Theorem 2 in the current paper I use SMA_κ instead of IP_κ. The relevant lemmas are these:

Lemma 1. *For arbitrarily large κ there are nonstandard models satisfying SMA_κ.*

Proof. See [18], §2. □

Lemma 2. *If $SMA_{\aleph_0}$ holds, and the (common) external cardinality of infinite internal sets is not cofinally measurable, then for every internal X and every internal $\mathcal{A} \subseteq {}^*\mathcal{P}(X)$ with $\kappa = \operatorname{card}(\mathcal{A})$ infinite, $\mathcal{A}$ is κ-terraced and $\operatorname{NMC}(\alpha, \kappa)$ for all α.*

Proof. This is Lemma 2.1 of [20]. □

Theorem 2 follows immediately from these two lemmas and Theorem 1.

The reason SMA_κ works in Lemma 2 is that it lets us stratify $\mathcal{A}$ in terms of the specializing chain: each $\mathcal{A}_i$ in the terracing is just the set of elements appearing in the ith model. IP_κ doesn't produce this sort of stratification. The question thus arises:

Question 1. *Suppose that $IP_{\aleph_0}$ holds, that X is internal and externally infinite with external cardinality κ. Is $\mathcal{P}(X)$ κ-terraced?*

An affirmative answer would mean that IP_κ is all that is necessary for the results here to hold; a negative answer would provide a reason for assuming SMA_κ in applied nonstandard analysis, and not just IP_κ.

4. Liftings and related corollaries

For this section (and the remainder of the paper) I will assume that the nonstandard model has the property that all Loeb spaces are ono.

Recall that a measurable function f from Ω to a topological space Y **has a lifting** F provided $F : \Omega \to {}^* Y$ is internal and $F(\omega) \approx f(\omega)$ μ_L-almost everywhere.

The seminal existence result for liftings is due to Robert Anderson ([3]; also [21], Theorem 2.1.4):

Lemma 3. *Every measurable function from a Loeb probability space to a second countable topological space has a lifting.*

(For $Y = \mathbb{R}$, this was shown rather earlier by Peter Loeb [13].)

Anderson's result is generalized in [17], where in addition an example is given of a Loeb-measurable function into a non-second-countable space which does *not* have a lifting. The question therefore arises:

Question 2. *For which topological spaces Y is it the case that every Loeb-measurable function into Y has a lifting?*

The argument used to prove the following standard result is well-known to anyone familiar with Ulam's original investigations into measurability of cardinals:

Lemma 4. *If $(X, \mathcal{B}, P)$ is ono, Y is metric, and $f : X \to Y$ is measurable, then there is a second-countable subset Y' of Y such that $P(f^{-1}(Y')) = 1$.*

Proof. The proof for the special case where $(X, \mathcal{B}, P)$ is Radon appears in ([5], Theorem 2B) and ([15], Theorem 4.1); for the more general case where $(X, \mathcal{B}, P)$ is compact, see ([14], Theorem 2); for the case where $(X, \mathcal{B}, P)$ is a Loeb space, see ([20], Corollary 2.3). These proofs are essentially the same, and all work for any ono space. □

Corollary 1. *Suppose Y is metric and $f : \Omega \to Y$ is Loeb measurable. Then f has a lifting.*

Proof. This is an immediate consequence of Lemma 4, Lemma 3, and the fact that $(\Omega, \mathcal{A}_L, \mu_L)$ is ono. □

For the next corollary, I assume the reader is familiar with the idea of the non-standard hull of a Banach space; see [8] for a reference.

Corollary 2. *Suppose $\hat{B}$ is the nonstandard hull of an internal Banach space B, and $f : \Omega \to \hat{B}$ is Loeb measurable. Then there is an internal $F : \Omega \to B$ such that $f(\omega) = F(\omega)/\mu(0)$ almost everywhere.*

Proof. This follows from Lemma 1 and ([17], Theorem 4.1; see also Example 4.4). □

Corollary 3. *Suppose Y is metric and $f : \Omega \to Y$ is Loeb measurable. Then the image measure $\mu_L \circ f^{-1}$ on Y is Radon.*

Proof. Let m be this image measure, and let F be a lifting of f. Pick an arbitrary Borel subset E of Y with $m(E) > 0$; it suffices to show that $m(E) = \sup\{m(K) \mid K \subseteq E, K \text{ compact }\}$. Fix $\epsilon > 0$, find an internal $A \subseteq f^{-1}(E)$ such that $f \approx F$ on A and $\mu_L(A) > m(E) - \epsilon$. Put $K = \mathrm{st}(F(A))$. It is easily verified that K is a compact subset of E and $m(K) \geq \mu_L(A) > m(E) - \epsilon$, as desired. □

5. Standard spaces with Loeb preimages

In this final section I discuss the extent to which the results above can be 'pushed down' to standard measure spaces.

Suppose that $(X, \mathcal{B}, P)$ is a standard probability space, that $(\Omega, \mathcal{A}_L, \mu_L)$ is as usual a Loeb space, and that there is a measurable, measure-preserving function $\phi : \Omega \to X$. In this case say that $(X, \mathcal{B}, P)$ has $(\Omega, \mathcal{A}_L, \mu_L)$ as a **preimage**, or less definitely that $(X, \mathcal{B}, P)$ **has a Loeb preimage**.

Probably the single most important question in the theory of Loeb spaces is:

Question 3. *Which (standard) probability spaces have Loeb preimages?*

The following is an easy exercise:

Lemma 5. *If $(X, \mathcal{B}, P)$ has an ono Loeb space $(\Omega, \mathcal{A}_L, \mu_L)$ as a preimage, then $(X, \mathcal{B}, P)$ is ono.*

If $(X, \mathcal{B}, P)$ is Lebesgue measure on $[0, 1]$ and $(\Omega, \mathcal{A}_L, \mu_L)$ is the Loeb measure generated from $^*(X, \mathcal{B}, P)$, then $(\Omega, \mathcal{A}_L, \mu_L)$ is the preimage of $(X, \mathcal{B}, P)$ via the standard part map; this was implicit in [13].

This result holds more generally for any Radon probability space, according to a result of Anderson [3].

Recently I showed that every compact measure space has a Loeb preimage. (The probability space $(X, \mathcal{B}, P)$ is **compact** provided P is inner–regular with respect to a compact family $\mathcal{K} \subseteq \mathcal{B}$.)

In all these cases the Loeb preimage can be constructed in a suitable nonstandard model, so can be assumed ono. Thus, Lemma 5 is stronger (strictly, by the result of Aldaz mentioned in the introduction) than the results of Fremlin, Prikry, et al.

I do not know whether other spaces of standard interest have ono Loeb space preimages.

There has recently been some effort to represent more general spaces by weakening the notion of 'Loeb space preimage' somewhat. For example, suppose that $(X, \mathcal{B}, P)$

and $(\Omega, \mathcal{A}_L, \mu_L)$ are as above, and $\phi : \Omega \to X$ satisfies
$$\overline{\mu_L}(\phi^{-1}(A)) = P(A) \text{ for every } A \in \mathcal{B}$$
where $\overline{\mu_L}$ is the outer measure induced on $\mathcal{P}(\Omega)$ by μ_L. We might in this case call $(\Omega, \mathcal{A}_L, \mu_L)$ a *semi-preimage* of $(X, \mathcal{B}, P)$ (or $(X, \mathcal{B}, P)$ a *semi-image* of $(\Omega, \mathcal{A}_L, \mu_L)$).

Recently Landers and Rogge [11] have proved that if $(X, \mathcal{B}, P)$ is regular and τ-smooth (the definition of which is not important here, except that it strictly generalizes 'Radon'), then it is a semi-image of a Loeb space (which we may take to be ono), where ϕ may be taken to be the standard part map. (Even more recently, Aldaz [2] has generalized this further, representing some nontopological spaces with an analogue of the inverse standard part map.)

At first glance, it would seem that this representation should suffice for a result like Lemma 5. After all, the inverse standard part map is still a measure-preserving function on $\mathcal{B}$ respecting arbitrary Boolean operations. However, the range of ϕ is no longer contained in $\mathcal{A}_L$, and in fact the 'push down' fails:

Lemma 6. *There is a τ-smooth regular probability measure which is not ono.*

Proof. Fremlin ([5], §3C) gives an example of a 'quasi-Radon' probability space $(X, \mathcal{B}, P)$ and a completely-measurable partition of X into nullsets. It suffices to note that quasi-Radon implies τ-smooth and regular, and that every partition is point-finite. □

6. Further comments

Section 1: The term 'ono' is not standard in the literature; it is a Hawaiian word which translates roughly to 'nice' or 'very good'. Fremlin only considers *disjoint* families $\mathcal{E}$ in his paper [5], and he calls probability spaces satisfying the corresponding property **c.m.p.p.**, for the *completely measurable partition property*. As far as I know, there is no other name in the literature for spaces satisfying the more general property, which is why I coined the term.

Section 4: Corollary 1 was already known to Anderson in the case where there are no measurable cardinals (or where any such are larger than the cardinality of Y); the reason is the standard result that in the absence of measurable cardinals all metric measure spaces are Radon. The proof of the latter result is the essential content of the proof of one case of Theorem 1.

If $(X, \mathcal{B}, P)$ is a topological probability space, then a function $f : X \to Y$ is **Lusin measurable** (or **almost continuous**) provided that whenever $0 < r < 1$ there is a $K \in \mathcal{B}$ such that $P(K) > r$ and such that the restriction of f to K is continuous. There is a very close connection between Lusin measurability and the existence of liftings; this connection is made precise in [17]. Fremlin's paper [5] raises the general question of which standard topological spaces Y have the property that if $(X, \mathcal{B}, P)$ is Radon and $f : X \to Y$ is measurable then f is automatically Lusin measurable. Call such spaces **Radon-Lusin targets**. Similarly, we might call a space Y a **Loeb-lifting target** if every Loeb measurable function into Y from an ono Loeb space has a lifting.

Lemma 7. *If Y is a Loeb-lifting target space then it is a Radon-Lusin target space.*

Proof. Suppose first that Y is a Loeb-lifting target space. Fix a Radon space $(X, \mathcal{B}, P)$ and a measurable function $f : X \to Y$. By Anderson's result, $(X, \mathcal{B}, P)$ has a Loeb preimage $(\Omega, \mathcal{A}_L, \mu_L)$ which we may take to be ono, where $\Omega =^* X$ and st is the measure-preserving map from Ω to X. The map $f \circ st$ is Loeb measurable, so has a lifting; it follows from Theorem 6.1 of [17] that f is Lusin measurable. □

Corollary 4. *If I is uncountable then $[0, 1]^I$ is not a Loeb-lifting target space.*

Proof. Fremlin ([5], Example 3F) shows that $[0, 1]^I$ is not a Radon-Lusin target. □

The obvious question, still open, is the converse to 7

Question 4. *Is every Radon-Lusin target space a Loeb-lifting target space?*

Section 5: The example of a noncompact Loeb measure given by Aldaz begins with the transfer of a $\{0, 1\}$-valued measure. The atomic nature of this space is essential to the construction. However, the Loeb spaces used most in practice are atomless. The following is therefore still open and interesting:

Question 5. *Are atomless Loeb spaces compact? In particular, is uniform Loeb measure on a hyperfinite set compact?*

References

[1] J. Aldaz, *On Compactness and Loeb measures*, Proc. Amer. Math. Soc., (to appear).
[2] J. Aldaz, *Representing abstract measures by Loeb measures: a generalization of the standard part map*, to appear.
[3] R. M. Anderson, *Star-finite representations of measure spaces*, Trans. Amer. Math. Soc. **271** (1982), pp. 667–687.
[4] C.C. Chang and H.J. Keisler, *Model Theory*, North Holland, Amsterdam, The Netherlands, 1973.
[5] D. H. Fremlin, *Measurable functions and almost continuous functions*, Manuscripta Math. **33** (1981), pp. 387–405.
[6] C.W. Henson, *The isomorphism property in nonstandard analysis and its use in the theory of Banach space*, The Journal of Symbolic Logic, **39** (1974), pp. 717—731.
[7] C.W. Henson, *Unbounded Loeb measures*, Proc. Amer. Math. Soc. **74** (1979), pp. 143—150.
[8] C. W. Henson, *Infinitesimals in functional analysis*, Nonstandard Analysis and its Applications (N. Cutland, ed.), Cambridge University Press, Cambridge, 1988, pp. 140–181.
[9] R. Jin, *The isomorphism property versus the special model axiom*, J. Symbolic Logic **57** (1992), pp. 975–987.
[10] R. Jin and S. Shelah, *The strength of the isomorphism property*, J. Symbolic Logic **59** (1994), pp. 292–301.
[11] D. Landers and L. Rogge, *Universal Loeb-measurability of sets and of the standard part map with applications*, Trans. Amer. Math. Soc. **304** (1987), pp. 229–243
[12] T. Lindstrom, *An invitation to nonstandard analysis*, Nonstandard Analysis and its Applications (N. Cutland, ed.), Cambridge University Press, Cambridge, 1988, pp. 1–105.
[13] P. A. Loeb, *Conversion from nonstandard to standard measure space and applications in probability theory*, Trans. Amer. Math. Soc. **211** (1975), pp. 113—122.

[14] G. Koumoullis and K. Prikry, *The Ramsey property and measurable selections*, J. Lond. Math. Soc. **28** (1983),pp. 203–210.

[15] J. Kupka and K. Prikry, *The measurability of uncountable unions*, Amer. Math. Monthly **91** (1984),pp. 85–97.

[16] D. A. Ross, *Measurable Transformations in Saturated Models of Analysis*, Ph.D. thesis, Univ. of Wisconsin–Madison, 1983.

[17] D. A. Ross, *Lifting theorems in nonstandard measure theory*, Proc. Amer. Math. Soc. **109** (1990),pp. 809–822.

[18] D. A. Ross, *The special model axiom in nonstandard analysis*, J. Symbolic Logic **55** (1990), pp. 1233-1242.

[19] D. A. Ross, *Compact measures have Loeb preimages*, Proc. Amer. Math. Soc. **115** (1992), pp. 365–370.

[20] D. A. Ross, *Unions of Loeb Nullsets*, Proc. Amer. Math. Soc., accepted pending revision.

[21] K. D. Stroyan and J. M. Bayod,*Foundations of Infinitesimal Stochastic Analysis*,North Holland/Elsevier Science Publishers, Amsterdam, The Netherlands, 1986.

Department of Mathematics
University of Hawaii
Honolulu, HI 96822
USA

E-mail ross@math.hawaii.edu

Siu-Ah Ng

Gradient lines and distributions of functionals in infinite dimensional Euclidean spaces

1. Introduction

Let us begin with the following elementary observation. Consider a differentiable function $f : \mathbb{R} \to \mathbb{R}$. Then the condition $f' \neq 0$ a.e. is both necessary and sufficient to insure that the measure on $\mathbb{R}$ induced by f is absolutely continuous with respect to the Lebesgue measure. Similar result holds also for distributions of functionals on finite dimensional Euclidean space $\mathbb{R}^n$.

Using methods from nonstandard analysis, we can define Euclidean spaces of the form $^*\mathbb{R}^N$, where $^*\mathbb{R}$ is the nonstandard version of the real line and where N is an infinite integer but finite in the sense of nonstandard analysis. In place of the Lebesgue measure, we consider the N-fold product of Gaussian measure of mean 0 and variance $\frac{1}{N}$. Then a functional $F : \, ^*\mathbb{R}^N \to \, ^*\mathbb{R}$ induces a nonstandard measure, the standard part of such measure can be extended to a unique σ-additive measure called the Loeb extension. We would like to study the distributions of such Loeb measures. To be more precise, we want to find sufficient conditions so that these measures are absolutely continuous with respect to the Lebesgue measure on the real line. To achieve this, we concentrate on functionals that are smooth in the nonstandard sense, then we consider the flow of gradient lines of these functionals. Each particular gradient line is a one-dimensional object, so we can apply the observation mentioned at the beginning. Then we piece local information together in a certain way, and draw conclusion about the global behaviour of the functional itself. The criteria involve natural notion of differentiation in infinite dimensional space. When applied to Wiener functionals, they will be associated to the derivation operator of the so-called Malliavin calculus.

This article is organized as follows. In Section 2 we briefly outline our framework, but the details of nonstandard analysis are referred to elsewhere. In Section 3, we show how to use classical differential geometry to define gradient lines. We show the connection of these gradient lines to the distribution of a functional. A theorem is proved for the criterion of absolute continuity with respect to Lebesgue measure. This criterion involves the Jacobian of the transformation along gradient lines. However, to make such result more usable, we need to formulate them in more naturally verifiable terms. This is done in Section 4, where we produce theorems using geometrical language. More variants of these results, including some for liftings, are proved in Section 5. In a future sequel to this paper, we will give concrete application in Wiener spaces and possibly in abstract Wiener spaces. But the main idea here is only to illustrate some geometric connection between objects in infinite dimensional spaces, and thus points to the possibility of developing a theory of infinite dimensional analysis based on differential geometry and nonstandard analysis.

We mention that our method here has some similarity with the fibering method discussed in [7]. Their method is based on the total integration formula and is quite different from the formula we use in Theorem 3.1. But if we can find a good way to assign weights to the gradient lines, then their method should yield interesting results. Yet another method can be found in [3], based on Federer's integration formula from geometric measure theory. (See [8].) It is not clear what are the connections between these approaches. More traditional approaches to the absolute continuity problem in the style of calculus of variation can be found in [13] and [12]. Another nonstandard approach can be found in [6].

2. Notation and preliminaries

We refer the reader to [AFHL], [9], [10] and [11] for the background of nonstandard analysis and its application in stochastic analysis.

Throughout this article, we fix $N \in {}^*\mathbb{N} \setminus \mathbb{N}$, write $\Delta t = N^{-1}$ and consider the hyperfinite dimensional Euclidean space ${}^*\mathbb{R}^N$ equipped with the product Gaussian measure γ, where $d\gamma(x) = (2\pi\Delta t)^{-\frac{N}{2}} \exp(-\frac{1}{2\Delta t}|x|^2)dx$, $x = (x_1, \cdots, x_N)$ and $|x| = (\sum_{0<i\leq N} x_i^2)^{\frac{1}{2}}$. We usually write $\Omega = {}^*\mathbb{R}^N$. We write $S^{N-1}(r)$ for the sphere centered at the origin with radius r, i.e. $\{x \in \Omega : |x| = r\}$.

When the hyperfinite time line $\{0, \Delta t, \cdots, 1\}$ is identified with $\{0, 1, \cdots, N\}$, the Loeb extension of γ represents the classical Wiener measure. (See [4].) If we consider the unit sphere $S^{N-1}(1) = \{x \in {}^*\mathbb{R}^N : |x| = 1\}$, with the uniform measure defined on it, we also obtain representation of the Wiener measure. (See [5].) Both models can be regarded as the *continuous analogue of Anderson's hyperfinite random walk model of Brownian motion, where x_i is intended to model the increment of a Brownian path at the time $i\Delta t$.

For a function $F : \Omega \to {}^*\mathbb{R}$, $F(\gamma)$ denotes the measure on ${}^*\mathbb{R}$ induced by F under γ. That is, $F(\gamma)(A) = \gamma(F^{-1}(A))$. Given two measure μ and ν, we say that μ is absolutely continuous with respect to ν, in symbols $\mu \ll \nu$, if whenever $\nu(A) = 0$, $\mu(A) = 0$. The Lebesgue measure is always written as Leb. For an internal measure μ, we use μ_L to denote the Loeb measure of μ. The following proposition and corollary are easy properties of absolute continuity. They show how the absolute continuity property can be transferred from the lifting.

Proposition 2.1. *Let $F : \Omega \to {}^*\mathbb{R}$ so that it lifts $f : \Omega \to \mathbb{R}$. Let ν be the measure on $\mathbb{R}$ induced by f. Then for each Lebesgue measurable $A \subset \mathbb{R}$ and standard $\epsilon > 0$, there is an internal $B \subset st^{-1}(A)$ such that B approximates $st^{-1}(A)$ within ϵ and $F^{-1}(B)$ approximates $f^{-1}(A)$ within ϵ. (Approximations with respect to the corresponding Loeb measures.)*

Proof. Let C be an internal subset of $f^{-1}(A)$ that approximates $f^{-1}(A)$ within ϵ. Let D be an internal subset of $st^{-1}(A)$ that approximates $st^{-1}(A)$ within ϵ. Then let $B = F(D) \cup D$ and the requirements are satisfied. □

Corollary 2.2. *In the above, if for all internal $A \subset \mathbb{R}$, *$Leb(A) \approx 0 \Rightarrow \gamma(F^{-1}(A)) \approx 0$, then $\nu \ll Leb$.* □

We say that F is spherical if whenever $x \neq 0$, $c \in$ *$\mathbb{R}$, $c \neq 0$, then $F(x) = F(cx)$. Quite often, we actually work on $\Omega \setminus \{0\}$, and when we say that F is smooth, we mean that F is smooth on $\Omega \setminus \{0\}$.

We say that a function F has Lipschitz constant c on a set B, if for all $y \in B$, $|F(x) - F(y)| \leq c|x - y|$.

Notions and elementary results on differential geometry are referred to [14].

The following easy proposition will be the basis of the absolute continuity results that we are going to prove. We will apply this proposition to each gradient line.

Proposition 2.3. *Let $f : \mathbb{R} \to \mathbb{R}$ be nondecreasing and continuously differentiable. Then for measurable $A \subset \mathbb{R}$,*

(1) $$\text{Leb}(A) \geq \int_{f^{-1}(A)} f'(u) du.$$

In particular, if $f' \neq 0$ a.e. then $f(\text{Leb}) \ll \text{Leb}$. □

3. Gradient lines and distributions

Let $F : \Omega \to$ *$\mathbb{R}$ be smooth and spherical. Then we define the flow of F as the gradient lines given by $\theta : \Omega \times$ *$\mathbb{R} \to \Omega$ as follows

(2) $$\begin{cases} \theta(x, 0) = x \\ d\theta(x, t)/dt = \nabla F(\theta(x, t)). \end{cases}$$

The existence of such gradient lines follows from the theory of ordinary differential equations. (See [14].) Note that by the independence of radius, for each x, $\nabla F(x) \cdot x = 0$, i.e. $\nabla F(x)$ is tangent to $S^{N-1}(|x|)$, so $\theta(x, t)$ always stays on $S^{N-1}(|x|)$. It should be also pointed out that together with the fact that $S^{N-1}(|x|)$ is compact, θ is parameterized by all $t \in$ *$\mathbb{R}$. (In general, it can be only parameterized by a subset of $\mathbb{R}$ depending on x.) The function θ is actually smooth in both variables.

We let M denote the Jacobian matrix

$$M(x, t) = [\frac{\partial \theta(x, t)_i}{\partial x_j}]_{1 \leq i, j \leq N}$$

and let J denote the Jacobian

$$J(x, t) = |\det M(x, t)|.$$

So $J(x, t)$ is also the density with respect to γ of the homeomorphism given by $\theta(\cdot, t)$. We remark that by F being spherical, it makes no difference whether the density is taken with respect to γ or the Lebesgue measure:

$$d\gamma(\theta(x, t)) = \frac{1}{(2\pi \Delta t)^{\frac{N}{2}}} \exp(-\frac{1}{2\Delta t}|\theta(x, t)|^2) d\theta(x, t)$$

$$= \frac{1}{(2\pi \Delta t)^{\frac{N}{2}}} \exp(-\frac{1}{2\Delta t}|x|^2) d\theta(x, t) = J(x, t) d\gamma(x).$$

So we have

(3) $$\frac{d\gamma(\theta(x,t))}{d\gamma(x)} = \frac{d\theta(x,t)}{dx} = J(x,t).$$

The following theorem gives a sufficient condition for the absolute continuity with respect to the Lebesgue measure of the distribution.

Theorem 3.1. *Let $F : \Omega \to {}^*\mathbb{R}$ be *smooth and spherical. Let J be given as above. Suppose $\int_{^*\mathbb{R}} J(x, s\Delta t) ds \not\approx 0$ a.s. and $0 \not\approx |\nabla F(\omega)|\sqrt{\Delta t} < \infty$ γ_L–a.s. Then $F(\gamma)_L \ll$ Leb.*

Proof. It suffices to show that for any *Lebesgue measurable $A \subset {}^*\mathbb{R}$, if $^*\text{Leb}(A) \approx 0$ then $\gamma(F^{-1}(A)) \approx 0$. (In fact we can even assume that A is *bounded and *closed, if one worries about the $^*L^1$ condition in the following calculations.)

We first fix $x \in \Omega$. Consider the function given by $(F \circ \theta)(x, \cdot) : {}^*\mathbb{R} \to {}^*\mathbb{R}$. By the property of gradient line, $(F \circ \theta)(x, \cdot)$ is nondecreasing and continuously differentiable. Moreover,

(4) $$\frac{d}{dt} F(\theta(x,t)) = \nabla F(\theta(x,t)) \circ \frac{d}{dt} \theta(x,t) = |\nabla F(\theta(x,t))|^2.$$

So by Proposition 2.3, we have for each *Lebesgue measurable $A \subset {}^*\mathbb{R}$,

$$^*\text{Leb}(A) \geq \int_{^*\mathbb{R}} 1_A(F(\theta(x,t)))|\nabla F(\theta(x,t))|^2 dt$$

for each fixed x. Now integrate over all $x \in \Omega$, we obtain:

$$\begin{aligned}
^*\text{Leb}(A) &\geq \int_\Omega \int_{^*\mathbb{R}} 1_A(F(\theta(x,t)))|\nabla F(\theta(x,t))|^2 dt \, d\gamma(x) \\
&= \int_\Omega \int_{^*\mathbb{R}} 1_A(F(x))|\nabla F(x)|^2 d\gamma(\theta(x,-t)) \, dt \\
&= \int_\Omega 1_A(F(x))|\nabla F(x)|^2 \int_{^*\mathbb{R}} J(x,t) dt \, d\gamma(x) \\
&= \int_{F^{-1}(A)} |\nabla F(x)|^2 \int_{^*\mathbb{R}} J(x,t) dt \, d\gamma(x) \\
&= \int_{F^{-1}(A)} |\nabla F(x)|^2 \Delta t \int_{^*\mathbb{R}} J(x, s\Delta t) ds \, d\gamma(x).
\end{aligned}$$

So it follows from the assumptions that if $^*\text{Leb}(A) \approx 0$ then $\int_{F^{-1}(A)} d\gamma(x) \approx 0$, i.e. $\gamma(F^{-1}(A)) \approx 0$, and the theorem is proved. $\square$

Now we give an intuitive argument that the condition $\int_\Omega J(x, s\Delta t) ds \not\approx 0$ a.s. holds under the following mild condition:

(5) a.a. x, if $t \approx 0$, then $|\nabla F(\theta(x,t))|\sqrt{\Delta t} \approx |\nabla F(x)|\sqrt{\Delta t}.$

Geometrically, a gradient line is perpendicular to level surfaces. Now consider an area element, say a cylinder with a base perpendicular to a gradient line and of height $\theta(x,\epsilon) - x$. Then if we transport this cylinder alone the gradient line by increasing the parameter by amount $t \approx 0$, then the new cylinder has height $\theta(x, \epsilon + t) - \theta(x, t)$, with little change to the base volume. So we have:

$$J(x,t) = \lim_{\epsilon \to 0} \frac{|\theta(x, \epsilon + t) - \theta(x, t)|}{|\theta(x, \epsilon) - x|} = \frac{|d\theta(x,t)/dt|}{|d\theta(x,0)/dt|}$$

$$= \frac{|\nabla F(\theta(x,t))|}{|\nabla F(x)|} \approx 1,$$

by using (5).

We hope to make the above argument rigorous in a sequel. A direct computation of the Jacobian $J(x,t)$ is difficult, and the geometrical criterion in the next section involves some ways to estimate this quantity.

4. A geometrical criterion for absolute continuity

Lemma 4.1. *Suppose that $I + A$ is an N-dimensional positive definite matrix. Then*

$$(\det(I + A))^{-\frac{1}{2}} = \int_{*\mathbb{R}^N} \exp(-\frac{1}{2\Delta t}(x, Ax)) d\gamma(x).$$

So if we write $A = [a_{ij}]_{1 \leq i,j \leq N}$, then

$$(\det(I + A))^{-\frac{1}{2}} = E_\Omega[\exp(-\frac{1}{2\Delta t} \sum_{i,j} a_{ij} x_i x_j)].$$

Proof. By [2] p.97, we have the following equation for any N-dimensional positive definite matrix M:

$$(\det(M))^{-\frac{1}{2}} = \pi^{-\frac{N}{2}} \int_\Omega \exp(-(x, Mx)) dx.$$

So we have the following

$$\begin{aligned}
(\det(I + A))^{-\frac{1}{2}} &= \pi^{-\frac{N}{2}} \int_\Omega \exp(-(x, Ax) - (x, x)) dx \\
&= \pi^{-\frac{N}{2}} \int_\Omega \exp(-(\frac{x}{\sqrt{2\Delta t}}, A \frac{x}{\sqrt{2\Delta t}})) \exp(-\frac{1}{2\Delta t}(x, x)) \frac{dx}{(2\Delta t)^{\frac{N}{2}}} \\
&= \int_\Omega \exp(-\frac{1}{2\Delta t}(x, Ax)) \exp(-\frac{1}{2\Delta t}|x|^2) \frac{dx}{(2\pi \Delta t)^{\frac{N}{2}}} \\
&= \int_\Omega \exp(-\frac{1}{2\Delta t}(x, Ax)) d\gamma(x).
\end{aligned}$$

□

We now follow the notations in Section 3. We think of the $\nabla^2 F(x)$ as the matrix $[\partial^2 F(x)/\partial x_i \partial x_j]_{1 \leq i,j \leq N}$.

Lemma 4.2.

(6) $$M(x,t) = {}^* \lim_{K \to \infty} \prod_{n < K} (I + \frac{t}{K} \nabla^2 F(\theta(x, \frac{n}{K}t))).$$

Proof. From equation (2), we have:

$$\theta(x,t) = x + \int_0^t \frac{d}{ds}\theta(x,s)ds = x + \int_0^t \nabla F(\theta(x,s))ds, \tag{7}$$

Where we write x, $\theta(x,\cdot)$ as column vectors. Then

$$\frac{\partial \theta(x,t)}{\partial x_j} = \frac{\partial x}{\partial x_j} + \int_0^t \nabla^2 F(\theta(x,s)) \frac{\partial \theta(x,s)}{\partial x_j} ds. \tag{8}$$

So we have

$$M(x,t) = I + \int_0^t \nabla^2 F(\theta(x,s)) M(x,s) ds. \tag{9}$$

Now discretize the interval $[0,t]$ into K points equally apart, we get the following approximation of (9):

$$\tilde{M}(x, \frac{n+1}{K}t) = I + \sum_{s=0,\frac{t}{K},\cdots,\frac{n}{K}t} \nabla^2 F(\theta(x,s)) \tilde{M}(x,s) \frac{t}{K}$$

$$= (I + \frac{t}{K}\nabla^2 F(\theta(x, \frac{n}{K}t)))\tilde{M}(x, \frac{n}{K}t),$$

and hence (6) follows from smoothness of M. $\square$

Corollary 4.3.

$$J(x,t) = {}^*\lim_{K\to\infty} \prod_{n<K} \det(I + \frac{t}{K}\nabla^2 F(\theta(x, \frac{n}{K}t))). \tag{10}$$

$\square$

We now briefly introduce some terminologies from differential geometry. (See [14] for more details.) Let p be a point on a smooth manifold, we use S_p to denote the tangent space at p. The Weingarten map $L_p : S_p \to S_p$ is given by $L_p(v) = -\nabla_v \mathbf{N} = -(\nabla_v N_1, \cdots, \nabla_v N_N)$, where $\mathbf{N}$ is the normal vector at p. The normal curvature of the surface at p in the tangent direction v is defined as $k(v) = L_p(v) \cdot v$. It is also called the second fundamental form of v at p. (Denoted by s_p.) We define the normal curvature of the surface at p as the maximum of such values. (The maximum exists, since we can identify s_p with a unit sphere.)

Lemma 4.4. *Let $p \in \Omega$ and $F : \Omega \to {}^*\mathbb{R}$ be a smooth function. Let κ_t be the normal curvature of the surface $F^{-1}(F(\theta(p,t)))$ at point $\theta(p,t)$. Assume that $\kappa_t/\sqrt{\Delta t} < \infty$ and $|\nabla F(\theta(p,t))|\sqrt{\Delta t} < \infty$ for $0 \le t \le \epsilon\Delta t$, where $\epsilon \approx 0$. Then ${}^\circ J(p, \epsilon\Delta t) \ge 1$.*

Proof. Let $q = \theta(p,t)$ for some $0 \le t \le \epsilon\Delta t$. First note that for any unit vector v tangent to q, the second fundamental form, the normal curvature, is

$$\nabla_v(|\nabla F(q)|^{-1}\nabla F(q)) \cdot v = (\nabla_v|\nabla F(q)|^{-1})\nabla F(q) \cdot v + |\nabla F(q)|^{-1}\nabla_v \nabla F(q) \cdot v$$
$$= |\nabla F(q)|^{-1}\nabla_v \nabla F(q) \cdot v,$$

by v being tangent. So

$$s_q(v) = -\frac{\nabla_v \nabla F(q) \cdot v}{|\nabla F(q)|} = -\frac{1}{|\nabla F(q)|} \sum_{1 \leq i,j \leq N} \frac{\partial^2 F(q)}{\partial x_i \partial x_j} v_i v_j, \tag{11}$$

For each $x \in \Omega$, we have the orthogonal decomposition $x = x_T \oplus x_N$, where x_N is parallel to $\nabla F(q)$.

Then $\nabla_x \nabla F(q) \cdot x = \nabla_x \nabla F(q) \cdot (x_T \oplus x_N) = \nabla_x \nabla F(q) \cdot x_T$, since $\nabla_x \nabla F(q) \perp x_N$. So $\nabla_x \nabla F(q) \cdot x = \nabla_{x_T} \nabla F(q) \cdot x_T + \nabla_{x_N} \nabla F(q) \cdot x_T$, but $\nabla_{x_N} \nabla F(q) \cdot x_T = \nabla_{x_T} \nabla F(q) \cdot x_N = 0$. (Since $\nabla_{x_T} \nabla F(q) \perp x_N$.) That is,

$$\nabla_x \nabla F(q) \cdot x = \nabla_{x_T} \nabla F(q) \cdot x_T. \tag{12}$$

Note also by (11)

$$\nabla_{x_T} \nabla F(q) \cdot x_T = \frac{1}{|\nabla F(q)|} \nabla_u \nabla F(q) \cdot u \, |x_T|^2 |\nabla F(q)|$$
$$\text{(where } u = x_T/|x_T|\text{)}$$
$$= -s_q(u)|\nabla F(q)||x_T|^2.$$

So by (12) and assumptions, we have for all $x \in \Omega$,

$$|\nabla_x \nabla F(q) \cdot x| \leq C|x_T|^2 \tag{13}$$

for some finite constant C, where

$$C = \frac{\kappa_t}{\sqrt{\Delta t}} |\nabla F(\theta(p,t))| \sqrt{\Delta t}. \tag{14}$$

Or we can write

$$|(x, \nabla^2 F(q) \cdot x)| \leq C|x_T|^2 \tag{15}$$

Now consider large enough $K \in {}^*\mathbb{N}$ such that $I + \frac{\epsilon}{NK} \nabla^2 F(\theta(p,t))$ is positive definite for each $0 \leq t \leq \epsilon z \Delta t$. Then apply Lemma 4.1, we have

$$\det(I + \frac{\epsilon}{NK} \nabla^2 F(q))^{-\frac{1}{2}} = \int_\Omega \exp(-\frac{\epsilon}{2K} \nabla_x \nabla F(q) \cdot x) d\gamma(x)$$
$$\leq \int_\Omega \exp(\frac{C\epsilon}{2K}|x_T|^2) d\gamma(x)$$
$$\leq \int_\Omega \exp(\frac{C\epsilon}{2K}|x|^2) d\gamma(x)$$
$$= \int_\Omega \exp(-\frac{1}{2\Delta t}(1 - \frac{C\epsilon}{K}\Delta t)|x|^2)) \frac{dx}{(2\pi \Delta t)^{\frac{N}{2}}}$$
$$= (1 - \frac{C\epsilon}{NK})^{-\frac{N}{2}}$$

i.e.

$$\det(I + \frac{\epsilon}{NK} \nabla^2 F(q)) \geq (1 - \frac{C\epsilon}{NK})^N. \tag{16}$$

The above finite C from (14) depends on t, for some $0 \leq t \leq \epsilon \Delta t$. By using (14), we choose a finite C that works for all t on the closed interval $0 \leq t \leq \epsilon \Delta t$. Using Corollary 4.3, we have

$$\begin{aligned} J(p, \epsilon \Delta t) &= {}^*\lim_{K \to \infty} \prod_{n<K} \det(I + \frac{\epsilon \Delta t}{K} \nabla^2 F(\theta(p, \frac{n}{K}t))) \\ &\geq (1 - \frac{C\epsilon}{NK})^{NK} \\ &\approx e^{C\epsilon} \\ &\approx 1 \end{aligned}$$

$\square$

Now we are ready to give a geometric characterization of the absolute continuity with respect to Lebesgue measure of the Loeb extension of the induced measure of a functional.

Theorem 4.5. *Let $F : \Omega \to {}^*\mathbb{R}$ be *smooth and spherical. For $x \in \Omega$, let κ_x be the curvature of the surface $F^{-1}(F(x))$. Suppose for all $x \in \Omega$, $\kappa_x/\sqrt{\Delta t} < \infty$, and $0 \not\approx |\nabla F(x)|\sqrt{\Delta t} < \infty$. Then $F(\gamma)_L \ll$ Leb.*

Proof. By Theorem 3.1, we only need to show that $\int_{{}^*\mathbb{R}} J(x, s\Delta t)ds \not\approx 0$ a.s. But it follows from Lemma 4.4 that ${}^\circ J(x, \epsilon \Delta t) \geq 1$, so for some noninfinitesimal interval I, $\int_I J(x, s\Delta t)ds \not\approx 0$, and the result follows. $\square$

5. More criteria and application to liftings

We now modify the criterion for the absolute continuity obtained from the last section. This modification should make the result more applicable. Normally, one does not always begin with a spherical function and one way to obtain it is to apply the function to the projection of each point on the unit sphere. We study the relation between them and give a result concerning liftings. We hope to provide more concrete applications in a sequel to this paper.

Lemma 5.1. *Assume that for almost all x, $|\nabla F(x)|\sqrt{\Delta t} < \infty$ and for almost all x, on the gradient line containing x, $|\nabla F(x)|$ has Lipschitz constant $C/\Delta t$ for some finite C. Then it holds for a.a.x that $\forall \epsilon \approx 0$,*
(i) $\theta(x, \epsilon \Delta t) \approx x$, (in fact $|\theta(x, \epsilon \Delta t) - x| < \delta \sqrt{\Delta t}$ for some $\delta \approx 0$) and
(ii) $|\nabla F(\theta(x, \epsilon \Delta t))|\sqrt{\Delta t} \approx |\nabla F(x)|\sqrt{\Delta t}$.

Proof. We first recall a simple form of the Gronwall's Lemma: given $y(0) = 0$ and $dy/dt \leq g(t)y(t) + a$, where a is a constant, then

(17) $$y(t) \leq a \int_0^t \exp(-\int_t^s g(u)du)ds.$$

Now from the gradient line equations (2), we have for each fixed x,

$$\frac{d}{dt}(\theta(x,t) - x) = \nabla F(\theta(x,t)).$$

We apply to a fixed x that satisfies the given Lipschitz condition, then for some finite constant C, we obtain

(18) $$||\nabla F(\theta(x,t))| - |\nabla F(x)|| \le \frac{C}{\Delta t}|\theta(x,t) - x|.$$

Notice that for $t \ge 0$, $|\theta(x,t) - x|$ is locally a monotone function. From (18), we have

(19) $$\frac{d}{dt}|\theta(x,t) - x| \le \frac{C}{\Delta t}|\theta(x,t) - x| + |\nabla F(x)|.$$

Now apply (17) with $y(t) = |\theta(x,t) - x|$, $g(t) = C/\Delta t$ and $a = |\nabla F(x)|$, we obtain

$$\begin{aligned}
|\theta(x,t) - x| &\le |\nabla F(x)|\int_0^t \exp(-\int_t^s \frac{C}{\Delta t}du)ds \\
&= |\nabla F(x)|\int_0^t \exp((t-s)\frac{C}{\Delta t})ds \\
&= |\nabla F(x)|\frac{\Delta t}{C}(\exp(\frac{tC}{\Delta t}) - 1).
\end{aligned}$$

Assume that $|\nabla F(x)|\sqrt{\Delta t} < \infty$, then for all $\epsilon \approx 0$, we have

$$\begin{aligned}
|\theta(x,\epsilon\Delta t) - x| &\le |\nabla F(x)|\sqrt{\Delta t}\frac{\sqrt{\Delta t}}{C}(e^{\epsilon C} - 1) \\
&= \delta\sqrt{\Delta t},
\end{aligned}$$

for some $\delta \approx 0$. That is (i) holds.

Now using (18), for the above fixed x, we have from this estimate that

$$||\nabla F(\theta(x,t))|\sqrt{\Delta t} - |\nabla F(x)|\sqrt{\Delta t}| \le \delta,$$

so (ii) holds. $\square$

Now we give a variant of Theorem 4.5.

Theorem 5.2. *Let $F : \Omega \to {}^*\mathbb{R}$ be *smooth and spherical. For $x \in \Omega$, let κ_x be the normal curvature of the surface $F^{-1}(F(x))$. Suppose for all $x \in \Omega$, $\kappa_x/\sqrt{\Delta t} < \infty$. Suppose also for a.a.x, $0 \napprox |\nabla F(x)|\sqrt{\Delta t} < \infty$ and $|\nabla F(x)|$ has Lipschitz constant $C/\Delta t$ on the gradient line containing x for some finite C. Then $F(\gamma)_L \ll$ Leb.*

Proof. We modify the argument in Theorem 4.5. The only difference is that now we use Lemma 5.1 and get for a.a.x, $\forall \epsilon \approx 0$, $|\nabla F(\theta(x,\epsilon\Delta t))|\sqrt{\Delta t} < \infty$. Hence ${}^\circ J(x,\epsilon\Delta t) \ge 1$ for a.a.x, and the result follows. $\square$

We are also interested in internal functions that are liftings of some standard functions on Ω. Since there is a measure-preserving map from this space to the standard Wiener space, so this also give results about Wiener functionals.

We first need some technical lemmas. For the following, we assume that $G : \Omega \to {}^*\mathbb{R}$ is *smooth and define $F(x) = G(x/|x|)$.

Lemma 5.3. *For any* $r \in {}^*\mathbb{R}$ *and* $i, j < N$, *we have*
$$\frac{\partial F(rx)}{\partial x_i} = \frac{1}{r}\frac{\partial F(x)}{\partial x_i} \quad \text{and} \quad \frac{\partial^2 F(rx)}{\partial x_i \partial x_j} = \frac{1}{r^2}\frac{\partial^2 F(x)}{\partial x_i \partial x_j}.$$

Proof. The results follow immediately from the following calculations:

(20)
$$\frac{\partial F(x)}{\partial x_i} = \frac{1}{|x|}\frac{\partial G}{\partial x_i}\left(\frac{x}{|x|}\right) - \frac{x_i}{|x|^3}\sum_k x_k \frac{\partial G}{\partial x_k}\left(\frac{x}{|x|}\right),$$

$$\begin{aligned}\frac{\partial^2 F(x)}{\partial x_i \partial x_j} &= -\frac{x_i}{|x|^3}\frac{\partial G}{\partial x_j}\left(\frac{x}{|x|}\right) - \frac{x_j}{|x|^3}\frac{\partial G}{\partial x_i}\left(\frac{x}{|x|}\right) \\ &\quad - \frac{x_i}{|x|^4}\sum_k x_k \frac{\partial^2 G}{\partial x_j \partial x_k}\left(\frac{x}{|x|}\right) - \frac{x_j}{|x|^4}\sum_k x_k \frac{\partial^2 G}{\partial x_i \partial x_k}\left(\frac{x}{|x|}\right) \\ &\quad + \frac{1}{|x|^2}\frac{\partial^2 G}{\partial x_i \partial x_j}\left(\frac{x}{|x|}\right) - \frac{1}{|x|^3}\frac{\partial x_i}{\partial x_j}\sum_k x_k \frac{\partial G}{\partial x_k}\left(\frac{x}{|x|}\right)\end{aligned}$$

(21)
$$+3\frac{x_i x_j}{|x|^5}\sum_k x_k \frac{\partial G}{\partial x_k}\left(\frac{x}{|x|}\right) + \frac{x_i x_j}{|x|^6}\sum_{k,r} x_k x_r \frac{\partial^2 G}{\partial x_k \partial x_r}\left(\frac{x}{|x|}\right).$$

$\square$

Lemma 5.4. *Let* $|x| = 1$. *If*
$$\sum_i \left(\partial G(x)/\partial x_i\right)^2 \Delta t < \infty \quad \text{and} \quad \sum_{i,j}\left(\partial^2 G(x)/\partial x_i \partial x_j\right)^2 \Delta t^2 < \infty$$

then
$$\sum \left(\partial^2 F(x)/\partial x_i \partial x_j\right)^2 \Delta t^2 < \infty.$$

Proof. From (21), we get
$$\begin{aligned}\frac{\partial^2 F(x)}{\partial x_i \partial x_j} &= -x_i \frac{\partial G}{\partial x_j}(x) - x_j \frac{\partial G}{\partial x_i}(x) - x_i \sum_k x_k \frac{\partial^2 G}{\partial x_j \partial x_k}(x) - x_j \sum_k x_k \frac{\partial^2 G}{\partial x_i \partial x_k}(x) \\ &\quad + \frac{\partial^2 G}{\partial x_i \partial x_j}(x) - \frac{\partial x_i}{\partial x_j}\sum_k x_k \frac{\partial G}{\partial x_k}(x) + 3x_i x_j \sum_k x_k \frac{\partial G}{\partial x_k}(x) \\ &\quad + x_i x_j \sum_{k,r} x_k x_r \frac{\partial^2 G}{\partial x_k \partial x_r}(x).\end{aligned}$$

Now it is straight-forward to check that for each term $g(x_i, x_j)$ in the above equation, it satisfies $\sum_{i,j} g^2(x_i, x_j)\Delta t^2 < \infty$. So the result follows. $\square$

Lemma 5.5. *Let* θ *be the flow of* F *as given in* (2). *Then for each* $x \in \Omega$ *and* $t \in {}^*\mathbb{R}$, *there is* $z \in S^{N-1}(|x|)$ *on the gradient line containing* x *such that*
$$\|\nabla F(\theta(x,t))\| - \|\nabla F(x)\| \leq \frac{1}{\Delta t}\sqrt{\sum_{i,j}\left(\frac{\partial^2 F}{\partial x_i \partial x_j}(z)\right)^2 \Delta t^2} \,|\theta(x,t) - x|.$$

195

Proof. By Mean Value Theorem and that F being spherical, there is $z \in S^{N-1}(|x|)$ on the gradient line containing x so that

$$|\|\nabla F(\theta(x,t))\| - |\nabla F(x)\|| = |\sum_i \Big(\frac{\partial}{\partial x_i}|\nabla F(z)|\Big)(\theta(x,t)_i - x_i)|$$

$$\leq \sqrt{\sum_i \Big(\frac{\partial}{\partial x_i}|\nabla F(z)|\Big)^2} |\theta(x,t) - x|.$$

So we only need to show that

(22) $$\sum_i \Big(\frac{\partial}{\partial x_i}|\nabla F(z)|\Big)^2 \leq \sum_{i,j} \Big(\frac{\partial^2 F}{\partial x_i \partial x_j}(z)\Big)^2.$$

But $\partial|\nabla F(z)|/\partial x_i = |\nabla F(z)|^{-1} \sum_j (\partial F(z)/\partial x_j)(\partial^2 F(z)/\partial x_i \partial x_j)$, so we have

$$\sum_i \Big(\frac{\partial}{\partial x_i}|\nabla F(z)|\Big)^2 \leq \sum_i \frac{1}{|\nabla F(z)|^2} \sum_j \Big(\frac{\partial F}{\partial x_j}(z)\Big)^2 \sum_j \Big(\frac{\partial^2 F}{\partial x_i \partial x_j}(z)\Big)^2$$

$$= \sum \frac{1}{|\nabla F(z)|^2} |\nabla F(z)|^2 \sum_j \Big(\frac{\partial^2 F}{\partial x_i \partial x_j}(z)\Big)^2.$$

So (22) holds. □

The following theorem uses a localization method.

Theorem 5.6. *Let $G : \Omega \to {}^*\mathbb{R}$ be *smooth. Let $F(x) = G(\frac{x}{|x|})$. Suppose that F is a lifting of a Loeb integrable function. Suppose*

$$\sum_i \big(\partial G(x)/\partial x_i\big)^2 \Delta t < \infty, \sum_{i,j} \big(\partial^2 G(x)/\partial x_i \partial x_j\big)^2 \Delta t^2 < \infty$$

and $\kappa_x/\sqrt{\Delta t} < \infty$ a.a.$x \in S^{N-1}(1)$ with respect to the uniform measure.

If $0 \neq |\nabla F(x)|\sqrt{\Delta t} < \infty$ a.a.x, then $F(\gamma)_L \ll$ Leb. So by Corollary 2.2, ${}^\circ F(\gamma_L) \ll$ Leb.

Proof. Let $A \subset {}^*\mathbb{R}$ be *Lebesgue measurable, such that *Leb$(A) \approx 0$. We want to show that $\gamma(F^{-1}(A)) \approx 0$. By countable additivity and F being a lifting, we can assume that A is bounded in a finite standard interval.

Let $\Omega_n = \{x \in \Omega : |\sum_i \big(\partial G(x)/\partial x_i\big)^2| < n, |\sum_{i,j}\big(\partial^2 G/\partial x_i \partial x_j\big)^2 \Delta t^2| < n$ and $\kappa_x < n\sqrt{\Delta t}\}$. Define *smooth functions G_n, so that each of them agrees with G on Ω_n and takes constant value c a.e. outside Ω_n, where c is some number outside A. Let $F_n(x) = G_n(\frac{x}{|x|})$.

By Lemma 5.4 and assumptions of the Theorem, $\sum_{i,j}\big(\partial^2 F_n/\partial x_i \partial x_j\big)^2 \Delta t^2 < \infty$ a.a.$x \in S^{N-1}(1)$ with respect to the uniform measure. Then by Lemma 5.3, the same holds a.a.x with respect to γ. Now Lemma 5.5 shows that, a.a.x, on the gradient line containing x, $|\nabla F(x)|$ has Lipschitz constant $C/\Delta t$ for some constant C. Therefore by Theorem 5.2, $\gamma(F_n^{-1}(A)) \approx 0$. Moreover, $\lim_{n\to\infty} {}^\circ\gamma(F_n^{-1}(A)) = {}^\circ\gamma(F(A))$, so the Theorem holds. □

References

[1] S. Albeverio, J.-E. Fenstad, R. Høegh-Krohn, and T. Lindstrøm, *Nonstandard Methods in Stochastic Analysis and Mathematical Physics*, Academic Press, New York 1986.

[2] R. Bellman, *Introduction to Matrix Analysis*, McGraw-Hill, 1970.

[3] N. Bouleau & F. Hirsch, *Dirichlet Forms and Analysis on Wiener Space*, de Gruyter Studies in Mathematics, No. 14, de Gruyter, 1991.

[4] N.J. Cutland, Infinitesimals in action, *J. Lond. Math. Soc.* **35**(1987), 202-216.

[5] N.J. Cutland & S.-A. Ng, The Wiener sphere and Wiener Measure, *The Annals of Probablity* **21**(1993) No. 1, 1-13.

[6] N.J. Cutland & S.-A. Ng, *Infinitesimal Methods for Malliavin calculus*, monograph in preparation.

[7] Yu.A. Davydov and M.A. Lifshits, *fibering Methods in some Probabilistic Problems*, Journal of Soviet Mathematics, Vol. 31 (1985), 2796-2858.

[8] H. Federer, *Geometric Measure Theory*, Springer-Verlag, 1969.

[9] A.E. Hurd and P.A. Loeb, *An Introduction to Nonstandard Real Analysis*, Academic Press, New York 1985.

[10] H.J. Keisler, An infinitesimal approach to stochastic analysis, *Mem. Amer. Math. Soc.* **297**(1984).

[11] T. Lindstrøm, An invitation to nonstandard analysis. In *Nonstandard Analysis and its Applications* (N.J. Cutland ed.) 1-105, Cambridge University Press, Cambridge, 1988.

[12] D. Ocone, Malliavin's calculus and stochastic integral representation of functionals of diffusion processes, *Stochastics* **12**(1984), 161-185.

[13] I. Shigekawa, Derivatives of Wiener functionals and absolute continuity of induced measures, *J. Mat. Kyoto Univ.*, **20**-2(1980), 263-289.

[14] J.A. Thorpe, *Elementary Topics in Differential Geometry*, Springer-Verlag, 1977.

Instytut Matematyki i Fizyki
WSRP
08–110 Siedlce Poland

E-mail ngsa@plearn.bitnet

Sergio Albeverio and Jiang-Lun Wu

Nonstandard flat integral representation of the free Euclidean field and a large deviation bound for the exponential interaction

1. Introduction

The aim of this paper is to present a new nonstandard description of Euclidean quantum field theory as well as some of its applications. For a general discussion of nonstandard approaches to physics, see e.g.[A] and [AFHL]. In order to understand, at least at some intuitive level, the meaning of the heuristic expressions defining Euclidean quantum field measures from the physical point of view, let us begin with the simple and familiar case of Wiener measure. Let $C_0[0,1] = \{x \in C[0,1] : x(0) = 0\}$. $C_0[0,1]$ endowed with the sup norm is a Banach space. Let $\mathcal{B}(C_0[0,1])$ be the Borel σ-algebra of $C_0[0,1]$, then Wiener measure W is defined starting from the formula

$$(1.1) \quad W(C_B) = \prod_{j=1}^{n} \left[2\pi(t_j - t_{j-1})\right]^{-\frac{1}{2}} \int_B \exp\left\{-\frac{1}{2}\sum_{j=1}^{n}\left[\frac{x(t_j) - x(t_{j-1})}{t_j - t_{j-1}}\right]^2 (t_j - t_{j-1})\right\} \prod_{j=1}^{n} dx(t_j)$$

for a cylinder set $C_B \equiv \{x \in C_0[0,1] : (x(t_1), \ldots, x(t_n)) \in B\} \in \mathcal{B}(C_0[0,1])$, where $\prod_{j=1}^{n} dx(t_j)$ is Lebesgue $\prod_{j=1}^{n} dx(t_j)$ is Lebesgue measure on $\mathbb{R}^n$ and B is a Borel set of $\mathbb{R}^n$. From this one arrives in several ways (the first one having been given already by N.Wiener in 1923) to an expression for $W(A)$, for any $A \in \mathcal{B}(C_0[0,1])$. In some heuristic sense(made rigorous by nonstandard analysis, see [C1]), this corresponds to considering a partition $0 = t_0 < t_1 < \ldots < t_n = 1$ of $[0,1]$, with mesh $\delta = \max\{|t_j - t_{j-1}| : 1 \leq j \leq n\}$ tending to 0 and with $n \to \infty$. If we take this heuristic limit separately for the normalization constant $\prod_{j=1}^{n}\left[2\pi(t_j - t_{j-1})\right]^{-\frac{1}{2}}$, the integrand and the n-dimensional Lebesgue measure in (1.1), we get the following heuristic formula

$$(1.2) \quad W(A) = \kappa \int_A \exp\left\{-\frac{1}{2}\int_0^1 [\dot{x}(t)]^2\, dt\right\} \prod_{t \in [0,1]} dx(t)$$

which is called Donsker's "flat integral", $\prod_{t \in [0,1]} dx(t)$ being a translation invariant (infinite-dimensional) "Lebesgue" measure on $C_0[0,1]$ (see e.g. [Do]). (1.2) gives only a heuristic formula for Wiener measure, since the integral of the kinetic energy $I(x) = \frac{1}{2}\int_0^1 [\dot{x}(t)]^2 dt = +\infty$ for all $x \in C_0[0,1]$ which are not absolutely continuous, κ is also positive infinite and the "flat measure" $\prod_{t \in [0,1]} dx(t)$ does not exist. However, the suggestive expression (1.2)(already used in early mathematical work by Cameron and Martin) is heuristically used in much of the physics literature, e.g. see [GJ] and references therein.

Let $d \in \mathbb{N}$ be fixed and Λ be a bounded region of $\mathbb{R}^d$ with a non empty interior. The generalization of (1.2) which corresponds to the (Nelson's)(quantized) free (Euclidean) scalar field of mass $m \geq 0 (m > 0$ for $d = 1, 2)$ in the region Λ of the d space-time $\mathbb{R}^d$ is given by the following heuristic expression

$$(1.3) \qquad d\mu(\phi) = \kappa \exp\left\{-\frac{1}{2} \int_\Lambda \left(|\nabla \phi(x)|^2 + m^2 [\phi(x)]^2\right) dx\right\} \prod_{x \in \Lambda} d\phi(x)$$

where $\prod_{x \in \Lambda} d\phi(x)$ is an infinite dimensional "Lebesgue" measure. There are essentially two standard approaches to give a mathematical meaning to (1.3): one is using a "lattice construction" of μ analogous to that in formula (1.1) to give the expression (1.3) a meaning as a probability measure on the Schwartz distribution space $\mathcal{D}'(\Lambda)$, see [AH1, AH2, AH3] and [GRS]; and the other is to interpret (1.3) with the help of Bochner-Minlos theorem (see e.g. [GJ], [N] and [S]; see also [Mo], [Pi], [Wo] for this measure in other contexts). To explain the use of Bochner-Minlos theorem in this context, let us observe that by a heuristic integration by parts in the exponent in (1.3) we obtain the following expression

$$(1.4) \qquad d\mu(\phi) = \kappa \exp\left\{-\frac{1}{2} \left\langle \phi, (-\Delta_\Lambda + m^2)\phi \right\rangle_{L^2(\Lambda)}\right\} \prod_{x \in \Lambda} d\phi(x),$$

where $-\Delta_\Lambda$ is a self-adjoint realization (see [RS, Section X.3]) of the Laplacian $-\Delta$ in $L^2(\Lambda)$ with some classical boundary (i.e., free, Dirichlet, Neumann or periodic) condition (see, e.g. [GJ]. Analogous to the finite dimensional case, we can then define μ as the unique probability measure on $\mathcal{D}'(\Lambda)$ whose Fourier transform is given by

$$(1.5) \qquad \int_{\mathcal{D}'(\Lambda)} e^{i \langle \phi, f \rangle} d\mu(\phi) = \exp\left\{-\frac{1}{2} \left\langle f, (-\Delta_\Lambda + m^2)^{-1} f \right\rangle_{L^2(\Lambda)}\right\}, \quad f \in \mathcal{D}(\Lambda).$$

The existence and uniqueness of this measure is precisely assured by Bochner-Minlos theorem (see e.g. [HKPS], [Yan] or [GV] for Bochner-Minlos theorem).

The free Euclidean field is of rather limited direct physical interest since it does not involve any particle interaction. However, it is the starting object for the construction of some non-Gaussian random fields which incorporate interaction potentials and are defined with the help of multiplicative functionals of the free field. Let us shortly recall how this is achieved. Let $U_u(\phi)$ be a "local additive functional" of the free field, i.e.,

$$U_u(\phi) = \int_\Lambda : u(\phi(x)) : dx,$$

with u a suitable continuous function on $\mathbb{R}$, where $: :$ denotes the Wick ordering with respect to the free field measure μ. Assume that U_u is μ-measurable, $e^{-U_u} \in L^1(\mu)$ and $E_\mu(e^{-U_u}) > 0$, with E_μ the mathematical expectation with respect to μ. For the existence of such functionals, see e.g. [AH2] and [S]. The interacting field measure in the region Λ is then defined by

$$(1.6) \qquad d\nu(\phi) = \frac{\exp\{-U_u(\phi)\} d\mu(\phi)}{E_\mu(\exp\{-U_u\})}.$$

The interacting scalar field in the region Λ is just ϕ considered as a random variable with respect to ν. For such interacting fields and their limits as $\Lambda \uparrow \mathbb{R}^d$, yielding interacting Euclidean (quantized) scalar fields (and via a suitable analytic continuation local relativistic quantum fields), we refer to [AH1, AH2, AH3], [AZ], [BSZ], [GJ], [S], [JoLM], [AFHL] (also for surveys and references).

A nonstandard (mathematical) formulation of the free Euclidean field measure μ and the above interacting fields ν has been provided in [AFHL] by replacing the underlying continuum Λ by a hyperfinite lattice $\Lambda \cap \delta^* \mathbb{Z}^d$ (where δ is a positive infinitesimal), obtaining the associated "lattice fields", and then by "taking the ultraviolet limit", realized mathematically by considering the associated Loeb measures. On the other hand, Cutland [C1] expressed, nonstandardly, Donsker's "flat integral" (1.2) as the Loeb measure of Anderson's infinitesimal random walk (which is a nonstandard realization of Brownian motion, see Anderson [An]). This makes rigorous mathematical sense of Donsker's "flat integral" for Wiener measure. Going a step further, Cutland [C2, C3, C4, C5] obtained nonstandard flat integral representations of certain other Gaussian measures such as the measures associated with the Brownian bridge, the fractional Brownian motion (which includes Lévy Brownian motion), and Gaussian measures on separable Hilbert spaces.

We remark that Donsker's "flat integral" is heuristically the free scalar field measure (Wiener measure) in one space-time dimension with mass parameter $m = 0$ and the free field of mass $m > 0$ in one space-time dimension corresponds to the one-dimensional Ornstein-Uhlenbeck process. Hence, it is interesting to extend such a type of nonstandard flat integral to the general free Euclidean field as well as to some interacting fields as the ones described above.

In this paper, we give an alternative nonstandard representation of the free lattice field, which can be used to obtain a flat integral realization of the free Euclidean field ϕ. This is a more concrete realization than the one for Gaussian measures in [C5]. From our result we also derive a flat integral representation of the above mentioned interacting fields. As an application, we give an upper large deviation bound for the measure corresponding to quantum fields with exponential interaction, i.e., the measure for Høegh-Krohn's model (cfr. [AH1, AH2] and [S]), in two dimensional space-time.

2. Preliminaries

Let us first outline the standard probabilistic construction of the free Euclidean field in a bounded region Λ of $\mathbb{R}^d$. Assuming $m > 0$ if $d = 1, 2$ and $m \geq 0$ if $d \geq 3$, let $\mathcal{D}_p(\Lambda), p \in \mathbb{R}$, be the real Sobolev-Hilbert space defined as the completion of $\mathcal{D}(\Lambda)$ with respect to the inner product

$$\langle f_1, f_2 \rangle_p := \langle f_1, (-\Delta_\Lambda + m^2)^p f_2 \rangle_{L^2(\Lambda)}, \quad f_1, f_2 \in \mathcal{D}(\Lambda).$$

Then the free Euclidean field of mass m in Λ (see [N] ,[AFHL], [GJ] [S] and also [Mo], [Pi] , [Wo]) can be defined as follows

Definition 2.1. Let $(\Omega, \mathcal{F}, P)$ be a probability space. The free Euclidean field of mass m (on $(\Omega, \mathcal{F}, P)$ in Λ) is a generalized Gaussian process $\phi : \mathcal{D}_{-1}(\Lambda) \times (\Omega, \mathcal{F}, P) \to \mathbb{R}$ satisfying the following conditions: (i) ϕ is linear on $\mathcal{D}_{-1}(\Lambda)$; (ii) $\phi(f, \cdot)$ is a Gaussian random variable with mean zero for each $f \in \mathcal{D}_{-1}(\Lambda)$; (iii) $\int_\Omega \phi(f_1, \omega)\, \phi(f_2, \omega)\, dP(\omega) = \langle f_1, f_2 \rangle_{-1}$, for all $f_1, f_2 \in \mathcal{D}_{-1}(\Lambda)$.

By Minlos theorem we can take Ω to be the Schwartz space $\mathcal{D}'(\Lambda)$, and we can also define the free field by the following

Definition 2.1.1. Let $\mathcal{B}$ be the σ-algebra generated by the weak topology (or the "cylinder sets") of Schwartz space $\mathcal{D}'(\Lambda)$. The free field measure is the probability measure μ on $(\mathcal{D}'(\Lambda), \mathcal{B})$ such that

$$(2.1) \qquad \int_{\mathcal{D}'(\Lambda)} e^{i <f,\omega>} d\mu(\omega) = \exp\left\{-\frac{1}{2}\langle f, f \rangle_{-1}\right\}, \quad f \in \mathcal{D}(\Lambda).$$

The generalized random process $\phi : \mathcal{D}(\Lambda) \times (\mathcal{D}'(\Lambda), \mathcal{B}, \mu) \to \mathbb{R}$ defined by

$$\phi(f, \omega) = <f, \omega>, \qquad f \in \mathcal{D}(\Lambda), \omega \in \mathcal{D}'(\Lambda)$$

is called the free (Euclidean) field (of mass m).

It was proved in [AW] that for $p > \frac{d}{4}$, $\mu(\mathcal{D}_{-2p+1}(\Lambda)) = 1$, i.e., $\mathcal{D}_{-2p+1}(\Lambda)$ is a support of μ. For various studies of support properties of the free field measure μ, see e.g. [Ca], [CoLa] and [RR].

The lattice formulation of the free Euclidean field is as follows. Let δ be a fixed positive real number and define the lattice Λ_δ with spacing δ by $\Lambda_\delta = \Lambda \cap \{z\delta : z \in \mathbb{Z}^d\}$. The free Euclidean field of mass m has the covariance operator $(-\Delta_\Lambda + m^2)^{-1}$, and the kernel of $(-\Delta_\Lambda + m^2)^{-1}$ is given by the Green function (fundamental solution) G of the operator $\Delta_\Lambda - m^2$, i.e.,

$$((-\Delta_\Lambda + m^2)^{-1} f)(x) = \int_\Lambda G(x - y)\, f(y)\, dy, \quad f \in L^2(\Lambda)$$

The discretization of the operator $(-\Delta_\Lambda + m^2)^{-1}$ on ℓ^2-space $\ell^2(\Lambda_\delta)$ is given by an $|\Lambda_\delta| \times |\Lambda_\delta|$-matrix, where $|\Lambda_\delta|$ denotes the cardinal number of Λ_δ, as follows

$$(2.2) \qquad ((-\Delta_\delta + m^2)^{-1} f)(z\delta) = \sum_{z'\delta \in \Lambda_\delta} \delta^d G_\delta(z\delta - z'\delta)\, f(z'\delta), \quad f \in \ell^2(\Lambda_\delta),$$

where Δ_δ is the discrete Laplace operator of Δ_Λ and G_δ is the discretization of G. Now by introducing the matrix C by the definition

$$C_{z\delta,\, z'\delta} = \delta^{-d} G_\delta(z\delta - z'\delta), \qquad z\delta,\, z'\delta \in \Lambda_\delta,$$

we define the free lattice field measure by

$$(2.3) \qquad d\mu^\delta(q) = (2\pi)^{-\frac{|\Lambda_\delta|}{2}} (\det C)^{-\frac{1}{2}} \exp\left\{-\frac{1}{2} \sum_{z\delta,\, z'\delta \in \Lambda_\delta} C^{-1}_{z\delta,\, z'\delta} q(z\delta)\, q(z'\delta)\right\} \prod_{z\delta \in \Lambda_\delta} dq(z\delta),$$

where $q \in \mathbb{R}^{\Lambda_\delta}$. The free field of mass m in Λ_δ is the map $\psi : \Lambda_\delta \times \mathbb{R}^{\Lambda_\delta} \to \mathbb{R}$ defined by $\psi(z\delta, q) = q(z\delta)$, $z\delta \in \Lambda_\delta$, $q \in \mathbb{R}^{\Lambda_\delta}$ (cfr. [GRS] , [S]).

To see that the free lattice field μ^δ approximates the free Euclidean field μ, we now pass to the nonstandard framework. We work with polysaturated models of nonstandard analysis(see [AFHL] and [L]) and make the notational convention that to each standard object S, we denote by $*S$ its nonstandard extension. Let δ be a positive infinitesimal and set $\Lambda_\delta = \Lambda \cap \{z\delta : z \in {}^*\mathbb{Z}^d \cap [-\delta^{-2}, \delta^{-2}]^d\}$. We denote by ${}^*\mathbb{R}^{\Lambda_\delta}$ the collection of all internal mappings $q : \Lambda_\delta \to {}^*\mathbb{R}$. By the transfer principle, we get the internal measure μ^δ from (2.3) on the internal algebra $\mathcal{A}({}^*\mathbb{R}^{\Lambda_\delta})$ of ${}^*\mathbb{R}^{\Lambda_\delta}$. The nonstandard lattice field is then the random field $\Phi : \Lambda_\delta \times ({}^*\mathbb{R}^{\Lambda_\delta}, \mathcal{A}({}^*\mathbb{R}^{\Lambda_\delta}), \mu^\delta) \to {}^*\mathbb{R}$ defined by

(2.4) $$\Phi(z\delta, q) = q(z\delta), \quad z\delta \in \Lambda_\delta, \ q \in {}^*\mathbb{R}^{\Lambda_\delta}.$$

Furthermore, we define for $f \in \mathcal{D}(\Lambda)$

(2.5) $$\psi(*f, q) = \sum_{z\delta \in \Lambda_\delta} \delta^d \, *f(z\delta) \, \Phi(z\delta, q), \quad q \in {}^*\mathbb{R}^{\Lambda_\delta}.$$

Let μ_L be the Loeb measure uniquely associated with the internal measure μ^δ defined by (2.3), then we have that $\psi(*f, q)$ is nearstandard on an internal subset of ${}^*\mathbb{R}^{\Lambda_\delta}$ with Loeb measure μ_L one, and the standard part st$[\psi(*f, q)]$, for $f \in \mathcal{D}(\Lambda)$ and $q \in {}^*\mathbb{R}^{\Lambda_\delta}$, is the free Euclidean field $\phi_{(L)}$ on the Loeb space $({}^*\mathbb{R}^{\Lambda_\delta}, \mathcal{A}_L({}^*\mathbb{R}^{\Lambda_\delta}), \mu_L)$, see Theorem 7.4.9 in [AFHL].

Finally in this section, let us define an S-white noise on the internal measure space $(\Lambda_\delta, \mathcal{A}(\Lambda_\delta), \delta^d)$. Let $(\Omega, \mathcal{A}(\Omega), P)$ be a given internal probability space. Let $\{\eta_{z\delta}(\omega) : z\delta \in \Lambda_\delta, \omega \in \Omega\}$ be an internal family of i.i.d. $N(0, \delta^d)$-random variables $\eta_{z\delta} : \Omega \to {}^*\mathbb{R}$ on $(\Omega, \mathcal{A}(\Omega), P)$.

Definition 2.2.

(i): For $A \in \mathcal{A}(\Lambda_\delta)$, define

(2.6) $$\eta_A(\omega) = \sum_{z\delta \in A} \eta_{z\delta}(\omega), \quad \omega \in \Omega.$$

The mapping $\eta : \mathcal{A}(\Lambda_\delta) \to {}^*\mathbb{R}$ is called S-white noise on $(\Lambda_\delta, \mathcal{A}(\Lambda_\delta), \delta^d)$ with respect to $(\Omega, \mathcal{A}(\Omega), P)$;

(ii): For $D \in \mathcal{A}_L(\Lambda_\delta)$, we define the standard part

$$\xi_D(\omega) = {}^\circ(\eta_A(\omega)), \quad A \in \{B \in \mathcal{A}(\Lambda_\delta) : |(D - B) \cap (B - D)| \delta^d \simeq 0\}.$$

We denote by $d\xi_x(\omega)$ the random measure such that

$$\int_D d\xi_x(\omega) = \xi_D(\omega),$$

for any $D \in \mathcal{A}_L(\Lambda_\delta)$.

Let $S\ell^2(\Lambda_\delta)$ denote the collection of all square S-integrable internal functions on Λ_δ. For $F \in S\ell^2(\Lambda_\delta)$, the stochastic integral of F with respect to η is defined as follows

(2.7) $$\left(\int F\, d\eta\right)(\omega) = \sum_{z\delta \in \Lambda_\delta} F(z\delta)\, \eta_{z\delta}(\omega).$$

Let us denote by $°$ the standard part map. We have the following results, the proof of which are in Propositions 2.1 and 2.2 of [AW] :

Proposition 2.1. *Let $(\Lambda_\delta, \mathcal{A}_L(\Lambda_\delta), (\delta^d)_L)$ and $(\Omega, \mathcal{A}_L(\Omega), P_L)$ be the Loeb spaces associated to the internal measure spaces $(\Lambda_\delta, \mathcal{A}(\Lambda_\delta), \delta^d)$ and $(\Omega, \mathcal{A}(\Omega), P)$, respectively. Then the following holds*

  (i): *the standard family $\{\xi_D(\omega) : D \in \mathcal{A}_L(\Lambda_\delta)$ and $°(|D|\delta^d) < \infty\}$ is a one-dimensional (standard) white noise on $(\Lambda_\delta, \mathcal{A}_L(\Lambda_\delta), (\delta^d)_L)$ with respect to $(\Omega, \mathcal{A}_L(\Omega), P_L)$.*
  (ii): *For every $f \in L^2(\Lambda)$, if $F \in S\ell^2(\Lambda_\delta)$ is a lifting of f, then*

(2.8) $$°\left(\int F\, d\eta\right)(\omega) = \int_\Lambda f(x)\, d\xi_x(\omega), \qquad P_L - a.s.$$

Moreover, the random variable $°\left(\int F\, d\eta\right)(\omega)$ is a Gaussian random variable with mean zero and covariance $\int_\Lambda [f(x)]^2\, dx$.

3. Nonstandard flat integral representation

In this section, we present a nonstandard formulation of the free lattice field in any dimension in terms of S-white noise. This provides a flat integral presentation of the free Euclidean field. The following result corresponds to the "square root lemma" in [RS, Theorem VI.9]:

Lemma 3.1.
  (i): *There exists a unique hyperfinite, positive, symmetric matrix $B = (B_{z\delta, z'\delta})_{z\delta, z'\delta \in \Lambda_\delta}$ such that*

(3.1) $$\left(G_\delta(z\delta - z'\delta)\right)_{z\delta, z'\delta \in \Lambda_\delta} = \delta^d \left(B_{z\delta, z'\delta}\right)^2_{z\delta, z'\delta \in \Lambda_\delta},$$

  i.e.,

(3.1)' $$G_\delta(z\delta - z'\delta) = \sum_{y\delta \in \Lambda_\delta} \delta^d\, B_{z\delta, y\delta}\, B_{y\delta, z'\delta}, \qquad z\delta, z'\delta \in \Lambda_\delta.$$

  (ii): *For $F \in S\ell^2(\Lambda_\delta)$,*

(3.2) $$\left((-\Delta_\delta + m^2)^{-\frac{1}{2}} F\right)(z\delta) = \sum_{z'\delta \in \Lambda_\delta} \delta^d\, B_{z\delta, z'\delta}\, F(z'\delta), \qquad z\delta \in \Lambda_\delta.$$

Let us now define $\Phi : \Lambda_\delta \times (\Omega, \mathcal{A}(\Omega), P) \to \mathbb{R}$ by

(3.3) $$\Phi(z\delta, \omega) = \sum_{z'\delta \in \Lambda_\delta} B_{z\delta, z'\delta}\, \eta_{z'\delta}(\omega), \qquad z\delta \in \Lambda_\delta,\ \omega \in \Omega,$$

where η is the S-white noise on $(\Lambda_\delta, \mathcal{A}(\Lambda_\delta), \delta^d)$ with respect to the given internal probability space $(\Omega, \mathcal{A}(\Omega), P)$. Furthermore, for $F \in {}^*\mathbb{R}^{\Lambda_\delta}$, we define

$$\Phi(F, w) = \sum_{z\delta \in \Lambda_\delta} \delta^d F(z\delta) \Phi(z\delta, w). \tag{3.4}$$

We have the following regularity result, which is proved by Lemmas 3.3–3.5 in [AW]:

Lemma 3.2.

(i): $\Phi(F, w)$ is S-continuous in $F \in {}^*\mathcal{D}_{2p-1}(\Lambda)$ for P_L - almost all $w \in \Omega$.

(ii): Let $Ns({}^*\mathcal{D}_{2p-1}(\Lambda))$ be the collection of all ${}^*\|\cdot\|_{2p-1}$-nearstandard elements of ${}^*\mathcal{D}_{2p-1}(\Lambda)$. If $F \in Ns({}^*\mathcal{D}_{2p-1}(\Lambda))$, then $\Phi(F, w)$ is nearstandard for P_L - almost all $w \in \Omega$ and

$$\mathrm{st}\,[\Phi(F,w)] = \int_{\Lambda_\delta} \mathrm{st}\left[\left((-\Delta_\delta + m^2)^{-\frac{1}{2}} F\right)(z\delta)\right] d\xi_{z\delta}(w), \quad P_L\text{-a.s.} \tag{3.5}$$

Let Γ be the internal measure on ${}^*\mathbb{R}^{\Lambda_\delta}$ induced by (3.3), namely, for $A \in \mathcal{A}({}^*\mathbb{R}^{\Lambda_\delta})$,

$$\begin{aligned}
\Gamma(A) &= P\{w : \Phi(\cdot, w) \in A\} \\
&= P\{w = (\eta_{z\delta}(w) : z\delta \in \Lambda_\delta) \in B^{-1}A\} \\
&= \prod_{z\delta \in \Lambda_\delta} P\{w : \eta_{z\delta}(w) \in \mathrm{Proj}_{z\delta}(B^{-1}A)\} \\
&= \prod_{z\delta \in \Lambda_\delta} (2\pi\delta^d)^{-\frac{1}{2}} \int_{\{q \in {}^*\mathbb{R}^{\Lambda_\delta} : q_{z\delta} \in \mathrm{Proj}_{z\delta}(B^{-1}A)\}} \exp\{-\frac{1}{2}\delta^{-d}q_{z\delta}^2\} dq_{z\delta} \\
&= (2\pi\delta^d)^{-\frac{|\Lambda_\delta|}{2}} \int_{B^{-1}A} \exp\{-\frac{1}{2}\sum_{z\delta \in \Lambda_\delta}\delta^{-d}q_{z\delta}^2\} \prod_{z\delta \in \Lambda_\delta} dq_{z\delta} \\
&= \kappa \int_A \exp\left\{-\frac{1}{2}\sum_{z\delta, z'\delta \in \Lambda_\delta} G_\delta^{-1}(z\delta - z'\delta)q_{z\delta}q_{z'\delta}\right\} \prod_{z\delta \in \Lambda_\delta} dq_{z\delta},
\end{aligned} \tag{3.6}$$

where $\kappa = (2\pi)^{-\frac{|\Lambda_\delta|}{2}}[\det(G_\delta^{-1})]^{\frac{1}{2}}$ is the normalization constant. We remark that the path space of (3.4) is the collection of all internal S-continuous linear mappings from ${}^*\mathcal{D}_{2p-1}(\Lambda)$ to ${}^*\mathbb{R}$, which is ${}^*\mathcal{D}_{-2p+1}(\Lambda)$, since ${}^*\mathcal{D}_{-2p+1}(\Lambda)$ is the internal dual Hilbert space of ${}^*\mathcal{D}_{2p-1}(\Lambda)$ with respect to the dualization given by the internal scalar product in $Sl^2(\Lambda_\delta)$. For $A \in \mathcal{A}({}^*\mathcal{D}_{-2p+1}(\Lambda))$, we define

$$\hat{A} = \left\{ q \in {}^*\mathbb{R}^{\Lambda_\delta} : (F \in {}^*\mathcal{D}_{2p-1}(\Lambda) \longrightarrow \sum_{z\delta \in \Lambda_\delta} \delta^d F(z\delta) q_{z\delta} \in {}^*\mathbb{R}) \in A \right\}. \tag{3.7}$$

We have the following main result:

Theorem 3.1. *The free field measure is realized by*

$$\mu(D) = \Gamma_L(\widehat{\mathrm{st}^{-1}(D)}) \tag{3.8}$$

where $D \in \mathcal{B}(\mathcal{D}_{-2p+1}(\Lambda))$ and $\mathrm{st}^{-1}(D) = \{F \in {}^*\mathcal{D}_{-2p+1}(\Lambda) : \mathrm{st}(F) \in D\}$.

Proof By Theorem 3.1 in [AW], $\{\phi(st(F),\omega) : F \in Ns(^*\mathcal{D}_{2p-1}(\Lambda)), \omega \in \Omega\}$ is the free Euclidean field of mass m (in Λ) on the Loeb space $(^*\mathbb{R}^{\Lambda_\delta}, \mathcal{A}(^*\mathbb{R}^{\Lambda_\delta}), \mu)$, according to the Definition 2.1 by taking $(\Omega, \mathcal{F}, P)$ to be $(^*\mathbb{R}^{\Lambda_\delta}, \mathcal{A}(^*\mathbb{R}^{\Lambda_\delta}), \mu)$. Now (3.8) is derived by the above (3.6), Lemma 3.6 and the proof of Theorem 3.2 in [AW]. □

Remark 3.1 Notice that all Schwartz distributions can be pointwise defined in the nonstandard framework, viewing $^*\mathbb{R}^{\Lambda_\delta}$ as a nonstandard enlargement of $\mathcal{D}_{-2p+1}(\Lambda)$ and of $\mathcal{D}'(\Lambda)$. Thus (3.6) and (3.8) may be viewed as a way to make precise sense of the quantized free scalar field ϕ, determined by (1.3). This is because for $A \in \mathcal{B}(^*\mathcal{D}'(\Lambda))$, we have the following rigorous formula from (3.6)

$$\Gamma_L(A) = \kappa \int_A \exp\left\{-\frac{1}{2} \sum_{z\delta \in \Lambda_\delta} \delta^d(|\nabla_\delta q_{z\delta}|^2 + m^2 q_{z\delta}^2)\right\} \prod_{z\delta \in \Lambda_\delta} dq_{z\delta}.$$

where ∇_δ is the discretization of ∇ such that $<\nabla_\delta f, \nabla_\delta g>_{l^2(\Lambda)} = <-\Delta_\delta f, g>_{l^2(\Lambda)}$ for $f, g \in l^2(\Lambda)$

Finally in this section, let us turn to derive the nonstandard flat integral representation of interacting fields. An internal function $V : {}^*\mathbb{R} \to {}^*\mathbb{R}$ defines an internal interaction U by

(3.9) $$U(\Phi) = \lambda \sum_{z\delta \in \Lambda_\delta} \delta^d \; ^*g(z\delta) V(\Phi(z\delta)),$$

where $g \in \mathcal{D}(\Lambda)$ and $\lambda \in {}^*\mathbb{R}_+$ is a coupling constant. We assume for simplicity $V \geq 0$. In accordance with (1.6), we introduce the following internal probability measure on $^*\mathbb{R}^{\Lambda_\delta}$

(3.10) $$d\nu(\Phi) = \frac{\exp\{-U(\Phi)\}d\Gamma}{E_\Gamma(\exp\{-U\})},$$

where Γ is defined by (3.6) and $E_\mu(\cdot)$ is the expectation with respect to the measure Γ. We have the following result:

Theorem 3.2. *If the $L^2(^*\mathbb{R}^{\Lambda_\delta}, d\Gamma)$–norm for $U(\Phi)$ is finite and $st(E_\Gamma \exp\{-U\}) > 0$, then the Loeb measure ν_L associated with the internal measure ν defined by (3.10) is absolutely continuous with respect to the free field measure $\mu = \Gamma_L$ and the Radon-Nikodym derivative is given by*

(3.11) $$\frac{d\nu_L}{d\mu} = st\left[\frac{\exp\{-U(\Phi)\}}{E_\Gamma(\exp\{-U\})}\right].$$

ν_L *is a probability measure on* $^*\mathcal{D}_{-2p+1}$. *In other words,* $\{st(\Phi(F,\omega)): F \in Ns(^*\mathcal{D}_{-2p+1}), \omega \in {}^*\mathbb{R}^{\Lambda_\delta}\}$ *is a random field on the Loeb space* $(^*\mathbb{R}^{\Lambda_\delta}, \mathcal{A}_L(^*\mathbb{R}^{\Lambda_\delta}), \nu_L)$.

Proof By the assumption that the $L^2(^*\mathbb{R}^{\Lambda_\delta}, d\Gamma)$– norm of the interaction $U(\Phi)$ is finite, we conclude that $U(\Phi)$ is nearstandard and so is $\exp\{-U(\Phi)\}$. Since by assumption $\frac{st(E_\Gamma(\exp\{-U\})) > 0, \exp\{-U(\Phi)\}}{E_\Gamma(\exp\{-U\})}$ is nearstandard. The rest follows by the Internal Definition Principle. □

Remark 3.2 In [AFHL] (see Theorems 7.4.11 and 7.4.13) some concrete conditions for the assumptions of Theorem 3.2 have been given for $d = 2$. For example, one can take

$$U(\Phi) = \lambda \sum_{z\delta \in \Lambda_\delta} \delta^2 \ {}^*g(z\delta) \int : e^{\alpha \Phi(z\delta)} : d\theta(\alpha),$$

with λ finite and positive, and θ a finite positive measure supported in $(-2\sqrt{\pi}, 2\sqrt{\pi})$. By extending the setting to cover situations when $U(\Phi) \in L^{1+\varepsilon}({}^*\mathbb{R}^{\Lambda_\delta}, d\Gamma)$, for some $\varepsilon > 0$, and using results of Kusuoka [Ku], one can extend this result to the case where the support of θ is in $(-\sqrt{8\pi}, \sqrt{8\pi})$. In all these cases $\frac{d\nu_L}{d\mu}$ is a non-constant measurable function. " Triviality results" concerning the cases $d \geq 3$ or other supports for θ can be deduced from [AH1] and [AGH].

The following result is just an alternative description of Theorem 3.2 in terms of measures and gives a flat integral representation of the interacting scalar field:

Theorem 3.3. *The field measure defined by Theorem 3.2, which we denote by ν^I, can be represented by*

(3.12) $$\nu^I(D) = \nu_L(\widehat{st^{-1}(D)})$$

for $D \in \mathcal{B}(\mathcal{D}_{-2p+1}(\Lambda))$ where $Z = (E_\Gamma(\exp\{-U\}))^{-1} > 0$.

Proof By (3.6) we have, for $q \in {}^*\mathbb{R}^{\Lambda_\delta}$

$$d\Gamma(q) = \kappa \exp\{-\frac{1}{2} \sum_{z\delta \in \Lambda_\delta} G_\delta^{-1}(z\delta - z'\delta) q_{z\delta} q_{z'\delta}\} \prod_{z\delta \in \Lambda_\delta} dq_{z\delta}.$$

Thus for every $A \in \mathcal{A}({}^*\mathbb{R}^{\Lambda_\delta})$, by (3.10) and (3.9), we get

$$\begin{aligned}
\nu(A) &= Z \int_A \exp\{-U(q)\} d\Gamma(q) \\
&= \kappa Z \int_A \exp\{-\lambda \sum_{z\delta \in \Lambda_\delta} {}^*g(z\delta) V(q_{z\delta}) \delta^d - \frac{1}{2} \sum_{z\delta, z'\delta \in \Lambda_\delta} G_\delta^{-1}(z\delta - z'\delta) q_{z\delta} q_{z'\delta}\} \\
&\quad \prod_{z\delta \in \Lambda_\delta} dq_{z\delta}.
\end{aligned}$$

Now by (3.7) and Lemma 3.1, we obtain for $D \in \mathcal{B}(\mathcal{D}_{-2p+1}(\Lambda))$ that

$$\nu^I(D) = \nu_L(\widehat{st^{-1}(D)}).$$

□

4. The upper large deviation bound

The flat integral representation obtained in section 3 was used in [AW] to obtain by nonstandard methods (for connections with other methods see Remark 1.2 in [AW]) a Schilder type large deviation principle for the free field measure μ with the rate function given by

$$(4.1) \quad I_\Lambda(f) = \begin{cases} \frac{1}{2} \int_\Lambda \left[((-\Delta_\Lambda + m^2)^{\frac{1}{2}} f)(x) \right]^2 dx, & \text{for } f \in (-\Delta_\Lambda + m^2)^{-\frac{1}{2}} L^2(\Lambda); \\ +\infty, & \text{otherwise.} \end{cases}$$

We state it as follows:

Theorem 4.1. *Let $\mu_\delta(A) = \mu(\delta^{-1} A)$ for $\delta > 0$ and $A \in \mathcal{B}(\mathcal{D}_{-2p+1}(\Lambda))$. Let ∂A denote the boundary of A. Then*
(i): *for every open set $O \subset \mathcal{D}_{-2p+1}(\Lambda)$ with $\mu(\partial O) = 0$,*

$$\underline{\lim}_{\delta \to 0} \delta^2 \log \mu_\delta(O) \geq -\inf I_\Lambda(O);$$

(ii): *for every closed set $C \subset \mathcal{D}_{-2p+1}(\Lambda)$, with $\mu(\partial C) = 0$*

$$\overline{\lim}_{\delta \to 0} \delta^2 \log \mu_\delta(C) \leq -\inf I_\Lambda(C).$$

From this result one can obtain immediately an upper large deviation bound for the exponential interacting field in two dimensional space-time (Høegh-Krohn's model), constructed in [AH2] , see Remark 3.2 above. In a bounded Borel subset Λ of $\mathbb{R}^2$, it is defined in terms of

$$(4.2) \quad U(\phi) = \lambda \int_\Lambda \int_{(-2\sqrt{\pi}, 2\sqrt{\pi})} : e^{\alpha \phi(x)} : d\theta(\alpha) dx$$

where $\lambda \geq 0$ and θ is a σ-finite positive measure supported in $(-2\sqrt{\pi}, 2\sqrt{\pi})$ (this corresponds to the expression in Remark 3.2 with g replaced by χ_Λ). One has $U \geq 0$, $E_\mu(\exp\{-U\}) > 0$ (the latter is easily seen by Jensen's inequality). The associated measure ν_Λ^I, called the exponential interacting field measure in the region Λ, is given by

$$(4.3) \quad d\nu_\Lambda^I(\phi) = \frac{\exp\{-U(\phi)\} d\mu(\phi)}{E_\mu(\exp\{-U\})},$$

and has support on $\mathcal{D}_{-2p+1}(\Lambda)$. We have the following

Corollary 4.1. *For every closed set $C \subset \mathcal{D}_{-2p+1}(\Lambda)$, with $\mu(\partial C) = 0$*

$$\overline{\lim}_{\delta \to 0} \delta^2 \log \nu_\Lambda^I(\delta^{-1} C) \leq -\inf I_\Lambda(C).$$

Proof By construction, $0 < \exp\{-U\} \leq 1$ μ-a.s.. Then, by (4.3), for every Borel set $D \subset \mathcal{D}_{-2p+1}(\Lambda)$, we have (with μ_δ defined in Theorem 4.1):

$$\nu_\Lambda^I(\delta^{-1} D) = \int_{\delta^{-1} D} d\nu_\Lambda^I(\phi) \leq \frac{\mu_\delta(D)}{E_\mu(\exp\{-U\})}.$$

Hence

$$\log \nu_\Lambda^I(\delta^{-1} D) \leq \log \mu_\delta(D) - \log E_\mu(\exp\{-U\})$$
$$\leq \log \mu_\delta(D) + |\log E_\mu(\exp\{-U\})|.$$

Therefore

$$\overline{\lim}_{\delta \to 0} \delta^2 \log \nu_\Lambda^I(\delta^{-1} D) \leq \overline{\lim}_{\delta \to 0} \delta^2 \log \mu_\delta(D).$$

The Corollary then follows by Theorem 4.1.(ii). □

Remark 4.1 The same method shows that the inequality in Corollary 4.1 also holds with ν_Λ^I replaced by any measure ν^ρ on $\mathcal{D}_{-2p+1}(\Lambda)$ such that $d\nu^\rho(\phi) := \rho(\phi)d\mu(\phi)$, with $E_\mu(\rho) = 1$ and $0 \leq \rho \leq K$, μ-a.s., for some constant $K \geq 1$. On the other hand, if ρ satisfies $L \leq \rho$, μ-a.s., for some constant $0 < L \leq 1$, then we get by a similar procedure a lower large deviation bound for ν^ρ, as in Theorem 4.1.(i).

Remark 4.2 Under general assumptions, large deviation estimates for a measure $\tilde{\mu}$ are related to small diffusion limits for the corresponding "stochastic quantization" diffusion having $\tilde{\mu}$ as invariant measure. One can indeed use our measure ν^I to construct such a $\mathcal{D}'(\Lambda)$-valued diffusion (by the "Dirichlet form method" of [AR] or the "regularization method" of [JoLM]). In this light, the corresponding estimate for the small diffusion limit for the stochastic quantization equation associated with the measure ν^I extends to the exponential interaction part of the results established in [JoLM] in the polynomial case.

It was proved in [AH2] and [FrP] that the infinite volume limit $\Lambda \to \mathbb{R}^2$ exists in such a way that ν_Λ^I converges weakly to a probability measure on $(\mathcal{D}'(\mathbb{R}^2), \mathcal{B}(\mathcal{D}'(\mathbb{R}^2)))$ (and defines a Euclidean invariant field measure), which we denote by ν_∞^I, as Λ tends to $\mathbb{R}^2$ in the manner that finally it covers all bounded subsets of $\mathbb{R}^2$, where $\mathcal{B}(\mathcal{D}'(\mathbb{R}^2))$ is the σ-algebra generated by the cylinder sets. Now let τ_w be the weak topology on $\mathcal{D}'(\mathbb{R}^2)$ and $\mathcal{B}_w(\mathcal{D}'(\mathbb{R}^2))$ be the topological σ-algebra associated with τ_w. It is clear (see Remark 3.2 of [AW]) that $\mathcal{B}_w(\mathcal{D}'(\mathbb{R}^2)) = \mathcal{B}(\mathcal{D}'(\mathbb{R}^2))$ and $\mathcal{B}(\mathcal{D}_{-2p+1}(\Lambda)) = \mathcal{B}_w(\mathcal{D}'(\mathbb{R}^2)) \cap \mathcal{D}_{-2p+1}(\Lambda)$. For the measure ν_∞^I, we have an upper large deviation bound as follows

Corollary 4.2. *For every τ_w-closed set $C \subset \mathcal{D}_{-2p+1}(\Lambda)$ such that $\nu_\infty^I(\partial C) = 0$ (with ∂C the boundary of C), we have*

$$\overline{\lim}_{\delta \to 0} \delta^2 \log \nu_\infty^I(\delta^{-1}C) \leq -\inf I(C),$$

where I is the rate function defined in terms of (4.1) replacing Λ by $\mathbb{R}^2$.

Proof By Corollary 4.1, the weak convergence of ν_Λ^I to ν_∞^I and the assumption $\nu_\infty^I(\partial C) = 0$, we have

$$\begin{aligned}
& \overline{\lim}_{\delta \to 0} \delta^2 \log \nu_\infty^I(\delta^{-1}C) \\
&= \overline{\lim}_{\delta \to 0} \lim_{\Lambda \to \mathbb{R}^2} \delta^2 \log \nu_\Lambda^I(\delta^{-1}(C \cap \mathcal{D}_{-2p+1}(\Lambda))) \\
&= \lim_{\Lambda \to \mathbb{R}^2} \overline{\lim}_{\delta \to 0} \delta^2 \log \nu_\Lambda^I(\delta^{-1}(C \cap \mathcal{D}_{-2p+1}(\Lambda))) \\
&\leq \lim_{\Lambda \to \mathbb{R}^2} (-\inf I_\Lambda(C \cap \mathcal{D}_{-2p+1}(\Lambda))) \\
&= -\inf I(C).
\end{aligned}$$

□

Acknowledgements.

We are very grateful to Professors Nigel Cutland and Tom Lindstrøm for stimulating discussions. The second named author gratefully acknowledges the financial support by Alexander von Humboldt Foundation and the kind invitation by Professors Vitor Neves, A.Franco Oliveira and J.Sousa Pinto to the Aveiro Conference.

References

[A]   S. Albeverio, *Nonstandard analysis in mathematical physics*, in: Nonstandard Analysis and its Applications (ed. by N.J. Cutland), 182–220, LMS Student Text 10, Cambridge, 1988.

[AFHL] S. Albeverio, J.E. Fenstad, R. Høegh–Krohn and T. Lindstrøm, Nonstandard Methods in Stochastic Analysis and Mathematical Physics, Academic Press, New York, 1986.

[AGH] S. Albeverio, G. Gallavotti and R. Høegh–Krohn, *Some results for the exponential interaction in two or more dimensions*, Commun. Math. Phys. $\underline{70}$(1979), 187–192.

[AH1] S. Albeverio and R. Høegh–Krohn, *Martingale convergence and the exponential interaction in $\mathbb{R}^n$*, in: Quantum Fields – Algebras, Processes (ed. by L. Streit), 331–353, Springer-Verlag, Berlin, 1980.

[AH2] S. Albeverio and R. Høegh–Krohn, *The Wightman axioms and the mass gap for strong interactions of exponential type in two-dimensional space-time*, J.Funct.Anal. $\underline{16}$(1974), 39–82.

[AH3] S. Albeverio and R. Høegh–Krohn, *Uniqueness of the physical vacuum and the Wightman functions in the infinite volume limit for some non-polynomial interactions*, Commun. Math. Phys. $\underline{30}$ (1973), 171–200.

[AR]   S. Albeverio and M. Röckner, *Stochastic differential equations in infinite dimension: solution via Dirichlet forms*, Prob. Th. Rel. Fields $\underline{89}$ (1991), 347–368.

[AW]   S. Albeverio and J.-L. Wu, *A mathematical flat integral realization and a large deviation result for the free Euclidean field*, SFB 237-Preprint Nr. 214, Bochum 1994, subm. to Acta Appl. Math.

[AZ]   S. Albeverio and B. Zegarlinski, *Global Markov property in quantum field theory and statistical mechanics: a review on results and problems*, in: R. Høegh–Krohn's Memorial Volume "Ideas and Methods in Quantum and Statistical Physics", Vol.2(edts. S.Albeverio, J.E.Fenstad, H.Holden and T.Lindstrøm),331-369, Cam. Univ. Press, Cambridge, 1992.

[An]   R.M. Anderson, *A non–standard representation for Brownian motion and Itô integration*. Israel J. Math. $\underline{25}$ (1976), 15–46.

[BSZ] J.C. Baez, I.E. Segal and Z. Zhou, Introduction to Algebraic and Constructive Quantum Field Theory, Princeton University Press, Princeton, NJ, 1992.

[Ca]   J. Cannon, *Continuous sample paths in quantum field theory*, Commun. Math. Phys. $\underline{35}$ (1974), 215–234.

[CoLa] P. Colella and O. E. Lanford, *Sample field behavior for the free Markov random field*, in: Constructive Quantum Field Theory (ed. by G. Velo and A. S. Wightman), 44–70, Lect. Notes Phys. vol 25, Springer-Verlag, Berlin/Heidelberg/New York, 1973.

[C1]   N.J. Cutland, *Infinitesimals in action*, J. London Math. Soc. $\underline{35}$ (1987), 202–216.

[C2]   N.J. Cutland, *The Brownian bridge as a flat integral*, Math. Proc. Cam. Phil. Soc. $\underline{106}$ (1989), 343–354.

[C3]   N.J. Cutland, *An action functional for Lévy Brownian motion*, Acta Appl. Math. $\underline{18}$ (1990), 261–281.

[C4]   N.J. Cutland, *On large deviations in Hilbert space*, Proc. Edinburgh Math. Soc. $\underline{34}$ (1991), 487–495.

[C5]   N.J. Cutland, *Nonstandard representation of Gaussian measures*, in: White Noise – Mathematics and Applications (ed. by T. Hida, H.-H. Kuo, J. Potthoff and L. Streit), 73–92, World Scientific, Singapore, 1990.

[Do]   M.D. Donsker, *On function space integrals*, in: Analysis in Function Space (ed. by W.T. Martin and I.E. Segal), 17–30, MIT Press, Cambridge, MA, 1964.

[FrP]   J. Fröhlich and Y.M. Park, *Remarks on exponential interactions and the quantum Sine-Gordon equation in two space-time dimensions*, Helv. Phys. Acta $\underline{50}$ (1977), 315–329.

[GV]   I.M. Gelfand and N. Ya. Vilenkin, Generalized Functions, vol.4.Applications of Harmonic Analysis, Academic Press, New York, 1964.

[GJ]   J. Glimm and A. Jaffe, Quantum Physics: A Functional Integral Point of View (2nd edn), Springer–Verlag, New York/Heidelberg/Berlin, 1987.

[GRS]   F. Guerra, L. Rosen and B. Simon, *The $P(\phi)_2$ Euclidean quantum field theory as classical statistical mechanics*, Ann. Math. $\underline{101}$ (1975), 111–259.

[HKPS]   T. Hida, H.-H. Kuo, J. Potthoff and L. Streit, White Noise: An Infinite Dimensional Calculus, Kluwer Academic Publishers, Dordrecht/Boston/London, 1993.

[JoLM]   G. Jona–Lasinio and P.K. Mitter, *Large deviation estimates in the stochastic quantization of φ_2^4*, Commun. Math. Phys. $\underline{130}$ (1990), 111–121.

[Ku]   S. Kusuoka, *Høegh–Krohn's model of quantum fields and the absolute continuity of measures*, in: R. Høegh–Krohn's Memorial Volume "Ideas and Methods in Quantum and Statistical Physics", Vol.2(edts. S.Albeverio, J.E.Fenstad, H.Holden and T.Lindstrøm), 405–424, Cam. Univ. Press, Cambridge, 1992.

[L]   T. Lindstrøm, *An invitation to nonstandard analysis*, in: Nonstandard Analysis and its Applications (ed. by N.J. Cutland), 1–105, LMS Student Text 10, Cambridge, 1988.

[Mo]   G. Molchan, *Characterization of Gaussian fields with Markovian property*, Soviet Math. Dokl. $\underline{12}$ (1971), 563–567.

[N]   E. Nelson, *The free Markov field*, J. Funct. Anal. $\underline{12}$ (1973), 211–227.

[Pi]   L. Pitt, *A Markov property for Gaussian processes with a multidimensional parameter*, Arch. Rat. Mech. Anal. $\underline{43}$ (1971), 367–391.

[RR]   M. Reed and L. Rosen, *Support properties of the free measure for boson fields*, Commun. Math. Phys. $\underline{36}$ (1974), 123–132.

[RS]   M. Reed and B. Simon, Methods of Mathematical Physics I: Functional Analysis (revised and enlarged edition), Academic Press, New York, 1980; II:Fourier Analysis, Self-Adjointness, Academic Press, New York, 1975.

[S]   B. Simon, The $P(\phi)_2$ Euclidean (Quantum) Field Theory, Princeton University Press, Princeton, NJ, 1974.

[Str]   D. W. Stroock, An Introduction to the Theory of Large Deviations, Springer–Verlag, New York, 1984.

[Wo]   E. Wong, Stochastic Processes in Information and Dynamical Systems, McGraw–Hill, New York, 1971.

[Yan]   J.- A. Yan, *Generalizations of Gross' and Minlos' theorems*, in: Séminaire de Probabilités. XXII (ed. by J. Azema, P.A. Meyer, M. Yor), 395–404, Lect. Notes Math. vol 1372, Springer–Verlag, Berlin/Heidelberg/New York, 1989.

SERGIO ALBEVERIO
SFB 237 Essen-Bochum-Düsseldorf
BiBoS-Research Centre
D-33615 Bielefeld
CERFIM Locarno Switzerland
E-mail:
sergio.albeverio@ruba.ruhr-uni-bochum.de

JIANG-LUN WU
Alexander von Humboldt research fellow;
on leave from Probability Centre
Institute of Applied Mathematics
Academia Sinica
Beijing 100080 PR China
E-mail jang-lun.wu@ruba.rz.ruhr-uni-bochum.de

Michael Benedikt
Nonstandard analysis in selective universes

1. Introduction

A major goal in foundational research in nonstandard analysis has been to pinpoint properties of nonstandard models beyond transfer and ω_1-saturation that are useful in applications. Research in this vein includes the work of Henson on the Isomorphism Property [10], and the work of Ross and Jin on the Special Model Axiom [20, 13]. However, most of the work in this direction focuses on 'big' models: models with lots of saturation. A natural counterpart to these investigations is to look at models that are in some sense as 'small' as possible. A nonstandard universe formed using a selective ultrafilter is definitely as small as an ω_1-saturated nonstandard universe can get: no nonstandard universe can be properly embedded in it [7, p. 291]. We would therefore expect these universes to have very distinctive properties. The purpose of this paper is to show that this is indeed the case. We will outline some unique features of nonstandard universes formed using selective ultrafilters and ultrafilters satisfying various weakenings of selectivity. In addition to giving new results on these universes, we will try to give a unified approach to results already in the literature that make use of such universes. We will give proofs only sporadically: to show how the main theorems are applied, and to show the basic tools for proving results on selective ultrafilters. Proofs can be found in the references (the 'default' reference being [2]).

Organization: Section two outlines the basic definitions and some facts about ultrafilters that are made use of in this paper. The next two sections discuss distinguishing features of selective ultrafilters with respect to topology and measure theory, respectively. Section three discusses nonstandard topology in a selective universe, focusing on the S-topology. Section four discusses some properties of the Loeb measure in selective universes.

2. Preliminaries

Let U be an ultrafilter on ω:

U is **selective** if for every $f{:}\omega \to \omega$, either f is constant on a set in U or f is one-to-one on a set in U.

U is **semi-selective** if for every sequence $\{X_n\}_{n\in\omega}$ with each X_n in U there is an $S \in U$ with $s_n \in X_n$, where $s_n = n^{th}$ element of S.

U is **rapid** if for every function f from ω to ω, there is $S \in U$ with $s_n > f(n)$.

Acknowledgments. The author wishes to thank H.J. Keisler for assistance with many different aspects of this work. Thanks are also due to Carlos Ortiz and Curtis Tuckey, for reviewing the manuscript.

All selective ultrafilters are semi-selective [5], and all semi-selective ultrafilters are rapid [19].

An important property of the selective ultrafilters, due to Kunen, is that they have the **Ramsey** property: that is,

Theorem 1. [5] *An ultrafilter U is selective iff for any $S \subseteq [\omega]^2$ either $\exists X \in U$ with $[X]^2 \subseteq S$ or $\exists X \in U$ with $[X]^2 \subseteq S^c$*

Let $[\omega]^\omega$ be given the subspace topology inherited from 2^ω.

The Ramsey property for an ultrafilter U implies the following stronger partition property (due to Mathias).

Theorem 2. [17] *Let $S \subseteq [\omega]^\omega$ be analytic. Then either $\exists X \in U$ with $[X]^\omega \subseteq S$ or $\exists X \in U$ with $[X]^\omega \subseteq S^c$*

For what follows, it will be convenient to use the following version of the Ramsey property, which follows directly from Kunen's Theorem. Following the notation of Blass say $T \subseteq \omega^{<\omega}$ is a U-**tree** for an ultrafilter U if it is closed under initial segments, and $\forall s \in T$ the set $\{n : s \frown n \in T\}$ is in U. We then have,

Proposition 1. [4] *If U is selective, then for all U-trees T, $\exists X \in U$ such that each initial segment of X (considered as a finite sequence) is in T. Such an X is called a* **path** *through T.*

The existence of selective or even semi-selective ultrafilters is dependent on axioms of set theory that go beyond ZFC. On the one hand, it can be shown that MA implies that each of these classes is nonempty. On the other hand, it is shown by Kunen [15] that there is a model of ZFC where there are no semi-selective ultrafilters, and by Miller [18] that there may be no rapid ultrafilters.

We assume familiarity with the construction of nonstandard universes from ultrafilters, and with the Loeb measure construction (see [1]). A nonstandard universe formed using a selective ultrafilter will be referred to as a **selective universe**. If μ is a standard measure, we will write $L\mu$ for the Loeb measure generated by $^*\mu$. Finally, the following notation is used throughout this paper: For a set S in the nonstandard universe $^*V$, we denote by $\sigma(S)$ the standard elements of S, that is:

$$\sigma(S) \stackrel{\text{def}}{=} \{x \in V : {^*x} \in S\}.$$

3. Topological properties of selective universes

We wish to identify the properties that distinguish nonstandard topology in a selective universe from topology in run-of-the-mill nonstandard universes. We will take as our point of departure the following result from nonstandard arithmetic:

Theorem 3. [19]*In a selective universe, for any two nonstandard integers there is a standard map taking one to the other.*

The intuition this gives us is the following: *If we are in a situation where there are many nonstandard objects realizing some type, then there exists a standard one realizing that type as well.* In this section we will give a topological theorem capturing the above fact. That is, we will show that *Sets that are topologically big (in terms of hitting many standard sets) must contain many standard elements.* This intuition is captured in the Density Theorem, Theorem 4, and is reiterated in Section 4 in Theorem 17. This is in sharp contrast to the situation in 'big' universes. In a c^+-saturated universe, there is a *-cofinite set in *$\mathbb{R}$ having no standard elements in it. This shows that in such a universe there is no relationship between the thickness of a set within the nonstandard universe and the thickness of its standard elements within the standard world. In a selective universe, if $S \in {}^*\mathcal{P}(\mathbb{R})$ has no standard elements, then Theorem 4 shows that there are many standard intervals I such that S is *-meager in I. The density theorem can be used to derive a number of useful topological properties of selective universes. We will give a couple of quick examples of how the Density Theorem is used. For more details, the reader can check [2].

Just as nonstandard integers can be represented in especially simple forms in selective universes, topological objects tend to have particularly nice representations. The second half of the section is devoted to two such representation theorems, one for *-compact sets and one for S-continuous functions. We show how these can be used to derive topological separation principles for the S-topology that are specific to selective universes. *Throughout this section we assume that the nonstandard universe is formed via a selective ultrafilter U.*

3.1. A Density Theorem

We wish to prove that in a standard topological space, nonstandard open sets that are dense within the standard open sets must contain standard elements. This is a strengthening of the Baire Category Theorem, and so will only hold in Baire spaces. We actually need a property slightly stronger than Baire, outlined in the next paragraph.

Let $\langle X, \mathcal{T} \rangle$ be a topological space. The two-player game $G(\mathcal{T})$ is played as follows: Player 1 chooses a nonempty open set in $\mathcal{T}$, Player 2 responds by playing a nonempty open $U_2 \subset U_1$, Player 1 responds with a nonempty open subset of U_2, and so on. A full play of the game is a sequence of alternating moves by 1 and 2. Player 2 wins a full play $S = \langle U_i \rangle_{i \in \omega}$ if $\bigcap_i U_i \neq \emptyset$. A space $\langle X, \mathcal{T} \rangle$ is called an *α-favorable Space* or a *Choquet Space* if Player 2 has a winning strategy for the game above. We will use the shorter term Choquet Space for this section. A Choquet space is a Baire Space, and the class of Choquet spaces captures most of the natural examples of Baire Spaces, such as locally compact T_2 spaces, complete metric spaces, Čech-complete spaces.

Given any topological space $\mathcal{T}$ on a set X, the S-*topology* is the topology on *X whose basic open sets are $\{{}^*A : A \in \mathcal{T}\}$. We will use S-dense to mean dense in the S-topology, S-interior (A) to mean the interior of A in the S-topology, etc.

Theorem 4. *Let $\langle X, \mathcal{T} \rangle$ be a Choquet space. If O is an internal set that is *-open and S-dense for *$\mathcal{T}$, then $\sigma(O)$ is dense in $\mathcal{T}$ and nonmeager.*

Proof: We first show that $\sigma(O)$ is nonempty. Since O is *-open, $O = \langle O_i : i \in \omega \rangle_U$ where each O_i is open in $\mathcal{T}$. Since O is nonempty, we can take each O_i to be nonempty.

We will construct a U-tree T such that every path through T corresponds to an R such that $\bigcap_{i \in R} O_i \neq \emptyset$. The nth level of the tree (i.e. $T \cap \omega^n$) will be denoted by T_n. For this proof, our tree will consist solely of increasing tuples, so we will identify these tuples with finite subsets (e.g., we will say things like $s \cup \{j\}$, where s is a tuple and $j > \max(\text{range}(s))$).

We will build T inductively. Let F be a winning strategy for Player 2 in $G(\mathcal{T})$. With each $s = \{s_1 \ldots s_n\}$ we associate a proposed play $P(s)$ in $G(\mathcal{T})$ as follows: Player 2 will play according to F at all times, Player 1's i^{th} move will be to play $O_{s_i} \cap \{\text{previous move of player 2}\}$. The partial play ends with the nth move of Player 2. This proposed play is a legal play if

$$\forall i \leq n \; O_{s_i} \cap \{\text{previous move of player 2}\} \neq \emptyset.$$

Say that s is playable if $P(s)$ is a legal play of the game, and for playable s denote the last move in $P(s)$ (made by Player 2) by O_s. We will assume inductively that $\forall s \in T_n$, s is playable. Let $T_0 = \omega$. The nonemptiness of the O_i's assures that every singleton is playable.

So suppose we have already constructed T_n. For each $s \in T_n$, we have, as O is S-dense, the set $\{j : O_j \cap O_s \neq \emptyset\} \in U$. Let $X_s = \{j : j > \max(s), O_j \cap O_s \neq \emptyset\}$. Now we can define:

$$T_{n+1} = \bigcup_{s \in T_n} \bigcup_{j \in X_s} s \cup \{j\}$$

Since each X_s is in U, T is a U-tree. Since we have defined T_{n+1} so that $\forall s \in T_n \; (s \cup \{j\} \in T_{n+1} \Rightarrow O_s \cap O_j \neq \emptyset)$, we get that $P(s \cup \{j\})$ is defined, so the inductive hypothesis is preserved. One can check inductively that for every $s \in T$, $O_s \subset \bigcap_{n \in s} O_n$. If S is a path through T and $\langle s_i \rangle_{i \in \omega}$ enumerates the initial segments of S, then $\langle O_{s_i} \rangle_{i \in \omega}$ is a sequence of consecutive moves by 2 in $G(\mathcal{T})$, so we have $\bigcup_{n \in S} O_n \neq \emptyset$ for every path through T. Proposition 1 now assures that there is a path through T in U, hence $\exists X \in U$ with $\bigcap_{i \in X} O_i \neq \emptyset \Rightarrow \sigma(O) \neq \emptyset$.

By applying the above paragraph within any standard open set, we can see that $\sigma(O)$ is dense. □

Example
Let $\langle X, \mathcal{T} \rangle$ be the Cantor space 2^ω. For any *-finite set $S \subset {}^*\mathbb{N} \setminus \mathbb{N}$ and any partial function $\rho \in 2^S$, the basic clopen set $[\rho] = \{F \in {}^*2^\omega : F \text{ extends } \rho\}$ is *-open and S-dense, and hence contains standard elements.

From Theorem 4 one can derive some basic facts about category in selective universes.

Theorem 5. *If $\mathcal{T}$ is Choquet and B is an internal *-Baire set, then $\sigma(B)$ is meager $\Leftrightarrow \exists D$ S-nowhere dense and *-meager E such that $B \subset D \cup E$.*

Proposition 2. *If $\langle X, \mathcal{T} \rangle$ is a Choquet space, C is a *-closed set in *X with $\sigma(C) = X$ then S-interior(C) is dense in the S-topology of *$\mathcal{T}$.*

A consequence of this proposition is the following strengthening of the Uniform Boundedness Principle:

Proposition 3. *If Γ is a *-bounded linear functional on *B, for some (standard) Banach Space B, and $(\forall x \in B, \|x\| = 1 \Rightarrow \Gamma(^*x)$ is finite$)$ then $\|\Gamma\|$ is finite.*

Proof: Otherwise let $\|\Gamma\| = K \in {}^*\mathbb{N} \setminus \mathbb{N}$, and let $H = K/\log K$. The set $C = \{x \in {}^*B : \|\Gamma(x)\| \leq H \cdot \|x\|\}$, is *-closed, and contains all standard elements of *B, since the hypothesis on Γ implies that Γx has finite norm for all standard x. So, by Proposition 2, C contains an S-open ball. This implies (by taking an S-closed ball contained in the S-open ball) that $\exists b_0 \in B$ $\epsilon > 0$ with $\{b \in {}^*B : \|b - {}^*b_0\| \leq \epsilon\} \subset C$. Then for any $x \in {}^*B$ with $\|x\| = 1$, we get: Let $\eta = 1/\epsilon$

$$\|\Gamma(x)\| = \eta \cdot \|\Gamma\left(\frac{x}{\eta}\right)\| \leq \eta \cdot (\|\Gamma(\frac{x}{\eta} + {}^*b_0)\| + \|\Gamma({}^*b_0)\|)$$

$$\leq \eta \cdot (H \cdot (\|\frac{x}{\eta}\| + \|{}^*b_0\|) + \|\Gamma({}^*b_0)\|)$$

$$= \eta \cdot H \cdot \|\frac{x}{\eta}\| + \eta \cdot H \cdot \|{}^*b_0\| + \|\Gamma({}^*b_0)\|$$

$$= H + \eta \cdot \|b_0\| \cdot H + \|\Gamma({}^*b_0)\|$$

So $\|\Gamma(x)\| \leq H + (\text{finite}) \cdot H + (\text{finite}) \leq (\text{finite}) \cdot H \leq \frac{K}{2}$.
So, $\sup\{\|\Gamma(x)\| : \|x\| = 1\} \leq \frac{K}{2} < K$, which contradicts the fact that $\|\Gamma\| = K$. $\square$

A disadvantage of Theorem 4 is that it applies only to internal sets, and generally the sets of interest in nonstandard analysis are external. For an arbitrary external set S it seems impossible to give a topological property which implies the existence of standard elements in S. It is, however, possible to say something when X is not too external.

Definition. *A set S is Π_1^0 if $S = \bigcap_i S_i$, where each S_i is internal.*

Theorem 6. *Let $\mathcal{T}$ be a Choquet space, and let S be a Π_1^0 set with the property that: For each standard open O, there is a *-open P such that $O \cap S \supset P$*
  Then $\sigma(S) \neq \emptyset$.

Proof: Similar to 4.

We give a few consequences of Theorems 4 and 6.

Proposition 4. *If $\langle a_i \rangle_{i \in {}^*\omega}$ is an internal sequence of *-reals with $\lim_{i \in {}^*\omega} a_i = 0$ and $\forall i \in \omega$ $a_i \approx 0$, then there is an $X \in [\omega]^\omega$ with $\sum_{i \in {}^*X} a_i \approx 0$.*

215

Proof: Let $\mathcal{T}$ be the Ellentuck Topology on $[\omega]^\omega$ namely, the topology whose basis consists of the sets
$$(s, S) = \{X \in [\omega]^\omega : s \subset X \subset s \cup S\}, \quad \text{where } s \in [\omega]^{<\omega}, S \in [\omega]^\omega.$$
The hypotheses on the sequence guarantee that the set
$$A = \{X \in {}^*[\omega]^\omega : \sum_{i \in X} a_i \approx 0\}$$
meets every S-open set in a *-open set. Also, we have A is Π_1^0, since
$$A = \bigcap_n \left\{X \in {}^*[\omega]^\omega : \sum_{i \in {}^*X} a_i \leq \frac{1}{n}\right\}.$$
Theorem 6 will now guarantee that there is an $X \in \sigma(A)$, since one can easily show that the Ellentuck topology is Choquet. □

We close this section with a result due to Burzyk, which we prove using Proposition 4.

Definition. X is an N-space if it is a normed linear space which also satisfies: If $\langle x_n \rangle_{n \in \omega}$ is a sequence in X converging to 0, then there is a subsequence $\langle x_{n_i} \rangle_{i \in \omega}$ such that $\lim_m \sum_{i=1}^m x_{n_i}$ exists in X.

A complete normed linear space is clearly an N-space.

Theorem 7 (MA). [6] *There is an N-space that is not complete.*

Proof: Start with the complete normed linear space $l_2(\omega)$, and let $\|\cdot\|$ be the usual norm on $l_2(\omega)$. Let $H \in {}^*\mathbb{N} \setminus \mathbb{N}$.
Define a second *-norm $\|\cdot\|_H$ on $*l_2(\omega)$ by:
$$\|x\|_H = |x_H| \cdot 2^H$$
Our N-space will be $M = \{x \in l_2(\omega) : \|{}^*x\|_H \approx 0\}$, with the linear and norm structure inherited from $l_2(\omega)$.

Since $\|\cdot\|_H$ is a *-norm, and the infinitesimals are closed under addition and scalar multiplication by a standard scalar, M is a linear space, and hence a normed linear space.

To see that M is not complete, let $x \in l_2(\omega)$ be such that $\forall i \ |x(i)| > 1/2^i$, and let x_n be defined by
$$\begin{cases} x_n(i) = 0 & \text{if } i \geq n \\ x_n(i) = x(i) & \text{if } i < n. \end{cases}$$
Then $\forall n, \ x_n \in M, \ x_n \to x$ in M, but $x \notin M$.

To see that M is an N-space, let $y_n \in M$ with $y_n \to 0$. By going to a subsequence, we can assume that $\sum_n \|y_n\| < \infty$. Consider the set $G = \{X : \sum_{J \in {}^*X} \|y_J\|_H \approx 0\}$. By applying Proposition 4 to $a_n = \|y_n\|_H$, we get $G \neq \emptyset$. Let $X = \{x_i\}_{i \in \omega}$ be an element of G.

216

Since $l_2(\omega)$ is complete and $\sum_n \|y_n\| < \infty$, the sequence $z_n = \sum_{i=1}^{n} y_{x_i}$ converges to some $z \in l_2(\omega)$.

We complete the proof by showing that $z \in M$:
$$\|^*z\|_H \leq \sum_{J \in {}^*\omega} \|^*y_{x_J}\|_H,$$
since $\|\cdot\|_H$ is a *-norm, and
$$\sum_{J \in {}^*\omega} \|^*y_{x_J}\|_H \approx 0,$$
since $X \in G$. This gives $\|^*z\|_H \approx 0$, which means $z \in M$. □

3.2. Separation Properties of Selective Universes

It is a trivial fact of point set topology that in a metric space points can be separated by open sets, and disjoint closed sets can be separated as well. Unfortunately, the same does not normally hold for the S-topology over a *-metric space, even a standard one. The goal of this section is to show that in a selective universe we can still obtain some separation theorems for the S-topology. We show that it separates points, and that there is a class of *-closed sets such that any two which are disjoint can be separated by S-open sets. We will also prove an analog of the Tietze Extension Theorem for S-continuous functions. The main tool in these results will be getting canonical representations of *-compact sets and S-continuous functions.

Here is what a *-compact set looks like in a selective universe:

Lemma 8. *Let C be a *-compact set in a (standard) complete metric space. Then C can be represented as $\langle C_i : i \in \omega \rangle_U$ with C_i satisfying:*

If $S \in [\omega]^\omega$ and $\forall j \in S\ x_j \in C_j$ and $\langle x_j \rangle_{j \in S}$ converges, then $\lim_j x_j \in st(C)$.

Proof: Start with any representation of C as $\langle C_i : i \in \omega \rangle_U$, with each C_i compact. For each k let $J_k = \{x : d(x, st(C)) \geq \frac{1}{k}\}$. Then J_k is closed, hence for each i the set $C_i \cap J_k$ is compact. Fix k, and let $D_i = C_i \cap J_k$. Since D_i is compact, it is totally bounded, and hence can be covered by a finite set of open balls $\langle B(x_j^i, \frac{1}{i}) \rangle_{j \leq n_i}$, with $x_j^i \in D_i$. Let
$$P_k = \{S \in [\omega]^\omega : \exists f : \omega \to \omega \ni f(i) \leq n_i \text{ and } \langle x_{f(i)}^i \rangle_{i \in S} \text{ converges}\}.$$

Claim 1: P_k is an analytic subset of $[\omega]^\omega$.

One can prove Claim 1 by tracing through the definitions.

Claim 2: $\exists S \in U$ such that $[S]^\omega \subset P_k^c$.

If not, then by Theorem 2, $\exists S \in U$ with $[S]^\omega \subset P_k$, so in particular, there is an $S \in U \cap P_k$. Fix an $f \in \omega^\omega$ such that $\langle x_{f(i)}^i \rangle_{i \in S}$ converges. Then $x = \lim_i x_{f(i)}^i$ is in J_k, since each $x_{f(i)}^i$ is in J_k and J_k is closed. So $x \notin st(C)$. But $^*x \approx \langle x_{f(i)}^i : i \in S \rangle_U \in \langle C_i : i \in S \rangle_U = C$, so x is in $st(C)$ and we have a contradiction.

Fix $S_k \in U$ with $[S_k]^\omega \subset P_k^c$.

Claim 3: If $T \subset S_k$, $x_j \in C_j$ for $j \in T$, and $\lim_{j \in T} x_j = x$ then $d(x, st(C)) \leq \frac{2}{k}$.

If x is a counterexample to the claim, then $d(x, st(C)) > \frac{2}{k}$ and there is a $T \subset S_k$ and $x_i \in C_i$ for $i \in T$ such that $x = \lim x_i$. Since $d(x, st(C)) > \frac{2}{k}$, x has a neighborhood that is in J_k, so WLOG all the x_i's are in J_k as well. Then we have $x_i \in D_i$ for $i \in T$, and it follows that there is $f \in \omega^\omega$ with $x_i \in B\left(x^i_{f(i)}, \frac{1}{i}\right)$. But then $x = \lim_{i \in T} x^i_{f(i)}$, which implies $T \in P_k$, a contradiction.

Since U is a P-point, we can get $S \in U$ with $|S \setminus S_k| < \omega$ for each k. Since $st(C)$ is closed, for this S we have:

If $T \subset S$, $x_j \in C_j$ for $j \in T$, and $\lim_{j \in T} x_j = x$, then $x \in st(C)$

Then $\langle C_i : i \in S \rangle_U$ is the desired representation of C. □

We now apply Lemma 8 to determine when *-compact sets can be separated by standard open sets.

Theorem 9. *If K^0 and K^1 are disjoint *-compact sets in a complete metric space X then*

(1) *There are disjoint open sets O^0 and O^1 with $K^0 \subset {}^*O^0$ and $K^1 \subset {}^*O^1$ $\iff$ $K^0 \cap {}^*st(K^1) = \emptyset$ and $K^1 \cap {}^*st(K^0) = \emptyset$*
(2) *$\exists$ a continuous $f : X \to \mathbb{R}$ with $K^0 \subset {}^*f^{-1}(\{0\})$ and $K^1 \subset {}^*f^{-1}(\{1\})$ $\iff$ $K^0 \cup {}^*st(K^0)$ is disjoint from $K^1 \cup {}^*st(K^1)$.*

Proof:

(1)

($\Rightarrow$)
If there are such open sets O^0 and O^1, and ${}^*st(K^1) \cap K^0 \neq \emptyset$, then $st(K^1) \cap O^0 \neq \emptyset$, so there is an $x \in X$ and $y \in {}^*X$ with $x \in O^0$, $y \in K^1$ and $y \approx {}^*x$. But then $y \in {}^*O^0$, which contradicts the fact that ${}^*O^0$ is disjoint from K^1. So ${}^*st(K^1) \cap K^0 = \emptyset$, and a symmetric argument shows ${}^*st(K^0) \cap K^1 = \emptyset$.

($\Leftarrow$)
Let $K^0 = \langle K^0_i : i \in \omega \rangle_U$ and $K^1 = \langle K^1_i : i \in \omega \rangle_U$. We can assume that $K^0_i \cap K^1_i = \emptyset$ and that $\forall i\ K^0_i \cap st(K^1) = \emptyset$, $K^1_i \cap st(K^0) = \emptyset$.

Applying Lemma 8, we can also arrange it so that:

$\forall j \in \{0,1\}$ if $S \in [\omega]^\omega$, $x_i \in K^j_i$ for $i \in S$ and $x_i \to x$, then $x \in st(K^j)$. (∗)

We wish to further restrict the K^j_i's to $S \in U$ so that

$$\forall i_1, i_2 \in S \ K^0_{i_1} \cap K^1_{i_2} = \emptyset \qquad (**)$$

By the Ramsey property of selective ultrafilters, if (∗∗) is not possible, then

$$\exists S \in U \ \forall i_1 \neq i_2 \in S \ K^0_{i_1} \cap K^1_{i_2} \neq \emptyset.$$

Then fixing i_1, we would have for every $j \in S$ an $x_j \in K^0_{i_1} \cap K^1_j$. But then there is a subsequence of the x_j's that converges to an $x \in K^0_{i_1}$ (since $K^0_{i_1}$ is compact), and this

x is in $K_{i_1}^0 \cap st(K^1)$, by (*) which is a contradiction. Therefore, we can take $S \in U$ to satisfy (**).

Letting $C_0 = \overline{\bigcup_{i \in S} K_i^0}$, we get $(*) \Rightarrow C_0 = \bigcup_i K_i^0 \cup st(K^0)$, so for each fixed i we have K_i^1 is disjoint from C_0 (using(**) and $K_i^1 \cap st(K^0) = \emptyset$).

For each i, we can take disjoint open sets $O_i^1 \supset K_i^1$ and $O_i^0 \supset C_0$. By intersecting, we can assume that $i < j \Rightarrow O_i^0 \supset O_j^0$.

Similarly, letting $C_1 = \overline{\bigcup_i K_j^1}$, we can get for each i disjoint open sets P_i^1 and P_i^0 with $P_i^0 \supset K_i^0$ and $P_i^1 \supset C_1$, and such that $i < j \Rightarrow P_i^1 \supset P_j^1$.

Let $O^0 = \bigcup_i O_i^0 \cap P_i^0$ and $O^1 = \bigcup_i O_i^1 \cap P_i^1$. Then O^0 covers $\bigcup_i K_i^0$, O^1 covers $\bigcup_i K_i^1$. To see that $O^0 \cap O^1 = \emptyset$:

If $x \in O^0 \cap O^1$, then $\exists i, j$ such that $x \in O_i^0 \cap P_i^0 \cap O_j^1 \cap P_j^1$. If $i \geq j$ then $O_i^0 \subset O_j^0$ which is disjoint from O_j^1, so we have a contradiction. If $i < j$ then $P_j^1 \subset P_i^1$ which misses P_i^0, so $P_j^1 \cap P_i^0 = \emptyset$, a contradiction.

Therefore, we have that $*O^0$ and $*O^1$ are disjoint sets covering K^0 and K^1.

(2)

$(\Rightarrow)$

If there is such an f, then the S-continuity of $*f$ implies $f''(st(K^0)) = \{0\}$ and $f''(st(K^1)) = \{1\}$, so $*st(K^0) \cup K^0 \subset f^1(\{0\})$ and $*st(K^1) \cup K^1 \subset f^{-1}(\{1\})$, so $*st(K^0) \cup K^0$ and $*st(K^1) \cup K^1$ are disjoint.

$(\Leftarrow)$

Get representations $\langle K_i^0 : i \in \omega \rangle_U$ and $\langle K_i^1 : i \in \omega \rangle_U$ as in **(1)**.

$$\text{Let } A = \bigcup_{i \in S} K_i^0 \cup st(K^0), \quad B = \bigcup_{i \in S} K_i^0 \cup st(K^1).$$

Then by (*), A and B are closed, and by (**) and the fact that $st(K^1) \cap st(K^0) = \emptyset$, A and B are disjoint, so there is a continuous $f : X \to \mathbb{R}$ mapping A to 0 and B to 1. Since $*A$ and $*B$ contain K^0 and K^1 respectively, f separates K^0 and K^1. $\square$

We now get a similar representation for S-continuous functions on a $*$-compact set:

For f S-continuous with domain K, let $st(f)$ be the (standard) function on $st(K)$ defined by:

$$st(f)(^\circ x) = {}^\circ f(x)$$

Lemma 10. *If X and Y are complete metric spaces, Y is separable and f is $*$-continuous and S-continuous from a $*$-compact subset of $*X$ to $*Y$, then f can be represented in the form $\langle f_i : i \in \omega \rangle_U$ with f_i satisfying:*

If $T \in [\omega]^\omega$, $x_i \in \text{domain}(f_i)$ for $i \in T$, and $x_i \to x$ in X, then $f_i(x_i) \to st(f)(x)$ in Y.

The representation theorem above can be used to give theorems concerning the separation of points by standard continuous functions.

Definition. *For $A \subset {}^*X$ we say A is **completely nonstandard** if ${}^*st(A) \cap A = \emptyset$ (the standard part map is w.r.t the topology on X).*

We can use Lemma 10 to derive the following version of the Tietze Extension Theorem:

Theorem 11. *Let X be a complete metric space and let D be a $*$-compact completely nonstandard subset of ${}^*X$. Let f be any $*$-continuous function from D into ${}^*\mathbb{R}$ that is also S-continuous. Then f extends to a standard continuous function: that is, there is an $F : X \to {}^*\mathbb{R}$ such that ${}^*F$ extends f.*

Note: A special case of the theorem is when D is a $*$-compact set containing no nearstandard elements: in this case S-continuity is satisfied vacuously, so any $*$-continuous function can be extended. The representation theorem can also be used to show

Theorem 12. *If X is a complete metric space, Y is a separable complete metric space, and f is a $*$-continuous, S-continuous function from a $*$-compact, completely nonstandard subset of X into Y, then there is a closed set $A \subset X$ and continuous function F on A such that ${}^*F$ extends f.*

If U is Σ_2^1-**selective** — that is, an ultrafilter satisfying Mathias's Theorem for all Σ_2^1 subsets of $[\omega]^\omega$ — then the hypothesis that Y is separable can be removed in Lemma 10 and Theorem 12. In [17] it is shown that the existence of measurable cardinals implies that any selective ultrafilter is Σ_2^1-selective.

Corollary 13. *If X is complete metric, $x \in {}^*X \setminus X$ and $y \in {}^*\mathbb{R} \setminus \mathbb{R}$, and either both or none of x or y are nearstandard, then there is a standard continuous f such that $f(x) = y$.*

Proof: In either case, the function on $\{x\}$ sending x to y is S-continuous and $*$-continuous on the completely nonstandard set $\{x\}$, and applying Theorem 11 to this function gives the desired f. □

Corollary 14. *If X is a complete metric space, the S-topology on ${}^*X$ separates points.*

Proof: If $x \neq y$ and x is standard, then ${}^*X \setminus \{x\}$ is a standard open set separating y from x. If both x and y are nonstandard, let f be the function on $\{x, y\}$ that sends x to 0 and y to some nonzero infinitesimal. Then f is S-continuous, and $\{x, y\}$ is completely nonstandard, so f extends to a standard continuous ${}^*F$ on ${}^*X$. the set ${}^*F^{-1}({}^*\mathbb{R} \setminus \{0\})$ is a standard open set separating y from x. □

4. Measure Theory in Selective Universes

One of the more striking properties of selective universes is a measure-theoretic version of the density theorem, Theorem 4, due to Henson and Wattenberg. We will state here a generalization of the Henson-Wattenberg Theorem, and give some consequences, the chief of which is a process for constructing extensions of measures. Extension measures were introduced in [2] and [9], and applications are given in both places. Here, we will show how extension measures are defined, and indicate how they inherit many of the important properties of Loeb measures.

Let $\langle X, \Sigma, \mu \rangle$ be a probability space. A family of sets F is said to be a **compact class** if every subfamily of F having the finite intersection property has nonempty intersection. A σ-algebra $\Gamma \subset \Sigma$ is said to be **compact** if there is a compact class F such that every set in Γ can be approximated from within (in measure) by sets in F. $\langle X, \Sigma, \mu \rangle$ is *perfect* if for every countably-generated sub-σ-algebra Γ of Σ, Γ is compact. Perfect measures are discussed in [21]. The class of perfect measures includes all Radon measures, and hence all Borel measures on a compact metric space.

The following result is due to Henson and Wattenberg. It can be proved by a U-tree argument similar to that used in Theorem 4.

Theorem 15. [11] *Let $\langle X, \Sigma, \mu \rangle$ be a perfect probability measure, and let our nonstandard universe be formed via a selective ultrafilter. If S is any $L\mu$-measurable set with positive Loeb measure, then $\sigma(S)$ is nonempty.*

Using results of Kunen (see notes in [17]), we have that Theorem 15 applies to all Borel measures on complete metric spaces as well.

Corollary 16. *For μ, S as in Theorem 15, we have*
$$\mu_{inner}(\sigma(S)) \leq L\mu(S) \leq \mu_{outer}(\sigma(S)).$$
In particular, $L\mu(S) = \mu(\sigma(S))$ whenever $\sigma(S)$ is μ-measurable.

Henson and Wattenberg also proved the following 'converse' to Theorem 15.

Proposition 5. [11] *Let the nonstandard universe be formed using a semi-selective ultrafilter, and let μ be an arbitrary complete measure in V, and S be an internal $L\mu$-measurable set with $L\mu(S) = 0$. Then $\sigma(S)$ is μ-measurable, and $\mu(\sigma(S)) = 0$.*

There is an extension of Proposition 5 to external sets, along the lines of Theorem 6.

Let $St(\Sigma) = \{{}^*S : S \in \Sigma\}$ and let $\mathcal{S}_{st}$ = the σ-algebra generated by $St(\Sigma)$. The function ${}^{\circ*}\mu$, a finitely-additive measure on $St(\Sigma)$, extends via the Loeb-Caratheodory extension procedure [16] to a σ-additive measure S on $\mathcal{S}_{st}$ that is inner- and outer-regular with respect to $St(\Sigma)$. Let S_μ denote the completion of this measure (the completion via S-nullsets, not $L\mu$ nullsets).

We now have the following result:

Theorem 17. *Suppose $\langle X, \Sigma, \mu \rangle$ is perfect, and A is an $L\mu$-measurable set that is Borel over the internal sets. Then*

(1) *If A has positive S_μ-outer measure, then $\sigma(A) \neq \emptyset$.*
(2) *$\sigma(A)$ is measurable $\Leftrightarrow A$ is S_μ measurable.*
(3) *If $\sigma(A)$ is measurable then $\mu(\sigma(A)) = S_\mu(A)$.*

We now apply the results above to the extension of perfect measures. Fix a perfect measure space $\langle X, \Sigma, \mu \rangle$ and, for the rest of this section, assume that the nonstandard universe is formed via a selective ultrafilter U. Then we define the **extension measure of μ formed via U**, $\hat{\mu}_U$ (or $\hat{\mu}$ when U is understood) by
$$domain(\hat{\mu}) = \{\sigma(A) : A \in {}^*V \text{ and } A \text{ is } L\mu\text{-measurable}\} \ ; \ \hat{\mu}(\sigma(A)) = L\mu(A)$$

Proposition 6. $\hat{\mu}$ *is a complete probability measure which extends μ. If μ is non-atomic, then $\hat{\mu}$ is a strict extension of μ (i.e. $domain(\hat{\mu})$ strictly contains Σ).*

The measure $\hat{\mu}$ inherits many of the properties of the Loeb measure $L\mu$, but lives on the original sample space X. In particular, for every $L\mu$ measurable function F from ${}^*X$ into some Polish space M, we get a $\hat{\mu}$ measurable function $\hat{F}$ by restricting F to X. The correspondence $Hat : F \mapsto \hat{F}$ is a bijection from (a.e. equivalence classes of) $L\mu$-measurable M-valued functions to $\hat{\mu}$-measurable functions. Furthermore, $\hat{F}$ has the same distribution as F, and in the case $M = \mathbb{R}$, the map Hat preserves integrals. Similarly, given an adapted Loeb space S of the form $\langle {}^*X, L\Sigma, L\mu, \{F_t\}_{t\in[0,1]}\rangle$, we can get an adapted space $\hat{S}$ on X by pushing down $L\mu$ onto the appropriate filtration on X. The resulting space has the same adapted distribution as the Loeb space S, and will be saturated whenever S is (see [HK] for definitions of adapted distribution and saturation of an adapted space).

We close this section with the following result on universal limit functions:

Theorem 18. *Let Σ be a σ-algebra, and M_Σ be the class of all perfect probabilities on Σ. For $\mu \in M_\Sigma$, let $\bar{\mu}$ denote the completion of μ. If $\langle f_i \rangle_{i\in\omega}$ is a sequence of Σ-measurable functions, let $f_\infty = Hat({}^\circ f_K)$, for some $K \in {}^*\mathbb{N} \setminus \mathbb{N}$.*
(1) *f_∞ is $\hat{\mu}$-measurable for every $\mu \in M_\Sigma$.*
(2) *$\forall \mu \in M_\Sigma$ such that $\langle f_i \rangle$ converges in μ-measure, f_∞ is in fact $\bar{\mu}$-measurable, and $\langle f_i \rangle$ converges in $\bar{\mu}$-measure to f_∞.*

This follows directly from Proposition 5 and the usual nonstandard characterization of convergence in measure. This method of getting functions that are 'universal limits' is due to Mokobodski [8], and requires only that the nonstandard universe be formed via a rapid ultrafilter. The extension measure process defined above requires only that the ultrafilter satisfy Property M, discussed in [3].

References

[1] Albeverio, S., Fenstad, J., Haøegh-Krohn, R., and Lindstrøm, T. *Nonstandard Methods in Stochastic Analysis and Mathematical Physics*, Academic Press (1986).
[2] Benedikt, M. *Nonstandard Analysis and Special Ultrafilters*, Ph.D. Thesis, University of Wisconsin, 1993
[3] Benedikt, M. *Property M Ultrafilters*, To appear.
[4] Blass, A. *Selective Ultrafilters and Homogeneity*, Annals of Pure and Applied Logic (1989), pp. 252-285.
[5] Booth, D. *Ultrafilters on a countable set*, Ann. Math. Logic 2 (1970), 1-24.
[6] Burzyk, J. *An example of a Non-complete Normed N-space*, Bulletin of the Polish Academy of Sciences Mathematics 35 (1987), pp. 449-454.

[7] Chang, C.C. and Keisler, H.J. *Model Theory*, Third Edition, North Holland, Amsterdam (1991).
[8] Dellacherie, C. and Meyer, P. *Probabilities and Potential*, North-Holland, Amsterdam (1978).
[9] Fremlin, D.H. unpublished manuscript, (1987).
[10] Henson, C.W. *The isomorphism property in nonstandard analysis and its use in the theory of Banach Spaces*, Journal of Symbolic Logic 39 (1974), pp. 717-731.
[11] Henson, C.W. and Wattenberg, B. *Egoroff's Theorem and the Distribution of Standard Points in a Nonstandard Model*, Proceedings of the AMS 81 (1981), pp. 455-461.
[12] Hoover, D.N. and Keisler, H.J. *Adapted Probability Distributions*, Transactions of the AMS 286 (1984), pp. 159-201.
[13] Jin, R. *The isomorphism property versus the special model axiom*, Journal of Symbolic Logic 57 (1992), pp. 975-987.
[14] Keisler, H.J. *An infinitesimal approach to stochastic analysis*, Memoirs of the AMS, 297 (1984).
[15] Kunen, K. *Some points in $\beta(N)$*, Math. Proc. Camb.
[16] Loeb, P.A. *Conversion from Nonstandard to Standard Measure Spaces and Applications in Probability Theory*, Transactions of the AMS 81 (1975), pp. 113-122.
[17] Mathias, A.R.D. *Happy Families*, Ann. Math. Logic 12 (1977), pp. 59-111.
[18] Miller, A.W. *There are no Q-points in Laver's Model for the Borel Conjecture*, Proceedings of the AMS 78 (1980), pp. 103-106.
[19] Puritz, C. *Ultrafilters and standard functions in Nonstandard Arithmetic*, Proceedings of the London Mathematical Society 22 (1971), pp. 705-733.
[20] Ross, D. *The special model axiom in nonstandard analysis*, Journal of Symbolic Logic 55 (1990), pp. 1233-1242.
[21] Sazonov, V.V. *On perfect measures*, Izv Akad. Nauk SSSR Ser. Mat. 26 (1962), pp. 391-414, Translations of the AMS, vol. 48 ser 2, pp. 229-254.

AT&T Bell Laboratories

E-mail benedikt@research.att.com

J. M. ALDAZ
Lattices and monads

1. Introduction

The idea of applying the Loeb measure construction to abstract measure spaces has served as motivation to define monads (and thus, the standard part map) in terms of arbitrary lattices of sets, rather than topologies. Since lattices are closed under finite unions and finite intersections, saturation arguments will still apply. Furthermore, by considering the topology generated by any such lattice, the usual results in the topological case also hold. The basic difference between using an arbitrary lattice on which a set function (or a measure) is defined, and a topology, is that in the first case one considers which properties the set function must satisfy in order to obtain representation results, while in the second case one considers topological properties such as metrizability, regularity, and the like. These type of questions have been examined by several authors, in papers such as, for instance, [Ro], [A2], and [R].

Next we introduce some terminology. A paving V on X is just a collection of subsets of the set X. We assume that the paving V is a lattice, i.e., it is closed under finite unions and finite intersections, and also that V contains X and $\emptyset$; in this case we say that V is a $[\emptyset, X, \cup f, \cap f]$-paving. The symbol $T(V)$ denotes the topology generated by V. Terms referring to topological notions, such as compactness or tightness, are understood in relation to $T(V)$. Thus, when we say that a set is compact, we mean compact with respect to $T(V)$. Given a measure space $(X, \mathcal{A}, \mu)$, where $\mathcal{A}$ is an algebra and μ a finite, finitely additive measure on $\mathcal{A}$, we always assume that $V \subset \mathcal{A}$.

If V does not separate points, then the standard part map is defined as a set valued map, following an idea of W. A. J. Luxemburg (see [Lux], page 61). Since in an algebra there may be several lattices which could be used to define the monads, we make this dependency explicit as follows. For each x in $^*X$, the monad of x with respect to V, denoted by $m(V, x)$, is the set $m(V, x) = \cap\{^*O : x \in {^*O}$ and $O \in V\}$. The set of near standard points of $^*X$ with respect to V, $ns(V, {^*X})$, is the union of all monads (with respect to V) of standard points in $^*X$; if y is near standard, $st(V, y)$ is the set of all standard points x such that $y \in m(V, x)$; and for each subset B of X, the standard part inverse with respect to V of the set B is $st^{-1}(V, B) := \cup\{m(V, x) : x \in B\}$.

As we mentioned above, by considering V as a base for the topology $T(V)$, it follows from the usual topological case that if O is open with respect to $T(V)$, then $st^{-1}(V, O) \subset {^*O}$; if C is closed, then $ns(V, {^*X}) \cap {^*C} \subset st^{-1}(V, C)$; and if K is compact, then $^*K \subset st^{-1}(V, K)$. One can easily check that the Hausdorff condition is not needed in the proofs of these facts.

We shall work with a κ-saturated nonstandard model, where κ is larger than the cardinality of the power set of any standard set being considered. All the measures or set functions under consideration are assumed to be bounded.

Given a real-valued set function μ defined on a lattice V, it is possible to define the class of "Loeb measurable sets" $A \subset {}^*X$ (see, for instance, [A1]) as those for which $\inf\{{}^{\circ*}\mu(O) : O \in {}^*V \text{ and } A \subset O\} = \sup\{{}^{\circ*}\mu(O) : O \in {}^*V \text{ and } O \subset A\}$. Then the "Loeb measure" of A, $L({}^*\mu)(A)$, is defined to be this common value.

Consider now the following example. Let $(X, \mathcal{A}_\backslash, \mathcal{P}_\backslash)_{\backslash \in \mathbb{N}}$ be the filtration obtained by choosing X to be the interval $[0,1)$, choosing $\mathcal{A}_\backslash$ to be the algebra generated by the intervals $[\frac{i}{2^n}, \frac{i+1}{2^n})$, $i = 0, 1, ..., 2^n - 1$, and choosing P_n to be the probability measure that assigns to each interval in $\mathcal{A}_\backslash$ its length. If for each $n \in \mathbb{N}$ we let $V_n = \mathcal{A}_\backslash$, then it is easy to see that $L({}^*P_n)(st^{-1}(V_n, \cdot)) = P_n(\cdot)$. Given a collection of measure spaces, one is frequently interested in finding another measure space which in some sense extends all the previous ones. This is often the case in the theory of stochastic processes, whence the use of infinite products and projective limits in Probability Theory. In the next section we examine a general version of the following method: Fix the space $({}^*X, \mathcal{A}_\mathcal{H}, \mathcal{P}_\mathcal{H})$ corresponding to some infinite index H in ${}^*[(X, \mathcal{A}_\backslash, \mathcal{P}_\backslash)_{\backslash \in \mathbb{N}}]$, and show that for all $n \in \mathbb{N}$, $L(P_H)(st^{-1}(V_n, \cdot)) = P_n(\cdot)$. We shall define the projective hull to be X_H modulo a suitable equivalence relation, and give an extension result along the preceding lines. We will also show that for a projective system of compact Hausdorff spaces, the projective hull is homeomorphic to the projective limit of the system.

In Section 3 we give a nonstandard treatment to the problem of extending a tight content on the compact sets of a topological space to a Borel measure.

2. The projective hull

Whenever

(i) D is a directed set,
(ii) for each $\alpha \in D$, $(X, \mathcal{T}_\alpha)$ is a topological space, and
(iii) for all $\alpha, \beta, \gamma \in D$ with $\alpha \leq \beta \leq \gamma$ there exists continuous, onto maps $f_{\alpha\beta} : X_\beta \to X_\alpha, f_{\beta\gamma} : X_\gamma \to X_\beta$ and $f_{\alpha\gamma} : X_\gamma \to X_\alpha$ such that $f_{\alpha\gamma} = f_{\alpha\beta} \circ f_{\beta\gamma}$, and for all $\eta \in D$, $f_{\eta\eta}$ is the identity on X_η,

$(X_\alpha, \mathcal{T}_\alpha, \{_{\alpha\beta}\})_{\substack{\alpha,\beta \in D \\ \alpha \leq \beta}}$ is said to be a *projective system* (or *inverse system*) of topological spaces. The *projective limit* of such a system is the topological space $(X, \mathcal{T})$, where X consists of all elements (x) of the cartesian product $\prod_\alpha X_\alpha$ satisfying $f_{\alpha\beta}(x_\beta) = x_\alpha$ for all $\alpha, \beta \in D$ with $\alpha \leq \beta$, and the topology $\mathcal{T}$ is the weakest topology on X making all the projections $f_\alpha : X \to X_\alpha$ continuous ($f_\alpha((x)) = x_\alpha$). It is well known that X is a closed subspace of $\to \prod_\alpha X_\alpha$.

A projective system of measure spaces $(X_\alpha, \mathcal{A}_\alpha, \{_{\alpha\beta}, P_\alpha\})_{\substack{\alpha,\beta \in D \\ \alpha \leq \beta}}$ is defined analogously, with the difference that the functions $f_{\alpha\beta}$, instead of being continuous and onto, are required to be measurable and measure preserving. Here $\mathcal{A}_\alpha$ is an algebra of subsets of X_α, and P_α a measure on $\mathcal{A}_\alpha$ (we will take P_α always to be a probability). The projective limit X of such a projective system is the same set as in the topological case. Projective systems of measure spaces were introduced by Bochner (cf. [Bo]), as a generalization of products of measure spaces.

Consider $^*[(X_\alpha, \mathcal{A}_\alpha, \{_{\alpha\beta}, \mathcal{P}_\alpha)_{\alpha,\beta \in \mathcal{D}}]$ and choose $H \in {}^*D$ such that $\alpha \leq H$ for all $\alpha \in D$. The space X_H has distinct advantages over the projective limit. It is well known that the projective limit X of a projective system $(X_\alpha, \mathcal{T}_\alpha, \{_{\alpha\beta}\}_{\alpha,\beta \in \mathcal{D}})$ of topological spaces may be empty even if $X_\alpha \neq \emptyset$ for each $\alpha \in D$ (cf. [Bou], p. 77). In the measure case, Bochner had to introduce the "sequential maximality condition" to prevent $X = \emptyset$ from happening. However, if $X_\alpha \neq \emptyset$ for each $\alpha \in D$, then $X_H \neq \emptyset$, by κ-saturation. In addition, extension theorems are easy to prove for X_H, so working with the Loeb space of X_H may be a useful alternative to the projective limit.

In the context of Loeb measures, the first application of this type was given by P. A. Loeb (see [L]), as a model for coin tossing. Here $X_n = \{0,1\}^n$; for $m \leq n$, $f_{mn} : X_n \to X_m$ is the projection function; and P_n, the uniform probability on X_n. In [Li], T. Lindstrom used X_H to prove the well known Prohorov's Theorem (apparently due to Kisynski, cf. [K]), which characterizes when there is a Radon measure on the projective limit X, extending the cylindrical measures. Lindstrom "pushed down" the measure $L(P_H)$ to X by utilizing the internal projection $f_H : {}^*X \to X_H$ and the standard part map $st : {}^*X \to X$. Here we are interested in the space X_H by itself, independently of whether the projective limit exists or not (i.e., whether or not it is empty). Next we present an example.

2.1. Example Given a projective system of measure spaces $(X_\alpha, \mathcal{A}_\alpha, \{_{\alpha\beta}, \mathcal{P}_\alpha)_{\alpha,\beta \in \mathcal{D}}}$, typical extension theorems tell us when there is a countably additive measure P on some σ-algebra of subsets of the projective limit X, such that each projection map f_α is measurable and $P_\alpha = P f_\alpha^{-1}$ for all $\alpha \in D$. The following example (see [S], pages 163–164) shows that such P may fail to exist even if X is nonempty. For all $n \in \mathbb{N}$, let $X_n = (0,1]$ and for $m \leq n$, let f_{mn} be the identity map, so we can identify the projective limit X with $(0,1]$. The algebra $\mathcal{A}_\setminus$ on X_n is generated by the sets $(0, 1/n], (1/n, 1/(n-1)], \ldots, (1/2, 1]$, and the measure P_n, defined on $\mathcal{A}_\setminus$, is given by $P_n(A) = 1$ if $(0, 1/n] \subset A$, and $P_n(A) = 0$ otherwise. No countably additive measure P that extends all the P_n's can exist on $(0,1]$, for then we would have $0 = P(\cap_n (0, 1/n]) = \lim_n P((0, 1/n]) = 1$. Let V_n be $\mathcal{A}_\setminus$; because each $\mathcal{A}_\setminus$ is finite, each P_n is automatically inner regular with respect to the compact sets (in the $T(V_n)$ topology, since $T(V_n) = V_n = \mathcal{A}_\setminus$). Also for $A \in \mathcal{A}_\setminus$ we have ${}^*A = st_{X_n}^{-1}(V_n, A)$. Hence, if $A \in \mathcal{A}_\setminus$ then $f_{nH}^{-1}\left(st_{X_n}^{-1}(V_n, A)\right)$ is $L(P_H)$ measurable (recall that in this case $D = {}^*\mathbb{N}$, so H is an infinite natural number).

On X_H, P_H is the internal probability measure which assigns mass 1 to the interval $(0, 1/H]$. Furthermore, since $f_{nH} : X_H \to {}^*X_n$ is the identity if follows that

$$L(P_H)\left[f_{nH}^{-1}(st_{X_n}^{-1}(V_n, A))\right]$$

equals one if $A \in \mathcal{A}_\alpha$ contains $(0, 1/n]$, and zero otherwise. Thus

$$L(P_H)\left[f_{nH}^{-1}\left(st_{X_n}^{-1}(V_n, \cdot)\right)\right] = P_n.$$

A measure is *tight* if it is inner regular with respect to the compact sets. We show next that the preceding result generalizes as follows: If all the measures P_α belonging to a projective system of measure spaces are tight then $L(P_H)$ extends them, in the sense that

$$L(P_H)\left[f_{\alpha H}^{-1}\left(st_{X_\alpha}^{-1}(V_\alpha,\cdot)\right)\right] = P_\alpha.$$

Let $\mathcal{A}$ be an algebra of subsets of X. We define an equivalence relation $\sim$ on X by setting $x \sim y$ if for every $A \in \mathcal{A}$, either both x and y belong to A, or both belong to its complement. To denote the atom to which x belongs, we use the symbol $[x]$. Let $V \subset \mathcal{A}$ be a lattice. We say that V *separates* $(X, \mathcal{A})$ if given two different atoms of X, $[x]$ and $[y]$, there are disjoint sets A, B in V such that $[x] \in A$ and $[y] \in B$. We will always assume that $V_\alpha \subset \mathcal{A}_\alpha$ is a $[\emptyset, X, \cup f, \cap f]$-paving which separates $(X_\alpha, \mathcal{A}_\alpha)$.

We denote by $\mathcal{B}$ the collection of sets

$$\mathcal{B} := \bigcup_{\alpha \in D}\{\{_{\alpha H}^{-\infty}(\int \sqcup_{X_\alpha}^{-\infty}(V_\alpha, \mathcal{A}_\alpha)) : \mathcal{A}_\alpha \in V_\alpha\},$$

and by $\sigma(\mathcal{B})$, the σ-algebra it generates.

2.2. Theorem

Let $(X_\alpha, \mathcal{A}_\alpha, \{_{\alpha\beta}, P_\alpha\})_{\alpha,\beta \in D, \alpha \leq \beta}$ be a projective system of probability measure spaces, and assume that $V_\alpha \subset \mathcal{A}_\alpha$ separates $(X_\alpha, \mathcal{A}_\alpha)$. If each P_α is tight with respect to the topology generated by V_α, then $\sigma(\mathcal{B}) \subset \mathcal{L}(\mathcal{A}_H)$, and for every subset A of $\mathcal{A}_\alpha$,

$$L(P_H)[f_{\alpha H}^{-1}(st_{X_\alpha}^{-1}(V_\alpha, A))] = P_\alpha(A).$$

Proof It is enough to show that every $B \in \mathcal{B}$, is $L(P_H)$-measurable, and has the "correct" measure. Now $B = f_{\alpha H}^{-1}\left(st_{X_\alpha}^{-1}(V_\alpha, A_\alpha)\right)$ for some $A_\alpha \in \mathcal{A}_\alpha$. Fix a standard $\epsilon > 0$. By tightness, there exist sets $I_1, I_2 \in {}^*\mathcal{A}_\alpha$ such that $I_1 \subset st_{X_\alpha}^{-1}(V_\alpha, A_\alpha) \subset I_2$ and $P_\alpha(A_\alpha) - \epsilon < {}^{o*}P_\alpha(I_1) \leq P_\alpha(A_\alpha) \leq {}^{o*}P_\alpha(I_2) < P_\alpha(A_\alpha) + \epsilon$. Since $f_{\alpha H}$ is measurable and measure preserving, $P_\alpha(A_\alpha) - \epsilon < {}^\circ P_H\left(f_{\alpha H}^{-1}(I_1)\right) \leq L(P_H)\left(f_{\alpha H}^{-1}(st_{X_\alpha}^{-1}(V_\alpha, A_\alpha))\right) \leq {}^\circ P_H\left(f_{\alpha H}^{-1}(I_2)\right) < P_\alpha(A_\alpha) + \epsilon$. Thus B is $L(P_H)$-measurable and $L(P_H)(B) = P_\alpha(A_\alpha)$. □

Let $(X_H, L(\mathcal{A}_H), \mathcal{L}(\mathcal{P}_H))$ be the Loeb space of $(X_H, \mathcal{A}_H, \mathcal{P}_H)$; since X_H is not in general the star of a standard space, we cannot speak of near standard points. We will consider a different notion, the set of points $nsbm(X_H)$ that are nearstandard under some bonding map $f_{\alpha H}$. More precisely,

$$nsbm(X_H) := \bigcup_{\alpha \in D} f_{\alpha H}^{-1}(ns(V_\alpha, {}^*X_\alpha)).$$

The *remote* points of X_H are those belonging to $r(X_H) := X_H \setminus nsbm(X_H)$. Let S be the collection of sets

$$S := \{r(X_H)\} \cup \left[\bigcup_{\alpha \in D}\{f_{\alpha H}^{-1}(st_{X_\alpha}^{-1}(V_\alpha, O_\alpha)) : O_\alpha \in V_\alpha\}\right],$$

and define an equivalence relation on X_H:

$$x \sim_S y \text{ if for all sets } A \in S, x \in A \text{ implies } y \in A.$$

We will call the space $\widehat{X} := X_H / \sim_S$ the *projective hull* of the projective system of measure spaces $(X_\alpha, \mathcal{A}_\alpha, \{_{\alpha\beta}, \mathcal{P}_\alpha)_{\alpha,\beta \in D \atop \alpha \leq \beta}$. In the topological case the definition of S (and hence of $\widehat{X}$) is the same, but with $O_\alpha \in \mathcal{T}_\alpha$ instead of $O_\alpha \in V_\alpha$. Let $q : X_H \to \widehat{X}$ be the canonical surjection. If in addition $(X_\alpha, \mathcal{T}_\alpha, \{_{\alpha\beta}\})_{\alpha,\beta \in D \atop \alpha \leq \beta}$ is a projective system of compact Hausdorff spaces, then we give to $\widehat{X}$ the weakest topology that makes all the maps $st_{X_\alpha}(f_{\alpha H}(q^{-1}(\cdot)))$ continuous.

Next we show that $\sim_S$ is indeed an equivalence relation, and then that the projective hull is homeomorphic to the projective limit of a projective system of compact Hausdorff spaces. The following Lemma is a straightforward generalization of the corresponding result for Hausdorff spaces, so the proof is omitted.

2.3. Lemma
Let $(X, \mathcal{A})$ be separated by V. If $x \in ns(V, {}^*X)$ then $st(V, x)$ is an atom of $(X, \mathcal{A})$. Furthermore, if $x, y \in {}^*X$ and $x \notin m(V, y)$, then $m(V, x) \cap m(V, y) = \emptyset$.

2.4. Proposition
$\sim_S$ is an equivalence relation on X_H.

Proof Reflexivity and transitivity are clearly satisfied, so only symmetry needs to checked. Let $x \sim_S y$, and choose $O \in S$ with $y \in O$. We show that $x \in O$. If $O = r(X_H)$, the result is clear, so assume $O \neq r(X_H)$. By the definition of S, there exist an $\alpha \in D$ and a set $O_\alpha \in V_\alpha$ such that $O = f_{\alpha H}^{-1}\left(st_{X_\alpha}^{-1}(V_\alpha, O_\alpha)\right)$. Suppose $x \notin O$; then $f_{\alpha H}(x) \in \neg st_{X_\alpha}^{-1}(V_\alpha, O_\alpha)$, whence $f_{\alpha H}(x) \notin m(f_{\alpha H}(y))$. Since $(X_\alpha, \mathcal{A}_\alpha)$ is separated by V_α, the points $f_{\alpha H}(x)$ and $f_{\alpha H}(y)$ have disjoint monads, so by κ-saturation there are sets O_1 and O_2 in V_α, with $f_{\alpha H}(x) \in st_{X_\alpha}^{-1}(V_\alpha, O_1) \subset {}^*O_1$, $f_{\alpha H}(y) \in st_{X_\alpha}^{-1}(V_\alpha, O_2) \subset {}^*O_2$, and ${}^*O_1 \cap {}^*O_2 = \emptyset$. Therefore $x \in f_{\alpha H}^{-1}(st_{X_\alpha}^{-1}(V_\alpha, O_1))$, while $y \notin f_{\alpha H}^{-1}(st_{X_\alpha}^{-1}(V_\alpha, O_1))$, contradicting $x \sim_S y$. □

In the coin tossing example mentioned above ($X_n = \{0,1\}^n$; $f_{mn} : X_n \to X_m$ is the projection and P_n the uniform probability on X_n), it is clear, since $X_n = {}^*X_n$, that $X_H / \sim_S$ is homeomorphic to the product (which in this case coincides with the projective limit) $X = \{0,1\}^{\mathbb{N}}$. The following Theorem tells us that the same is true for any projective system of compact Hausdorff spaces. Note that under the hypotheses given below, the set X_H has no remote points.

2.5. Theorem
Let $(X_\alpha, \mathcal{T}_\alpha, f_{\alpha\beta})_{\alpha,\beta \in D \atop \alpha \leq \beta}$ be a projective system of topological spaces, where each X_α is compact and Hausdorff and each $f_{\alpha\beta}$ continuous and onto. If V_α is any $[\emptyset, X_\alpha, \cap f, \cup f]$-base for the topology of X_α and

$$S = \bigcup_{\alpha \in D} \left\{ f_{\alpha H}^{-1}\left(st_{X_\alpha}^{-1}(O_\alpha)\right) : O_\alpha \in V_\alpha \right\},$$

then the projective limit X of $(X_\alpha, \mathcal{T}_\alpha, \{_{\alpha\beta}\})_{\alpha,\beta \in \mathcal{D} \atop \alpha \le \beta}$ is homeomorphic to $X_H/\sim_S = \hat{X}$, the projective hull.

Proof Given $(x) \in X$, set $A_{(x)} = \{z \in X_H : \text{for all } \alpha \in D, st_{X_\alpha}(f_{\alpha H}(z)) = f_\alpha((x))\}$. First we show $A_{(x)}$ is an atom of the partition of X_H determined by the equivalence relation $\sim_S$; select $w \in A_{(x)}$ and $y \in [w]_S$. If we prove that $st_{X_\alpha}(f_{\alpha H}(y)) = st_{X_\alpha}(f_{\alpha H}(w))$ for all $\alpha \in D$, then $[w]_S \subset A_{(x)}$ follows. Suppose there is an α such that $st_{X_\alpha}(f_{\alpha H}(y)) \ne st_{X_\alpha}(f_{\alpha H}(w))$; we can find disjoint V_α sets O_1 and O_2 with $m(f_{\alpha H}(y)) \subset st_{X_\alpha}^{-1}(O_1)$ and $m(f_{\alpha H}(w)) \subset st_{X_\alpha}^{-1}(O_2)$. But then

$$y \in f_{\alpha H}^{-1}\left(st_{X_\alpha}^{-1}(O_1)\right),$$

$$w \in f_{\alpha H}^{-1}\left(st_{X_\alpha}^{-1}(O_2)\right) \text{ and}$$

$$f_{\alpha H}^{-1}\left(st_{X_\alpha}^{-1}(O_1)\right) \cap f_{\alpha H}^{-1}\left(st_{X_\alpha}^{-1}(O_2)\right) = \emptyset.$$

This contradicts the fact that $y \in [w]_S$. Therefore $[w]_S \subset A_{(x)}$. Next we show that $A_{(x)} \subset [w]_S$. Select $f_{\alpha H}^{-1}\left(st_{X_\alpha}^{-1}(O)\right) \in S$ with $w \in f_{\alpha H}^{-1}\left(st_{X_\alpha}^{-1}(O)\right)$, and choose $z \in A_{(x)}$. Since $st_{X_\alpha}(f_{\alpha H}(z)) = st_{X_\alpha}(f_{\alpha H}(w)) \in O$, it follows that $z \in f_{\alpha H}^{-1}\left(st_{X_\alpha}^{-1}(O)\right)$ and hence $z \in [w]_S$.

Let $q : X_H \to \hat{X}$ be the canonical surjection; define $\varphi : X \to \hat{X}$ by setting $\varphi((x)) = A_{(x)}/\sim_S$. By the preceding argument, the map $st_{X_\alpha}(f_{\alpha H}(q^{-1}(\cdot)))$ is well defined, and clearly the diagram

$$\begin{array}{c} X \xrightarrow{\varphi} \hat{X} \\ f_\alpha \searrow \swarrow st_{X_\alpha}(f_{\alpha H}(q^{-1}(\cdot))) \\ X_\alpha \end{array}$$

commutes. Give $\hat{X}$ the weakest topology that makes all the maps $st_{X_\alpha}(f_{\alpha H}(q^{-1}(\cdot)))$ continuous. The function φ is both injective and surjective. To prove the continuity of φ it is enough to show that for some subbase, the inverse image of every set in it is open. The topology for $\hat{X}$ is generated by the subbase

$$\left\{q\left(f_{\alpha H}^{-1}\left(st_{X_\alpha}^{-1}(O_\alpha)\right)\right) : \alpha \in D \text{ and } O_\alpha \in \mathcal{T}_\alpha\right\}.$$

Fix O_α. Since $f_\alpha^{-1}(O_\alpha)$ is open and equals $\varphi^{-1}(q(f_{\alpha H}^{-1}(st_{X_\alpha}^{-1}(O_\alpha))))$, φ is continuous. To show that φ^{-1} is also continuous we use the same argument: $\{f_\alpha^{-1}(O_\alpha) : \alpha \in D \text{ and } O_\alpha \in \mathcal{T}_\alpha\}$ is a subbase for the topology of X, and $\varphi(f_\alpha^{-1}(O_\alpha)) = q\left(f_{\alpha H}^{-1}\left(st_{X_\alpha}^{-1}(O_\alpha)\right)\right)$, which is open in $\hat{X}$. Therefore X and $\hat{X}$ are homeomorphic. □

2.6 Example In the preceding theorem the assumption that the spaces X_α are Hausdorff is essential. It is not enough that $(X_\alpha, \mathcal{A}_\alpha)$ be separated by V_α, as we show next: Let $\{\mathcal{A}_\alpha\}$ be the collection of all finite algebras of subsets of $[0,1]$, ordered by inclusion, let X_α be $[0,1]$, let $f_{\alpha\beta}$ be the identity function, and let $V_\alpha = \mathcal{T}(V_\alpha) = \mathcal{A}_\alpha$. Trivially, the space $(X_\alpha, T(V_\alpha))$ is compact and separated by V_α. Select a free ultrafilter U on $[0,1]$, and define P_α on $\mathcal{A}_\alpha$ by setting $P_\alpha(A) = 1$ if $A \in U$, $P_\alpha(A) = 0$ otherwise. By Theorem 2.2, $L(P_H)$ extends all the P_α's, so if $\hat{X}$ were homeomorphic to the projective limit $[0,1]$, we would be able to produce a countably additive $0-1$ measure on all the subsets of $[0,1]$ which vanishes on the singletons. But this is clearly impossible, by a Bolzano-Weierstrass type argument: either $[0, 1/2)$ or $[1/2, 1]$ has measure one. Without loss of generality we may assume that $[1/2, 1]$ has full measure. Then so does either $[1/2, 3/4)$ or $[3/4, 1]$. Iterate to produce a singleton with mass one.

3. Contents and measures

Nonstandard methods have often been used to obtain extension results from finitely additive measures to countably additive measures, and from the Baire to the Borel sets (see, for instance, [AFHL] or [LR]). By a content we mean a monotone, subadditive and additive function defined on a lattice of sets, which vanishes on the empty set. That is, if V is a lattice of subsets of X, containing X and $\emptyset$, a *content* λ defined on V satisfies the following conditions:

i) $\lambda(\emptyset) = 0$
ii) If $A \subset B$ then $\lambda(A) \leq \lambda(B)$
iii) $\lambda(A \cup B) \leq \lambda(A) + \lambda(B)$
iv) If $A \cap B = \emptyset$ then $\lambda(A \cup B) = \lambda(A) + \lambda(B)$.

A content λ defined on the paving $\mathcal{K}$ of compact sets of a topological space is said to be *tight* if for every pair of compact sets K_1 and K_2 with $K_1 \subset K_2$, $\lambda(K_2) - \lambda(K_1) = \sup\{\lambda(K) : K \in \mathcal{K} \text{ and } K \subset K_2 \setminus K_1\}$.

The problem of extending contents to measures by nonstandard means was dealt with in [A1] and also in [R]. Here we present a nonstandard proof of Kisynski's extension Theorem (see [K]). We shall assume that X is Hausdorff and λ is bounded. While Kisynski did not need these assumptions, they are usually satisfied in applications (for instance in the proof of Prohorov's Theorem). Furthermore, it is easy to see how to extend the proof to the abstract case, where the content is defined on an arbitrary compact family of sets (along the lines of [Ro] or [A2]), rather than on the compact sets. We shall not pursue these elaborations here (for a standard proof, see [T]).

Let λ be a bounded content defined on the lattice of compact sets of X. If X is not compact, define $\lambda(X) := \sup\{\lambda(K) : K \in \mathcal{K}\}$. Recall that a measure defined on the completion of the Borel sets is *Radon* if it is inner regular with respect to the compact sets (Radon is the analogue of tight for the completion of the Borel sets).

3.1 Theorem
Every bounded and tight content on the compact subsets of a topological Hausdorff space can be extended to a Radon measure.

Proof Let λ be a bounded and tight content on the compact subsets of the Hausdorff space X. First we define a content on the topology $\mathcal{T}$ of X by setting $\mu(O) = \sup\{\lambda(K) : K \subset O \text{ and } K \in \mathcal{K}\}$ for every open set O. Clearly μ is finite and monotone, and if D and G are disjoint open sets, then $\mu(D \cup G) \geq \mu(D) + \mu(G)$. Hence, in order to prove that μ is a content, it suffices to show that subadditivity holds. This is a direct consequence of tightness: Let A and B be open sets, not necessarily disjoint. Fix $\epsilon > 0$ and select a compact set $K \subset A \cup B$ with $\mu(A \cup B) < \lambda(K) + \epsilon$. Then $K \setminus A$ is compact and contained in B. Choose $K_1 \subset K \setminus (K \setminus A)$ so that $\lambda(K_1) + \epsilon > \lambda(K) - \lambda(K \setminus A)$. Then $K_1 \subset A$, and

$$\mu(A \cup B) < \lambda(K) + \epsilon \leq \lambda(K_1) + \lambda(K \setminus A) + 2\epsilon \leq \mu(A) + \mu(B) + 2\epsilon$$

proving subadditivity.

Next we show that for every Borel subset B of X and any standard $\epsilon > 0$, there exists internal open sets O_1 and O_2 with $O_1 \subset st_X^{-1}(B) \subset O_2$ and $^*\mu(O_2) - ^*\mu(O_1) < \epsilon$. Let K be a compact subset of X with $\lambda(X) - \lambda(K) < \epsilon/3$. The symbol $\mathcal{K}|\mathcal{K}$ stands for the lattice of compact subsets of K. Denote the restriction of λ to $\mathcal{K}|\mathcal{K}$ also by λ. Now $B \cap K$ is a Borel subset of K, so by Theorem 3.6 of [A1], there are sets $K_1, K_2 \in {}^*(\mathcal{K}|\mathcal{K})$ such that $K_1 \subset st_K^{-1}(B \cap K) \subset K_2$ and $^*\lambda(K_2) - ^*\lambda(K_1) < \epsilon/3$. Note that for every open subset O of X with compact complement, $\mu(O) = \lambda(X) - \lambda(\neg O)$. Since K_2 is internally compact, there exists, by *tightness, a *compact set $K_3 \subset {}^*K \setminus K_2$ with $^*\lambda(^*K) - ^*\lambda(K_2) < ^*\lambda(K_3) + \epsilon/3$. Hence

$$^*\mu(\neg K_3) = {}^*\lambda(^*X) - {}^*\lambda(K_3) \leq {}^*\lambda(^*X) - {}^*\lambda(^*K) + {}^*\lambda(K_2) + \epsilon/3 < {}^*\lambda(K_1) + \epsilon.$$

Let O_1 be an internal open set such that $K_1 \subset O_1 \subset st_X^{-1}(B)$ (such a set exists by Theorem 2.2, i of [A1]), and set $O_2 = \neg K_3$. Then

$$O_1 \subset st_X^{-1}(B) \subset K_2 \cup \neg^*K \subset O_2$$

and $^*\lambda(K_1) \leq {}^*\mu(O_1)$, whence $^*\mu(O_2) - {}^*\mu(O_1) < \epsilon$, as desired.

Recall how the set function $L(^*\mu)st_X^{-1}$ is defined: on the standard sets A such that $st_X^{-1}(A)$ can be approximated from within and from without by internal sets (or equivalently, by internal open sets, see Theorem 2.2, i of [A1]),

$$L(^*\mu)st_X^{-1}(A) := inf\{{}^{\circ *}\mu(O) : O \text{ is an internal open set with } st_X^{-1}(A) \subset O\}$$
$$= sup\{{}^{\circ *}\mu(O) : O \text{ is an internal open set with } O \subset st_X^{-1}(A)\}.$$

Then $L(^*\mu)st_X^{-1}$ is a Radon measure, by arguments essentially identical to the proof of Theorem 3.3 in [A1] (so we will not repeat them). Finally we show that $L(^*\mu)st_X^{-1}$ extends λ. Clearly, $\lambda(X) = L(^*\mu)st_X^{-1}(X)$. If C is compact, then $^*C \subset st_X^{-1}(C)$, so $\lambda(C) \leq L(^*\mu)st_X^{-1}(C)$. To see why $\lambda(C) \geq L(^*\mu)st_X^{-1}(C)$, note that by tightness, for any $\epsilon > 0$ there is a compact set D with $D \subset \neg C$ and $\lambda(X) - \lambda(C) < \lambda(D) + \epsilon$. Hence

$$\lambda(C) + \epsilon > \lambda(X) - \lambda(D) \geq L(^*\mu)st_X^{-1}(C) + L(^*\mu)st_X^{-1}(D) - \lambda(D) \geq L(^*\mu)st_X^{-1}(C).$$

Since ϵ is arbitrary, we get $\lambda(C) \geq L(^*\mu)st_X^{-1}(C)$, and we have proven that $L(^*\mu)st_X^{-1}$ extends λ. $\square$

References

[A1] J. M. Aldaz *A characterization of universal Loeb measurability for completely regular Hausdorff spaces*, Can. J. Math., 1992 **44**, no 4, 673–690

[A2]   —— *On compactness and Loeb measures*, to appear, Proc. Amer. Math. Soc.
[AFHL]   S. Albeverio, J. E. Fenstad, R. Hoeg-Krohn and T. Lindstrøm *Nonstandard Methods in Stochastic Analysis and Mathematical Physics*, Academic Press, New York (1986)
[Bo]   S. Bochner *Harmonic Analysis and the Theory of Probability* University of California Press, 1955
[Bou]   N. Bourbaki *Theorie des Ensembles* Hermann, Paris, 1963
[K]   J. Kisynski *On the generation of tight measures* Studia Math. 1968 **30**, 141–151
[L]   P. A. Loeb *Conversion from nonstandard to standard measure spaces and applications in probability theory*, Trans. Amer. Math. Soc. **211**, 113–122 1975.
[Li]   T. Lindstrom *A Loeb-measure approach to theorems by Prohorov, Sazonov and Gross* Trans. Amer. Math. Soc. 1982 **269**, no 2, 521–534
[LR]   D. Landers and L. Rogge *Universal Loeb-measurability of sets and of the standard part map with applications* Trans. Amer. Math. Soc. 1987 **304** no 1 229–243
[Lux]   W. A. J. Luxemburg *A general theory of monads* Applications of Model Theory to Algebra, Analysis and Probability ed. W. A. J. Luxemburg, Holt, Rinehart and Winston, 18–69, 1969
[R]   H. Render *Pushing down Loeb measures* Math. Scan. **72**, 61–84 1993
[Ro]   D. Ross *Compact measures have Loeb preimages* Proc. Amer. Math. Soc. 1992 **115**, no 2, 365–370
[S]   A. N. Shiryayev *Probability, Statistics, and Random Processes*, Springer-Verlag, Berlin and New York, 1984
[T]   F. Topsøe *Topology and measure* Lecture Notes in Math. 133, Springer-Verlag, 1970

Departamento de Matemáticas
Facultad de Ciencias
Universidad Autónoma de Madrid
28049 Madrid Spain

E-mail aldaz@ccuam3.sdi.uam.es

H. Jerome Keisler
A neometric survey

1. Introduction

Nonstandard analysis is often used to prove that certain objects exist, i.e., that certain sets are not empty. In the literature one can find many existence theorems whose only known proofs use nonstandard analysis; see, for example, [AFHL].

This article will survey a new method for existence proofs, based on the concept of a neometric space. We shall state definitions and results (usually without proofs) from several other papers, and try to explain how the ideas from these papers fit together as a whole. The purpose of the neometric method is twofold: first, to make the use of nonstandard analysis more accessible to mathematicians, and second, to gain a deeper understanding of why nonstandard analysis leads to new existence theorems. The neometric method is intended to be more than a proof technique— it has the potential to suggest new conjectures and new proofs in a wide variety of settings. However, it bypasses the notion of an internal set and the lifting and pushing down arguments which are the main feature of many nonstandard existence proofs.

The central notion is that of a neocompact family, which is a generalization of the classical family of compact sets. A neocompact family is a family of subsets of metric spaces with certain closure properties. In applications, nonstandard analysis is needed at only one point—to obtain neocompact families which are countably compact. From that point on, the method can be used without any knowledge of nonstandard analysis at all.

This program grew out of earlier work on adapted probability distributions ([K2], [HK]) and a first approach to neocompactness in the paper [K3]. Various aspects of our program will appear in the papers [CK], [FK1], [FK2], [FK3], [FK4], [K4], and [K5]. In this article we shall give an overview of the entire program. We shall explain how the method can be painlessly applied, and discuss the relationship of the method to nonstandard practice and to adapted probability distributions.

Let's take an informal look at a common way of solving existence problems in analysis (or in a metric space): We want to show that within a set C there exists an object x with a particular property $\phi(x)$, that is, $(\exists x \in C)\phi(x)$. If we cannot find a solution x directly, we may proceed to find "approximate" solutions; we construct an object which is close to C, but perhaps not in C, and which almost has property ϕ. What is usually done is the following: define a sequence $\langle (x_n) \rangle$ of approximations which get better and better as n increases; if we do things right the sequence has a limit and that limit is the desired x.

The hard part is to show that the limit exists. In the classical setting, the most common way to do so is to show that the sequence x_n is contained in a compact set, and to use the fact that every sequence in a compact set has a convergent subsequence.

A simple example of an existence proof by approximation is Peano's existence theorem for differential equations: One first constructs a sequence of natural approximations (i.e Euler polygons). Then, using Arzela's theorem, a consequence of compactness that guarantees that under certain conditions a sequence of functions converges, one shows that the limit exists and is precisely the solution wanted. Written in symbolic form, the theorem is a statement of the form

$$(\exists x \in C)(f(x) \in D).$$

The approximation procedure gives us the following property:

$$(\forall \varepsilon > 0)(\exists x \in C^\varepsilon)(f(x) \in D^\varepsilon).$$

Here C^ε is the set $\{x : \rho(x, C) \leq \varepsilon\}$ with ρ the metric on the space where C lives, and similarly for D^ε. Then, if we choose a sequence ε_n approaching 0, we obtain a sequence of approximations. The compactness argument (Arzela's theorem) gives the existence of the limit.

The centerpiece of our method is a result (called the Approximation Theorem) which intuitively says "it is enough to approximate", or "if you can find approximate solutions then you can conclude that an exact solution exists without going through the convergence argument." In the above notation, the theorem states that:

(1) $\qquad$ If $(\forall \varepsilon > 0)(\exists \in C^\varepsilon)(f(x) \in D^\varepsilon)$ then $(\exists x \in C)(f(x) \in D)$.

The reader should have no problem showing that condition (1) holds in the following case: C is a compact subset of a complete separable metric space M, D is a closed subset of another complete separable metric space N, and f is a continuous function from M into N.

The main point of the neometric method is that our Approximation Theorem goes beyond the familiar case of convergence in a compact set. First, we work in metric spaces that are not necessarily separable. Second, we identify new families $\mathcal{C}, \mathcal{D}$ and $\mathcal{F}$ of sets C, D and functions f, such that (1) holds. These are the families of neocompact sets, neoclosed sets, and neocontinuous functions. The family of neocompact sets is much larger than the family of compact sets, and provides a wide variety of new opportunities for proving existence theorems by approximation.

2. Neometric Families

In this section we summarize the central notion of a neometric family from the paper [FK1], and state the main approximation theorem.

We use script letters

$$\mathcal{M} = (M, \rho), \mathcal{N} = (N, \sigma)$$

for complete metric spaces which are not necessarily separable. Given two metric spaces $\mathcal{M}$ and $\mathcal{N}$, the **product metric** is the metric space $\mathcal{M} \times \mathcal{N} = (M \times N, \rho \times \sigma)$ where

$$(\rho \times \sigma)((x_1, x_2), (y_1, y_2)) = \max(\rho(x_1, y_1), \sigma(x_2, y_2)).$$

The first notion we need is that of a neocompact family.

Definition 2.1. *Let* **M** *be a collection of complete metric spaces which is closed under finite products, and for each* $M \in \mathbf{M}$ *let* $\mathcal{B}(M)$ *be a collection of subsets of* M, *which we call* **basic sets**. *By a* **neocompact family** *over* $(\mathbf{M}, \mathcal{B})$ *we mean a triple* $(\mathbf{M}, \mathcal{B}, \mathcal{C})$ *where for each* $M \in \mathbf{M}$, $\mathcal{C}(M)$ *is a collection of subsets of* M *with the following properties, where* $M, \mathcal{N}$ *vary over* **M**:

- (a): $\mathcal{B}(M) \subset \mathcal{C}(M)$;
- (b): $\mathcal{C}(M)$ *is closed under finite unions; that is, if* $A, B \in \mathcal{C}(M)$ *then* $A \cup B \in \mathcal{C}(M)$;
- (c): $\mathcal{C}(M)$ *is closed under finite and countable intersections;*
- (d): *If* $C \in \mathcal{C}(M)$ *and* $D \in \mathcal{C}(\mathcal{N})$ *then* $C \times D \in \mathcal{C}(M \times \mathcal{N})$;
- (e): *If* $C \in \mathcal{C}(M \times \mathcal{N})$, *then the set*

$$\{x : (\exists y \in \mathcal{N})(x, y) \in C\}$$

belongs to $\mathcal{C}(M)$, *and the analogous rule holds for each factor in a finite Cartesian product;*

- (f): *If* $C \in \mathcal{C}(M \times \mathcal{N})$, *and* D *is a nonempty set in* $\mathcal{B}(\mathcal{N})$, *then*

$$\{x : (\forall y \in D)(x, y) \in C\}$$

belongs to $\mathcal{C}(M)$, *and the analogous rule holds for each factor in a finite Cartesian product.*

The sets in $\mathcal{C}(M)$ are called **neocompact sets**. The neocompact family $(\mathbf{M}, \mathcal{B}, \mathcal{C})$ induces a family of metric spaces with extra structure, $(M, \mathcal{B}(M), \mathcal{C}(M))$, which we call **neometric spaces**. A neometric space thus consists of a complete metric space $M \in \mathbf{M}$ and two families $\mathcal{B}(M)$ and $\mathcal{C}(M)$ of subsets of M. The properties (a)–(f) not only give conditions on single neometric spaces, but also on finite Cartesian products of neometric spaces.

We call $(\mathbf{M}, \mathcal{B}, \mathcal{C})$ the **neocompact family generated by** $(\mathbf{M}, \mathcal{B})$ if $\mathcal{C}(M)$ is the collection of all sets obtained by finitely many applications of the rules (a)–(f).

The classical example of a neocompact family is the family generated by $(\mathbf{S}, \mathcal{B})$ where $\mathbf{S}$ is the collection of all complete metric spaces, and for each $M \in \mathbf{S}$, $\mathcal{B}(M)$ is equal to the set of all compact subsets of M. It is not hard to see that the family of compact sets is closed under all of the rules (a)–(f). Thus the collection of neocompact sets $\mathcal{C}(M)$ generated by $(\mathbf{S}, \mathcal{B})$ is just $\mathcal{B}(M)$ itself, i.e. every neocompact set is compact.

It is easy to produce neocompact families by first choosing the basic sets and then closing them under the rules (a)–(f). The interesting neocompact families have an extra feature expressed in the following property, which is a familiar property of the family of compact sets in a topological space and plays a key role in the new theory of neometric spaces (see [FK1]).

Definition 2.2. *We say that a neocompact family* $(\mathbf{M}, \mathcal{B}, \mathcal{C})$ *is* **countably compact** *if for each* $M \in \mathbf{M}$, *every decreasing chain* $C_0 \supset C_1 \supset \cdots$ *of nonempty sets in* $\mathcal{C}(M)$ *has a nonempty intersection* $\bigcap_n C_n$ *(which, of course, also belongs to* $\mathcal{C}(M)$).

The classical neocompact family $(\mathbf{S},\mathcal{B},\mathcal{C})$ of compact sets is clearly countably compact. The interesting question is whether there are other, nontrivial, neocompact families which have it. The only examples we know are built using nonstandard analysis! (See [FK2]).

We now introduce notions for neocompact families analogous to familiar notions for metric spaces, and then introduce the slightly stronger notion of a neometric family.

Definition 2.3.

(a) A set $C \subset \mathcal{M}$ is **neoclosed** in $\mathcal{M}$ if $C \cap D$ is neocompact in $\mathcal{M}$ for every neocompact set D in $\mathcal{M}$.

(b) Let $D \subset \mathcal{M}$. A function $f : D \to \mathcal{N}$ is **neocontinuous** from $\mathcal{M}$ to $\mathcal{N}$ if for every neocompact set $A \subset D$ in $\mathcal{M}$, the restriction $f|A = \{(x, f(x)) : x \in A\}$ of f to A is neocompact in $\mathcal{M} \times \mathcal{N}$.

(c) A set A is said to be **neoseparable** in $\mathcal{M}$ if A is the closure of the union of countably many basic subsets of $\mathcal{M}$.

Definition 2.4. We call a neocompact family $(\mathbf{M},\mathcal{B},\mathcal{C})$ a **neometric family** if the distance functions in $\mathbf{M}$ and the projection functions for finite Cartesian products in $\mathbf{M}$ are neocontinuous. That is, the metric space $\mathbb{R}$ of reals is contained in some member $\mathcal{R}$ of $\mathbf{M}$, and for each $\mathcal{M} \in \mathbf{M}$ the distance function ρ of $\mathcal{M}$ is neocontinuous from $\mathcal{M} \times \mathcal{M}$ into $\mathcal{R}$. Moreover, for each $\mathcal{M}, \mathcal{N} \in \mathbf{M}$, the projection functions from $\mathcal{M} \times \mathcal{N}$ to $\mathcal{M}$ and to $\mathcal{N}$ are neocontinuous.

In the classical family $(\mathbf{S},\mathcal{B},\mathcal{C})$ a set is neoclosed if and only if it is closed, and neoseparable if and only if it is closed and separable, and a function is neocontinuous if and only if it is continuous. Since the distance and projection functions on any metric space are continuous, $(\mathbf{S},\mathcal{B},\mathcal{C})$ is a neometric family.

The following is a list of facts taken from [FK1]. Taken together, these facts show that the notions of neocompactness, neoclosedness, and neocontinuity behave in a manner analogous to the classical notions of compactness, closedness, and continuity.

Blanket Hypothesis 1. *For the rest of this section, we assume that* $\mathbf{M}$ *is a collection of complete metric spaces closed under finite Cartesian products, and* $(\mathbf{M},\mathcal{B},\mathcal{C})$ *is a countably compact neometric family such that for each* $\mathcal{M} \in \mathbf{M}$, $\mathcal{B}(\mathcal{M})$ *contains at least all compact sets in* $\mathcal{M}$.

Basic Facts 1.

1. Every neocompact set in $\mathcal{M}$ is neoclosed and bounded.

2. Every section of a neocompact set is neocompact. That is, if C is neocompact in $\mathcal{M} \times \mathcal{N}$ and $z \in \mathcal{N}$ then the set $\{x \in \mathcal{M} : (x, z) \in C\}$ is neocompact in $\mathcal{M}$.

3. If $f : D \to \mathcal{N}$ is neocontinuous from $\mathcal{M}$ to $\mathcal{N}$ and $A \subset D$ is neocompact in $\mathcal{M}$, then the set $f(A) = \{f(x) : x \in A\}$ is neocompact in $\mathcal{N}$.

4. If $f : C \to \mathcal{N}$ is neocontinuous from $\mathcal{M}$ to $\mathcal{N}$, C is neoclosed in $\mathcal{M}$, and D is neoclosed in $\mathcal{N}$, then $f^{-1}(D) = \{x \in C : f(x) \in D\}$ is neoclosed in $\mathcal{M}$.

5. Compositions of neocontinuous functions are neocontinuous.

6. Every closed separable subset of $\mathcal{M}$ is neoseparable in $\mathcal{M}$.

7. Every neoclosed set in $\mathcal{M}$ is closed in $\mathcal{M}$.

8. If $f : D \to \mathcal{N}$ is neocontinuous from $\mathcal{M}$ to $\mathcal{N}$, then f is continuous on D.

We now introduce one more property of a well behaved neometric family which is crucial for the deeper applications. This property is called closure under diagonal intersections.

Definition 2.5. A neometric family $(\mathbf{M}, \mathcal{B}, \mathcal{C})$ is said to be **closed under diagonal intersections** if the following holds. Let $M \in \mathbf{M}$, let $A_n \in \mathcal{C}(\mathcal{M})$ for each $n \in \mathbb{N}$, and let $\lim_{n \to \infty} \varepsilon_n = 0$. Then
$$A = \bigcap_n ((A_n)^{\varepsilon_n}) \in \mathcal{C}(\mathcal{M}).$$

The paper [FK1] has several consequences of closure under diagonal intersections. One example is a neometric analogue of Arzela's theorem. The most important consequence is the approximation theorem which was mentioned in the introduction.

Theorem 2.6. *(Approximation Theorem)* Suppose $(\mathbf{M}, \mathcal{B}, \mathcal{C})$ is closed under diagonal intersections. Let A be neoclosed in $\mathcal{M}$ and $f : A \to \mathcal{N}$ be neocontinuous from $\mathcal{M}$ to $\mathcal{N}$. Let B be neocompact in $\mathcal{M}$ and D be neoclosed in $\mathcal{N}$. Suppose that for each $\varepsilon > 0$, we have
$$(\exists x \in A \cap B^\varepsilon) f(x) \in D^\varepsilon.$$
Then
$$(\exists x \in A \cap B) f(x) \in D.$$

In the paper [A2], Anderson proved a form of the approximation theorem for the classical neometric family $(\mathbf{S}, \mathcal{B}, \mathcal{C})$, and gave several applications. To go further, we need other examples of neometric families which are countably compact and closed under diagonal intersections, and also need a library of useful neocompact sets and neocontinuous functions. In the next section we discuss a neometric family which has been studied in detail and was the was original motivation for our method, the family of neocompact sets in a rich adapted probability space. Other interesting neometric families will be discussed later on in this paper.

3. Rich Adapted Spaces

Anderson's construction of Brownian motion in [A1] and the lifting method for proving existence theorems for stochastic differential equations on an adapted Loeb space in [K1] were among the earliest applications of the Loeb measure construction in nonstandard analysis. These results were the primary motivation for both the adapted probability distributions in [K2] and [HK] and for the neometric method being discussed in this paper. In this section we shall review the neometric family on a rich adapted space which was introduced in the paper [FK1].

Let $\mathbf{B}$ be the set of dyadic rationals in $\mathbb{R}_+$. We say that $\Omega = (\Omega, P, \mathcal{G}, \mathcal{G}_t)_{t \in \mathbf{B}}$ is a **B-adapted (probability) space** if P is a complete probability measure on $\mathcal{G}$, $\mathcal{G}_t$ is a σ-subalgebra of $\mathcal{G}$ for each $t \in \mathbf{B}$, and $\mathcal{G}_s \subset \mathcal{G}_t$ whenever $s < t$ in $\mathbf{B}$. Let Ω be a $\mathbf{B}$-adapted probability space which will remain fixed throughout our discussion. For $s \in \mathbb{R}_+$ we let $\mathcal{F}_s$ be the P-completion of the σ-algebra $\bigcap\{\mathcal{G}_t : s < t \in \mathbf{B}\}$. Then the filtration $\mathcal{F}_s$ is right continuous, that is, for all $s < \infty$ we have $\mathcal{F}_s = \bigcap\{\mathcal{F}_t : s < t\}$. Each $\mathbf{B}$-adapted space $(\Omega, P, \mathcal{G}, \mathcal{G}_t)_{t \in \mathbf{B}}$ has an associated **right continuous adapted space** $(\Omega, P, \mathcal{F}, \mathcal{F}_t)_{t \in \mathbb{R}_+}$.

We say that P is **atomless** if any set of positive measure can be partitioned into two sets of positive measure, and that P is atomless on a σ-algebra $\mathcal{F} \subset \mathcal{G}$ if the restriction of P to $\mathcal{F}$ is atomless.

We let $M = (M, \rho)$ and $N = (N, \sigma)$ be complete separable metric spaces. We use the corresponding script letter $\mathcal{M} = L^0(\Omega, M)$ to denote the space of all P-measurable functions from Ω into M with the metric ρ_0 of convergence in probability,

$$\rho_0(x, y) = \inf\{\varepsilon : P[\rho(x(\omega), y(\omega)) \le \varepsilon] \ge 1 - \varepsilon\}.$$

(We identify functions which are equal P-almost surely). Note that the product metric $\mathcal{M} \times \mathcal{N}$ is topologically equivalent to the space $L^0(\Omega, M \times N)$.

The space of Borel probability measures on M with the Prohorov metric

$$d(\mu, \nu) = \inf\{\varepsilon : \mu(K) \le \nu(K^\varepsilon) + \varepsilon \text{ for all closed } K \subset M\}$$

is denoted by $\mathrm{Meas}(M)$. It is again a complete separable metric space, and convergence in $\mathrm{Meas}(M)$ is the same as weak convergence. Each measurable function $x : \Omega \to M$ induces a measure $\mathrm{law}(x) \in \mathrm{Meas}(M)$, and the function

$$\mathrm{law} : \mathcal{M} \to \mathrm{Meas}(M)$$

is continuous.

Definition 3.1. *Let $\Omega = (\Omega, P, \mathcal{G}, \mathcal{G}_t)_{t \in \mathbf{B}}$ be a $\mathbf{B}$-adapted space, and let $\mathbf{M}_\Omega$ be the family of all the metric spaces $\mathcal{M} = L^0(\Omega, M)$ where M is a complete separable metric space. A subset B of $\mathcal{M}$ will be called* **basic**, *$B \in \mathcal{B}_\Omega(\mathcal{M})$, if either*

- **(1):** B *is compact, or*
- **(2):** $B = \mathrm{law}^{-1}(C)$ *for some compact set $C \subset \mathrm{Meas}(M)$, or*
- **(3):** $B = \{x \in \mathrm{law}^{-1}(C) : x \text{ is } \mathcal{G}_t - \text{measurable}\}$ *for some compact $C \subset \mathrm{Meas}(M)$ and $t \in \mathbf{B}$.*

We say that a $\mathbf{B}$-adapted space Ω is **rich** if the measure P is atomless on $\mathcal{G}_0$, Ω admits a Brownian motion, and the neocompact family $(\mathbf{M}_\Omega, \mathcal{B}_\Omega, \mathcal{C}_\Omega)$ generated by $(\mathbf{M}_\Omega, \mathcal{B}_\Omega)$ is countably compact. The sets in $\mathcal{C}_\Omega(\mathcal{M})$ are said to be **neocompact for the $\mathbf{B}$-adapted space Ω**.

It is convenient to identify each complete separable metric space M with the set of all constant functions in $\mathcal{M} = L^0(\Omega, M)$. With this identification we get a notion of a neocontinuous function from M into $\mathcal{N}$, and a neocontinuous function from $\mathcal{N}$ into M.

The simpler notion of a **rich probability space** is defined in the same way as a rich $\mathbf{B}$-adapted space except that condition (3) is left out of the definition of the basic sets.

The paper [FK1] gives examples showing that the usual probability spaces and adapted spaces considered in the classical literature are not rich. Moreover, the universal projection condition (f) cannot be strengthened by allowing the set D to be neocompact rather than basic. We shall discuss the existence of rich probability and adapted spaces later on.

An extensive library of neocompact sets and neocontinuous functions for a rich B-adapted space is developed in [FK1], and is extended further in [CK]. Here is a sampling from this library.

Blanket Hypothesis 2. *For the rest of this section we assume that Ω is a rich B-adapted space and that M, N are complete separable metric spaces.*

Theorem 3.2. *The family of neocompact sets for Ω is a neometric family which is closed under diagonal intersections.*

Thus the distance and projection functions are neocontinuous.

Let $T > 0$. A stochastic process $x \in L^0(\Omega, L^0([0,T], M))$ or a continuous stochastic process $x \in L^0(\Omega, C([0,T], M))$ is said to be **adapted** if $x(\cdot, t)$ is $\mathcal{F}_t$-measurable for each $t \in [0, T]$.

Theorem 3.3. *The following sets are neocompact.*
 (a) The set $L^0(\Omega, C)$ where C is a compact subset of M.
 (b) The set of all $\mathcal{F}_t$-stopping times in $L^0(\Omega, [0,T])$.
 (c) The set of all Brownian motions on $\Omega \times [0,T]$, that is the set of continuous adapted processes on Ω with values in $\mathbb{R}$ whose law is the Wiener measure on $C([0,T], \mathbb{R})$.
 (d) The set of all $x \in L^0(\Omega, \mathbb{R})$ such that $E[|x|] \leq r$ where $r > 0$ is fixed.

The proof of (d) uses closure under diagonal intersections.

Theorem 3.4. *The following sets are neoclosed.*
 (a) The set of all $\mathcal{F}_t$-measurable $x \in \mathcal{M}$, where $t \in \mathbb{R}_+$ is fixed.
 (b) The set of adapted stochastic processes on $[0,T]$ with values in M.
 (c) The set of continuous adapted stochastic processes on $[0,T]$ with values in M.

Theorem 3.5. *The following functions are neocontinuous.*
 (a) (Randomization Lemma) The function $g : \mathcal{M} \to \mathcal{N}$ defined by $(g(x))(\omega) = f(x(\omega))$ where $f : M \to N$ is continuous.
 (b) The law function from $\mathcal{M}$ to $\operatorname{Meas}(M)$.
 (c) The stochastic integral function

$$(y, b) \mapsto \int_0^t y(\omega, s) db(\omega, s)$$

where r is finite, y belongs to the neoclosed set of adapted stochastic processes on Ω with values in $[-r, r]$, and b belongs to the neocompact set of Brownian motions.

Moreover, the range of the function (c) is contained in a neocompact set of continuous adapted stochastic processes.

Theorem 3.6. *The following functions are neocontinuous on each uniformly integrable subset of $L^1(\Omega, \mathbb{R})$.*
  (a) The expected value function $x \mapsto E[x(\omega)]$.
  (b) The conditional expectation function $x \mapsto E[x(\omega)|\mathcal{G}_t]$ where $t \in \mathbf{B}$.
  (c) The conditional expectation function $x \mapsto E[x(\omega)|\mathcal{F}.]$ where the value is a stochastic process.

The paper [FK1] has an example which complements Theorem 3.4 (a) and 3.6 (c). The example shows that for each t, the neocompact set of all $\mathcal{F}_t$-measurable x with law$(x) \in [0,1]$ cannot be basic, and the continuous function $x \mapsto E[x(\omega)|\mathcal{F}_t]$ cannot be neocontinuous from $L(\Omega, [0,1])$ to itself.

A variety of optimization and existence theorems for rich **B**-adapted spaces are proved in [FK1]. In [CK] the method is applied to obtain new optimization and existence theorems for stochastic Navier-Stokes equations. To give an idea of what can be done, we give two examples here.

Theorem 3.7. *Let Ω be a rich **B**-adapted space and let $T > 0$. For each continuous stochastic process $x \in L^0(\Omega, C([0,T], \mathbb{R}))$ there is a Brownian motion on $\Omega \times [0,T]$ whose ρ_0-distance from x is a minimum.*

Proof sketch: The function $f(b) = \rho_0(x, b)$ is neocontinuous from the neocompact set B of Brownian motions on $\Omega \times [0, T]$ into the reals. Therefore its range $f(B)$ is neocompact and hence closed and bounded in the reals, and thus has a minimum.

The next result was proved for adapted Loeb spaces by the lifting method in [K1]. It is a stochastic analogue of the Peano existence theorem, and improves the weak existence theorem of Skorokhod [Sk] for stochastic differential equations,

Theorem 3.8. *Let $T, r > 0$, let g be an adapted stochastic process on $\Omega \times [0, T]$ with values in $C(\mathbb{R}, [-r, r])$, and let b be a Brownian motion on $\Omega \times [0, T]$. Then there exists a continuous adapted stochastic process x such that*

$$x(\omega, t) = \int_0^t g(\omega, s)(x(\omega, s)) db(\omega, x).$$

Moreover, the set A of all such solutions x is neocompact, and hence any neocontinuous function f from A into the reals has a minimum (i.e. and optimal solution with respect to f).

Proof sketch: Our library of neocontinuous functions shows that the function

$$h(x, u) = \int_0^t g(\omega, s)(x(\omega, s - u)) db(\omega, s),$$

with the convention that $x(\omega, u) = 0$ when $u < 0$, is neocontinuous and its range is contained in a neocompact set D of continuous stochastic processes. Since A is the set of all x such that $h(x, 0) - x = 0$ and $A \subset D$, it follows that A is neocompact.

The set C of all $u \in [0, 1]$ such that $(\exists x \in D) h(x, u) = x$ is also neocompact. Thus C is a closed subset of the unit interval $[0, 1]$. By successively integrating over subintervals $[0, u], [u, 2u], \ldots$, for each $u \in (0, 1]$ we get an $x \in D$ such that $h(x, u) = x$.

It follows that $(0,1] \subset C$. Since C is closed, we have $0 \in C$, and therefore the set A of solutions is nonempty.

The above proof used the basic facts about neocontinuous functions and neocompact sets in a direct manner. To illustrate the use of the Approximation Theorem, we give a second proof.

Alternative Proof by Approximation: Let h and D be the neocontinuous function and neocompact set from the first proof. By successively integrating over subintervals we see that
$$(\exists (x,u) \in (D \times \{0\})^\varepsilon) h(x,u) - x \in \{0\}^\varepsilon.$$
By the approximation theorem we have
$$(\exists x \in D) h(x,0) - x = 0$$
as required.

The preceding examples are illustrations of a general approach to the discovery of new conjectures and proofs. In a wide variety of situations, one can ask whether a given set is neocompact, or whether a given function is neocontinuous. Neocompactness can often be proved by checking through the definition of the set to see that it can be constructed from basic sets using the operations (a)—(f). Once neocompactness and neocontinuity are established, the Approximation Theorem immediately suggests a way to prove an existence theorem by proving that approximate solutions exist. In many cases, such as in the preceding theorem, one can then go on to ask if the set of solutions itself is neocompact, and continue the process.

4. Saturated Adapted Spaces

In the paper [HK] the notions of an adapted distribution and of a saturated adapted probability space were introduced. The adapted distribution of a random variable on an adapted space (with values in a complete separable metric space) is the natural analogue of the distribution of a random variable on a probability space. The results of [HK] suggest that two stochastic processes on possibly different spaces may be considered alike if they have the same adapted distribution. For stochastic differential equations and a wide variety of other existence problems, every existence theorem which holds on some adapted space holds on a saturated adapted space. The relationship between saturated and rich adapted spaces was studied in the papers [K4] and [K5]. The key result was a quantifier elimination theorem showing that every neocompact set can be represented in a simple form by means of the adapted distribution.

We begin with the simple notion of a saturated probability space, and then take up the more complicated notions of a saturated **B**-adapted space and a saturated right continuous adapted space.

Definition 4.1. *A probability space Ω is* **saturated** *if for any random variable $x \in L^0(\Omega, M)$ and pair of random variables $\bar{x} \in L^0(\Gamma, M)$ and $\bar{y} \in L^0(\Gamma, N)$ on another probability space Γ such that $law(x) = law(\bar{x})$, there is a random variable $y \in L^0(\Omega, N)$ such that $law(x,y) = law(\bar{x}, \bar{y})$.*

The following theorem was proved in [K5].

Theorem 4.2. *A probability space is rich if and only if it is saturated.*

The main tool in the proof was a quantifier elimination theorem which is of interest in its own right.

Theorem 4.3. *(Quantifier Elimination for Probability Spaces) In a saturated probability space Ω, a set is neocompact in $\mathcal{M}$ if and only if it is of the form*
$$\{x \in \mathcal{M} : law(x, z) \in C\}$$
for some compact set $C \subset Meas(\mathcal{M} \times \mathcal{N})$ and some $z \in \mathcal{N}$.

We now present analogous results for **B**-adapted spaces. In order to state these results we need the notion of an adapted function, which was essentially introduced in [HK].

Definition 4.4. *Let $\mathcal{R} = L^0(\Omega, \mathbb{R})$. The class of **B-adapted functions** on $\mathcal{M}$ is the least class of functions from $\mathcal{M}$ into $\mathcal{R}$ such that:*

*(i) For each bounded continuous function $\phi : M \to \mathbb{R}$, the function $(\hat{\phi}(x))(\omega) = \phi(x(\omega))$ is a **B**-adapted function on $\mathcal{M}$;*

*(ii) If $f_1, \ldots, f_m$ are **B**-adapted functions on $\mathcal{M}$ and $g : \mathbb{R}^m \to \mathbb{R}$ is continuous, then $h(x) = g(f_1(x), \ldots, f_m(x))$ is a **B**-adapted function on $\mathcal{M}$;*

*(iii) If f is a **B**-adapted function on $\mathcal{M}$ and $t \in \mathbf{B}$, then $g(x)(\omega) = E[f(x)|\mathcal{G}_t](\omega)$ is a **B**-adapted function on $\mathcal{M}$.*

*Two random variables $x \in L^0(\Omega, M)$ and $\bar{x} \in L^0(\Gamma, M)$ have the same **B-adapted distribution**, in symbols $x \equiv_\mathbf{B} \bar{x}$, if $E[f(x)] = E[f(\bar{x})]$ for every **B**-adapted function f on $\mathcal{M}$.*

*A **B**-adapted space Ω is **saturated** if for every other **B**-adapted space Γ, every $x \in L^0(\Omega, M)$, and every pair $\bar{x} \in L^0(\Gamma, M), \bar{y} \in L^0(\Gamma, N)$ such that $x \equiv_\mathbf{B} \bar{x}$, there exists $y \in L^0(\Omega, N)$ such that $(x, y) \equiv_\mathbf{B} (\bar{x}, \bar{y})$.*

With this definition, the following theorem is proved in [K5] using results from [K4].

Theorem 4.5. *A **B**-adapted space Ω is rich if and only if it is saturated.*

The main tool for the implication from left to right is the following consequence of our library of neocontinuous functions.

Theorem 4.6. *If Ω is a rich **B**-adapted space then every **B**-adapted function for Ω is neocontinuous.*

The main tool for the other direction is the following quantifier elimination theorem which is again of interest in its own right.

Theorem 4.7. *(Quantifier Elimination for **B**-adapted Spaces) Let Ω be a rich **B**-adapted space. A set is neocompact in $\mathcal{M}$ if and only if it is the intersection of a set of the form*
$$\{x \in \mathcal{M} : law(x, z) \in C\}$$
and countably many sets of the form
$$\{x \in \mathcal{M} : E[f_n(x, z)] \in D_n\}$$
*where each f_n is a **B**-adapted function on $\mathcal{M} \times \mathcal{N}$, $z \in \mathcal{N}$, C is compact in Meas($M \times N$), and each D_n is compact in $\mathbb{R}$.*

We now turn to right continuous adapted spaces.

Definition 4.8. *The notion of an $\mathbb{R}_+$-adapted function is defined in exactly the same way as a **B**-adapted function except that times are taken from the set $\mathbb{R}_+$ and conditional expectations are taken with respect to the right continuous filtration $\mathcal{F}_t$.*

*We shall say that two random variables x and $\bar{x}$ on right continuous adapted spaces Ω and Γ have the same **adapted distribution**, in symbols $x \equiv \bar{x}$, if $E[f(x)] = E[f(\bar{x})]$ for each $\mathbb{R}_+$-adapted function f.*

*A right continuous adapted space Ω is **saturated** if for every other right continuous adapted space Γ, every $x \in L^0(\Omega, M)$, and every pair $\bar{x} \in L^0(\Gamma, M), \bar{y} \in L^0(\Gamma, N)$ such that $x \equiv \bar{x}$, there exists $y \in L^0(\Omega, N)$ such that $(x, y) \equiv (\bar{x}, \bar{y})$.*

It is shown in [K5] that rich right continuous adapted spaces do not exist, and that nontrivial $\mathbb{R}_+$-adapted functions of the form $E[f(x)|\mathcal{F}_t]$ for a right continuous adapted space can never be neocontinuous.

Here is the main result on right continuous adapted spaces which is proved in [K5].

Theorem 4.9. *For every rich **B**-adapted space, the associated right continuous adapted space is saturated.*

5. The Huge Neometric Family

Our discussion up to this point has not involved nonstandard analysis at all, but we have postponed the proof that rich **B**-adapted spaces exist until this section. Now it is time to enter the nonstandard world. We present the huge neometric family $(\mathcal{H}, \mathcal{B}, \mathcal{C})$ associated with each $\aleph_1$-saturated nonstandard universe, which was introduced in [FK2]. The huge neometric family is constructed by giving an explicit definition of basic and neocompact sets that captures the way internal sets are used in nonstandard probability practice. The idea is that basic sets should be standard parts of internal sets, and neocompact sets should be standard parts of countable intersections of internal sets. The huge neometric family lives up to its name and contains all neometric spaces studied so far.

We shall see that the huge neometric family contains the neometric family over a **B**-adapted Loeb space. It follows that **B**-adapted Loeb spaces are rich, and therefore that rich **B**-adapted spaces exist.

The huge neometric family is a generalization of the approach to neocompactness originally developed in [K3]. In that paper, the neometric family over a rich **B**-adapted space was introduced in a nonstandard setting. From our current viewpoint, this neometric family is a subfamily of the huge neometric family.

We fix an $\aleph_1$-saturated nonstandard universe. We shall use the notions of a *metric space and a *probability measure, which are obtained from the corresponding standard notions by transfer: a *metric space is a structure $(\bar{M}, \bar{\rho})$ where $\bar{M}$ is an internal set and $\bar{\rho}$ is an internal function $\bar{\rho} : \bar{M} \times \bar{M} \to {}^*\mathbb{R}$ which satisfies the transfer of the usual rules for a metric. We now quickly review the nonstandard hull construction.

If $X, Y \in \bar{M}$, we write $X \approx Y$ if $\bar{\rho}(X, Y) \approx 0$. The **standard part** of an element $X \in \bar{M}$ is the equivalence class

$$°X = \{Y \in \bar{M} : X \approx Y\}.$$

If $x = °X$, we say that X **lifts** x.

Definition 5.1. *Consider a* *metric space $(\bar{M}, \bar{\rho})$ and a point $c \in \bar{M}$. The **galaxy** of c is the set $G(\bar{M}, c)$ of all points $X \in \bar{M}$ such that $\bar{\rho}(X, c)$ is finite. By the **nonstandard hull** of $\bar{M}$ at c we mean the metric space $(\mathcal{H}(\bar{M}, c), \rho)$ where*

$$\mathcal{H}(\bar{M}, c) = \{°X : X \in G(\bar{M}, c)\}, \rho(°X, °Y) = st(\bar{\rho}(X, Y)).$$

Note that any two points $b, c \in \bar{M}$ such that $\bar{\rho}(b, c)$ is finite have the same galaxies and nonstandard hulls,

$$G(\bar{M}, b) = G(\bar{M}, c) \text{ and } \mathcal{H}(\bar{M}, b) = \mathcal{H}(\bar{M}, c).$$

The neometric spaces in our huge family **H** will be the closed subspaces of nonstandard hulls. We need more definitions.

Given a set $B \subset G(\bar{M}, c)$, the **standard part** of B is the set

$$°B = \{°X : X \in B\}$$

of standard parts of elements of B. In the opposite direction, for a set $A \subset \mathcal{H}(\bar{M}, c)$ the **monad** of A is the set

$$\text{monad}(A) = \{X : °X \in A\}.$$

By a Σ_1^0 (Π_1^0) set we mean the union (intersection) of countably many internal subsets of the galaxy $G(\bar{M}, c)$.

Observe that every countable subset of $G(\bar{M}, c)$ is Σ_1^0, and hence every countable subset of $\mathcal{H}(\bar{M}, c)$ is the standard part of a Σ_1^0 set.

For a set $B \subset G(\bar{M}, c)$ and a hyperreal $\varepsilon > 0$, we write

$$\bar{\rho}(X, B) = \inf\{\bar{\rho}(X, Y) : Y \in B\}, B^\varepsilon = \{X : \bar{\rho}(X, B) \leq \varepsilon\}.$$

Observe also that for each *metric space $\bar{M}$ and distinguished point $c \in \bar{M}$, the galaxy $G(\bar{M}, c)$ is a Σ_1^0 set, and the monad of the nonstandard hull $\mathcal{H}(\bar{M}, c)$ is the galaxy $G(\bar{M}, c)$.

Definition 5.2. *The* **huge neometric family** $(\mathbf{H}, \mathcal{B}, \mathcal{C})$ *is defined as follows.* $\mathbf{H}$ *is the class of all metric spaces* $(\mathcal{M}, \rho)$ *such that* $\mathcal{M}$ *is a closed subset of some nonstandard hull* $\mathcal{H}(\bar{M}, c)$. *For each* $\mathcal{M} \in \mathbf{H}$, *the collections of basic and neocompact subsets of* $\mathcal{M}$ *are*

$$\mathcal{B}(\mathcal{M}) = \{A \subset \mathcal{M} : A = {}^\circ B \text{ for some internal set } B \subset G(\bar{M}, c)\},$$

$$\mathcal{C}(\mathcal{M}) = \{A \subset \mathcal{M} : A = {}^\circ B \text{ for some } \Pi_1^0 \text{ set } B \subset G(\bar{M}, c)\}.$$

Note that the standard part of the union of two sets is the union of the standard parts, and therefore $\mathcal{B}(\mathcal{M})$ is closed under finite unions. Moreover, finite Cartesian products of basic sets are basic, and every finite subset of $\mathcal{M}$ is basic. On the other hand, the intersection of two basic sets need not be basic (see Example 3.6 in [FK2]).

The standard neometric family $(\mathbf{S}, \mathcal{B}, \mathcal{C})$ may be regarded as a subfamily of $(\mathbf{H}, \mathcal{B}, \mathcal{C})$. Given a standard complete metric space $(M, \rho) \in \mathbf{S}$, we may consider the *metric space $({}^*M, {}^*\rho)$. We abuse notation by identifying M with the set $\{{}^*x : x \in M\}$. Thus M is a closed subset of the nonstandard hull $\mathcal{H}({}^*M, x)$ where x is any element of M, and hence M itself belongs to the huge family $\mathbf{H}$.

Here is a list of facts about the huge neometric family taken from [FK2]

Basic Facts 2. *Let* $\mathcal{M}$ *and* $\mathcal{N}$ *belong to* $\mathbf{H}$.

1. $(\mathbf{H}, \mathcal{B}, \mathcal{C})$ *is a countably compact neometric family which is closed under diagonal intersections.*

2. Every compact set $C \subset \mathcal{M}$ *is basic.*

3. Let $C = {}^\circ(\bigcap_n C_n)$ *be a neocompact set in* $\mathcal{M}$ *where* $\langle C_n \rangle$ *is a decreasing chain of internal sets. Then*

$$monad(C) = \bigcap_n ((C_n)^{1/n}).$$

4. A set $B \subset G(\bar{M}, c)$ *is the monad of a neoseparable set if and only if* B *can be written in the form*

$$B = \bigcap_n \bigcup_m ((B_m)^{1/n})$$

where $\langle B_m \rangle$ *is an increasing chain of internal subsets of* $G(\bar{M}, c)$.

5. Let M *be a standard complete metric space, that is,* $M \in \mathbf{S}$. *A subset* C *of* M *is neocompact with respect to* $\mathbf{H}$ *if and only if it is compact, neoclosed with respect to* $\mathbf{H}$ *if and only if it is closed, and neoseparable with respect to* $\mathbf{H}$ *if and only if it is closed and separable. If* $C \subset M$ *is closed and* $\mathcal{N} \in \mathbf{H}$, *a function* $f : C \to \mathcal{N}$ *is neocontinuous with respect to* $\mathbf{H}$ *if and only if it is continuous.*

6. Let $\mathcal{M}$ *be neoseparable. A set* $C \subset \mathcal{M}$ *is neocompact in* $\mathcal{M}$ *if and only if* C *is neoclosed in* $\mathcal{M}$ *and any countable covering of* C *by neoopen sets in* $\mathcal{M}$ *has a finite subcovering.*

7. Let $C \subset \mathcal{M}$ *be neocompact and* $f : C \to \mathcal{N}$. *Then* f *is neocontinuous if and only if there is an internal function* F *such that* ${}^\circ F(X) = f({}^\circ X)$ *for all* $X \in monad(C)$.

We now look at adapted Loeb spaces within the huge neometric family. We first need some notation for internal probability spaces.

Definition 5.3. *Let $(\Omega, \bar{P}, \bar{\mathcal{G}})$ be a *probability space and let $(\Omega, P, \mathcal{G})$ be the corresponding Loeb probability space. $SL^0(\Omega, M)$ denotes the *metric space of all $\bar{\mathcal{G}}$-measurable functions $X : \Omega \to {}^*M$ with the *metric*
$$\bar{\rho}_0(X, Y) = {}^*\inf\{\varepsilon : \bar{P}[{}^*\rho(X(\omega), Y(\omega)) \geq \varepsilon] \leq \varepsilon\}.$$
*We say that $X \in SL^0(\Omega, M)$ is a **lifting** of a function $x : \Omega \to M$, in symbols ${}^\circ X = x$, if $X(\omega)$ has standard part $x(\omega) \in M$ for P-almost all $\omega \in \Omega$.*

By the fundamental result of Loeb, that a function $x : \Omega \to M$ is Loeb measurable if and only if it has a lifting, we may take $L^0(\Omega, M)$ to be a subset of the standard part of $SL^0(\Omega, M)$. We now introduce adapted Loeb spaces.

Definition 5.4. *By a **B-adapted Loeb space** we mean an **B**-adapted space $\Omega = (\Omega, P, \mathcal{G}_t)_{t \in \mathbf{B}}$ such that $(\Omega, P, \mathcal{G})$ is a Loeb probability space, $\mathcal{G}_s \subset \mathcal{G}_t$ whenever $s < t \in \mathbf{B}$, and each $\mathcal{G}_t$ is a σ-algebra generated by an internal subalgebra $\bar{\mathcal{G}}_t$ of $\bar{\mathcal{G}}$.*

The following theorem from [FK2] shows that **B**-adapted Loeb spaces are rich and hence that rich **B**-adapted spaces exist. As we have emphasized in the introduction, this is the one place where nonstandard analysis is needed in order to prove of existence theorems via neocompact sets.

Theorem 5.5. *Let Ω be an atomless **B**-adapted Loeb space.*
*(i) The set $L^0(\Omega, M)$ is neoseparable with respect to the *metric $\bar{\rho}_0$ on $SL^0(\Omega, M)$, and the metric space $\mathcal{M} = (L^0(\Omega, M), \rho_0)$ belongs to the huge neometric family $\mathbf{H}$.*
(ii) Every basic set, neocompact set, neoclosed set, neoseparable set, and neocontinuous function with neoclosed domain in $(\mathbf{M}_\Omega, \mathcal{B}_\Omega, \mathcal{C}_\Omega)$ is also basic, neocompact, neoclosed, neoseparable, or neocontinuous, respectively, in the huge neometric family $(\mathbf{H}, \mathcal{B}, \mathcal{C})$.
(iii) Ω is rich.

Corollary 5.6. *Every atomless Loeb probability space is rich.*

The paper [FK2] gave several other examples of natural neometric families within the huge neometric family. We mention three of them here.

Theorem 5.7. *Let M be a standard Banach space. In the huge neometric family, the nonstandard hull $\mathcal{H}({}^*M, 0)$ of the galaxy of 0 in ${}^*M$ is neoseparable, each closed ball in $\mathcal{H}({}^*M, 0)$ is neocompact, and the norm function $x \mapsto \|x\|$, the addition function $(x, y) \mapsto x + y$, and the scalar multiplication function $x \mapsto \alpha x, \alpha \in \mathbb{R}$, are neocontinuous.*

Theorem 5.8. *Let Ω be an atomless Loeb probability space. The set $L^1(\Omega, \mathbb{R})$ of Loeb integrable functions on Ω is neoseparable with respect to the *metric $\bar{\rho}_1(X, Y) = \bar{E}[\bar{\rho}(X(\cdot), Y(\cdot))]$, and the metric space $(L^1(\Omega, \mathbb{R}), \rho_1)$ belongs to the huge neometric family $\mathbf{H}$.*

The next example from [FK2] concerns Loeb integrable functions with values in a neoseparable space rather than in the separable space of reals. We first need some definitions.

Definition 5.9. *Let Ω be an atomless Loeb probability space and let $(\bar{M}, \bar{\rho})$ be a *metric space. Let $\bar{\rho}_1$ be the *metric on the set $SL^0(\bar{\Omega}, \bar{M})$ defined by $\bar{\rho}_1(X, Y) = \bar{E}[\bar{\rho}(X(\cdot), Y(\cdot))]$. For each $c \in \bar{M}$, let $SL^1(\bar{\Omega}, \bar{M}, c)$ be the set of all $X \in SL^0(\bar{\Omega}, \bar{M})$ such that $\bar{\rho}(X(\cdot), c)$ is S-integrable.*

Choose a point c in the monad of $\mathcal{M}$. We let $\mathcal{L}^1(\Omega, \mathcal{M})$ denote the metric space of all functions $x : \Omega \to \mathcal{M}$ such that x has a lifting in $SL^1(\bar{\Omega}, \bar{M}, c)$, with the metric ρ_1 such that $\rho_1(x, y) = {}^\circ\bar{\rho}_1(X, Y)$ whenever $X, Y \in SL^1(\bar{\Omega}, \bar{M}, c)$, X lifts x and Y lifts y.

Theorem 5.10. *Let $(\mathcal{M}, \rho)$ be neoseparable in the huge neometric family $\mathbf{H}$. The set $\mathcal{L}^1(\Omega, \mathcal{M})$ is neoseparable with respect to the *metric $\bar{\rho}_1$, and the metric space $(\mathcal{L}^1(\Omega, \mathcal{M}), \rho_1)$ belongs to $\mathbf{H}$.*

6. Forcing and Approximations

The paper [FK3] in this volume develops another approach to our program. It centers on the notion of a long sequence in the huge neometric family, that is, a sequence indexed by the hyperintegers. The notions of neocompactness, neoclosedness, and neocontinuity have natural characterizations in terms of long sequences, and long sequence arguments have a flavor much like the more traditional lifting and pushing down arguments in nonstandard analysis.

One of the central ideas in the paper [K3] was a notion of forcing for formulas in an infinitary language built from neocompact sets and neocontinuous functions. This notion of forcing is defined by an induction on formulas which is reminiscent of forcing in set theory. However, the "names" in the statements to be forced are sequences of elements, the "conditions" are infinite sets of natural numbers, and proofs by forcing resemble classical proofs by convergence. Forcing was applied in that paper to prove several existence theorems in stochastic analysis. Long sequences played an important role in the treatment of forcing.

The forthcoming paper [FK4] generalizes the treatment of forcing introduced in [K3], and introduces a second kind of forcing which applies only to positive bounded formulas but appears to be easier to use. The paper [FK4] also introduces a notion of an approximation for positive bounded formulas, and uses the results about forcing to generalize the approximation theorem 2.6 stated earlier in this paper. This approximation theorem is closely related to a theorem of Anderson [A2] in a classical compact setting, and the work of Henson [He] and Henson and Iovino [HI] in the setting of Banach space model theory. In this section we shall present the main notions and results from [FK4] on forcing and approximations of positive bounded formulas.

We begin by introducing the language **PB** of **positive bounded formulas**. We shall always work in the huge neometric family **H**.

Definition 6.1. *The language* **PB** *of positive bounded formulas has the following symbols:*

: *Infinitely many variables $u, v, \ldots$ of sort $\mathcal{M}$ for each neometric space $\mathcal{M} \in \mathbf{H}$,*
: *An n-ary function symbol for each total neocontinuous function*

$$f : \mathcal{M}_1 \times \cdots \times \mathcal{M}_n \to \mathcal{N},$$

: *A constant symbol for each element $c \in \mathcal{M}$,*
: *A unary predicate symbol of sort $\mathcal{M}$ for each neoclosed set C in $\mathcal{M}$.*

*Terms are built in the usual way by applying function symbols to variables and constants of the appropriate sorts. The atomic formulas of **PB** are $\tau(\vec{v}) \in C$ where τ is a term and C is a neoclosed set of the same sort. The formulas of **PB** are built from atomic formulas using finite and countable conjunctions, finite disjunctions, existential quantifiers of the form $(\exists v \in C)\phi$ where C is neocompact, and universal quantifiers $(\forall v \in D)\phi$ where D is neoseparable.*

Since the distance functions are neocontinuous, an equation $\tau(\vec{u}) = \pi(\vec{v})$ between two terms can be expressed by the **PB** formula $\rho(\tau(\vec{u}), \pi(\vec{v})) \in \{0\}$.

The next theorem says that every positive bounded formula defines a neoclosed set.

Theorem 6.2. *For every **PB** formula $\phi(\vec{v})$, the set $\{\vec{x} : \phi(\vec{x}) \text{ is true }\}$ is neoclosed in the sort space of $\vec{v}$.*

A sequence of k-tuples $\langle \vec{x}_n \rangle$ in $\mathcal{M}$ is said to be **neotight** if it is contained in a neocompact set in $\mathcal{M}$. By a **condition** we mean an infinite set $p \subset \mathbb{N}$. In the following, p, q, and r will denote conditions.

We now introduce positive bounded forcing.

Definition 6.3. *For each **PB** formula $\phi(\vec{v})$, neotight sequence $\langle \vec{x}_n \rangle$ of the same sort as $\vec{v}$, and condition p, the **forcing relation** $p \Vdash \phi(\langle \vec{x}_n \rangle)$ is defined inductively as follows:*

: $p \Vdash f(\langle \vec{x}_n \rangle) \in C$ *iff* $\lim_{n \in p} \rho(f(\vec{x}_n), C) = 0$.
: $p \Vdash \bigwedge_m \phi_m(\langle \vec{x}_n \rangle)$ *iff* $(\forall m) p \Vdash \phi_m(\langle \vec{x}_n \rangle)$.
: $p \Vdash (\forall v \in D)\phi(\langle \vec{x}_n \rangle, v)$ *iff* $(\forall \text{ neotight } \langle y_n \rangle \text{ in } D) p \Vdash \phi(\langle \vec{x}_n, y_n \rangle)$.
: $p \Vdash (\phi \vee \psi)(\langle \vec{x}_n \rangle)$ *iff*

$$(\forall q \subset p)(\exists r \subset q) r \Vdash \phi(\langle \vec{x}_n \rangle) \vee r \Vdash \psi(\langle \vec{x}_n \rangle).$$

: $p \Vdash (\exists v \in C)\phi(\langle \vec{x}_n \rangle, v)$ *iff*

$$(\forall q \subset p)(\exists r \subset q)(\exists \text{ neotight } \langle y_n \rangle \text{ in } C) r \Vdash \phi(\langle \vec{x}_n, y_n \rangle).$$

The main technical result about positive bounded forcing in [FK4] uses long sequences, which are defined in [FK3] in this volume. This theorem and its corollary capture the analogy between forcing and classical proofs by convergence.

Theorem 6.4. *Let $\phi(\vec{v})$ be a positive bounded formula, let $\langle \vec{x}_n \rangle$ be a neotight sequence in $\mathcal{M}$, and let $\langle \vec{x}_J \rangle$ be a long sequence which is an $\mathcal{M}$-extension of $\langle \vec{x}_n \rangle$. If $p \Vdash \phi(\langle \vec{x}_n \rangle)$ then $\phi(\vec{x}_J)$ is true for all sufficiently small infinite $J \in {}^*p$.*

Corollary 6.5. *Let $\phi(\vec{v})$ be a positive bounded formula and let $\lim_{n \in p} \vec{x}_n = \vec{x}$ in $\mathcal{M}$. If $p \Vdash \phi(\langle \vec{x}_n \rangle)$ then $\phi(\vec{x})$ is true.*

We now define the set of approximations of a **PB** formula.

Definition 6.6. *The set $\mathcal{A}(\phi)$ of **approximations** of a **PB** formula $\phi(\vec{v})$ is defined by induction on the complexity of ϕ as follows. For each neoseparable set D, let $\langle D_m \rangle$ be a chain of basic sets such that $\bigcup_m D_m$ is dense in D.*

$$\mathcal{A}(\tau(\vec{v}) \in C) = \{\tau(\vec{v}) \in C^{1/n} : n \in \mathbb{N}\}.$$

$$\mathcal{A}(\bigwedge_m \phi_m) = \{\bigwedge_{m \leq n} \psi_m : n \in \mathbb{N} \text{ and } \psi_m \in \mathcal{A}(\phi_m) \text{ for all } m \leq n\}.$$

$$\mathcal{A}(\phi \vee \psi) = \{\phi_0 \vee \psi_0 : \phi_0 \in \mathcal{A}(\phi) \text{ and } \psi_0 \in \mathcal{A}(\psi)\}.$$

$$\mathcal{A}((\exists v \in C)\phi) = \{(\exists v \in C^{1/n})\psi : \psi \in \mathcal{A}(\phi) \text{ and } n \in \mathbb{N}\}.$$

$$\mathcal{A}((\forall v \in D)\phi) = \{(\forall v \in D_m)\psi : \psi \in \mathcal{A}(\phi) \text{ and } m \in \mathbb{N}\}.$$

Note that each approximation of a **PB** formula ϕ is a consequence of ϕ. The approximations of ϕ are finite formulas but are not necessarily positive bounded, because if C is neocompact, the set $C^{1/n}$ is neoclosed but not necessarily neocompact.

The next theorem from [FK4] characterizes positive bounded forcing in terms of approximate truth.

Theorem 6.7. *Let $\phi(\vec{v})$ be a positive bounded formula and $\langle \vec{x}_n \rangle$ be neotight. The following are equivalent.*

(i) $p \Vdash \phi(\langle \vec{x}_n \rangle)$.

(ii) For all $\psi \in \mathcal{A}(\phi)$, $\psi(\vec{x}_n)$ is true for all but finitely many $n \in p$,

Corollary 6.8. *(Positive Bounded Approximation Theorem) Let $\phi(\vec{v})$ be a **PB** formula and $\vec{c}$ be a tuple of constants. Then $\phi(\vec{c})$ is true if and only if $\psi(\vec{c})$ is true for every approximation $\psi \in \mathcal{A}(\phi)$.*

In the case that Ω is an atomless **B**-adapted Loeb space, the Approximation Theorem 2.6 is a special case of the Positive Bounded Approximation Theorem. To see this, write the formula

$$(\exists x \in A \cap B) f(x) \in D$$

in the equivalent form

$$(\exists x \in B)[x \in A \wedge f(x) \in D].$$

The latter formula is positive bounded by Theorem 5.5.

Here are two further results about approximations from [FK4] which are analogous to theorems from [HI] and [A2].

249

Theorem 6.9. *(Perturbation Principle)* For each **PB** formula $\phi(\vec{v})$, neocompact set D, and approximation $\psi \in \mathcal{A}(\phi)$, there is a real $\delta > 0$ such that whenever $\vec{x}, \vec{y} \in D$, $\phi(\vec{x})$ holds, and $\rho(\vec{x}, \vec{y}) \leq \delta$, we have $\psi(\vec{y})$.

Theorem 6.10. *(Almost-Near Theorem)* Let $\phi(v)$ be a **PB** formula where v has sort $\mathcal{M}$, let C be a neocompact set in $\mathcal{M}$, and let D be a neoseparable set in $\mathcal{M}$ such that every $x \in C$ such that $\phi(x)$ is true belongs to D. Then for every real $\varepsilon > 0$ there is an approximation $\psi \in \mathcal{A}(\phi)$ such that every $x \in C$ such that $\psi(x)$ is true belongs to D^ε.

References

[AFHL] S. Albeverio, J. E. Fenstad, R. Hoegh-Krohn, and T. Lindstrøm. *Nonstandard Methods in Stochastic Analysis and Mathematical Physics*, Academic Press, New York (1986).

[A1] R. Anderson. *A Nonstandard Representation for Brownian Motion and Ito Integration*, Israel J. Math 25 (1976), 15-46.

[A2] R. Anderson. *Almost Implies Near*, Trans. Amer. Math. Soc. 296 (1986), pp. 229-237.

[CK] N. Cutland and H. J. Keisler. *Applications of Neocompact Sets to Navier-Stokes Equations*, to appear.

[FK1] S. Fajardo and H. J. Keisler. *Existence Theorems in Probability Theory*, to appear.

[FK2] S. Fajardo and H. J. Keisler. *Neometric Spaces*, to appear.

[FK3] S. Fajardo and H. J. Keisler. *Long Sequences and Neocompact Sets*, This volume.

[FK4] S. Fajardo and H. J. Keisler. *Neometric Forcing*, to appear.

[He] C. Ward Henson. *Nonstandard Hulls of Banach Spaces*, Israel J. Math. 25 (1976), 108-144.

[HI] C. Ward Henson and Jose Iovino. *Banach Space Model Theory I*, to appear.

[HK] D. N. Hoover and H. J. Keisler. *Adapted Probability Distributions*, Trans. Amer. Math. Soc. 286 (1984), 159-201.

[K1] H. J. Keisler. *An Infinitesimal Approach to Stochastic Analysis*, Memoirs Amer. Math. Soc. 297 (1984).

[K2] H. J. Keisler. *Probability Quantifiers*, Model Theoretic Languages, Springer-Verlag, pages 509-556 in Model Theoretic Logics, edited by J. Barwise and S. Feferman, 1985.

[K3] H. J. Keisler. *From Discrete to Continuous Time*, Ann. Pure and Applied Logic 52 (1991), 99-141.

[K4] H. J. Keisler. *Quantifier Elimination for Neocompact Sets*, to appear.

[K5] H. J. Keisler. *Rich and Saturated Adapted Spaces*, to appear.

[Sk] A. V. Skorokhod. *Studies in the Theory of Random Processes*, Addison-Wesley, 1965.

University of Wisconsin, Madison WI.

E-mail keisler@math.wisc.edu

Sergio Fajardo and H. Jerome Keisler

Long sequences and neocompact sets

1. Introduction

Users of non-standard methods in mathematics have always been interested in the following question: what does non-standard analysis offer the mathematical community? The issues raised by this question are neverending and there is a whole spectrum of possible answers.

The paper [HK] offered an explanation from the point of view of logic, explaining how the principles used in the superstructure approach to nonstandard analysis are related to standard mathematical practice. One difficulty is that the mathematical community is not agreed on what "standard mathematical practice" is. [HK] used mathematical logic to provide a formal framework where these issues can be discussed.

The question posed above was approached from a different point of view in the series of papers beginning with [K1] and continuing with [FK1], [CK], [FK2], [K2] and [K3]. This series of papers develops the notion of a neometric space, and the whole program is explained in the survey paper [K6] in this volume. The approach may be intuitively described as follows. Start from a part of mathematics, probability theory, where nonstandard methods have clearly offered new insights and enriched the field with new and interesting results. Then isolate and present in "standard terms" those features of nonstandard practice that have made this success possible. The results appeared in [FK1] and [FK2] where the notions of neocompact sets and neometric families were presented, and the basic mathematical theory around these new concepts was developed.

A few words about these two papers will help to explain our reason for writing the present paper. In [FK1], entitled "Existence Theorems in Probability Theory", we developed a standard theory which captured the key elements from nonstandard analysis that made it possible to prove new existence theorems in stochastic analysis (see [AFHL], [K4] and [K5]). Using neocompact sets and neometric spaces we introduced a new class of probability spaces called "Rich Probability Spaces" and then proceeded to show that in those spaces the results obtained using nonstandard methods are true. The main new ingredient is that these results, in the new setting, are proved within standard mathematical practice. We just asked our readers to accept the existence of such spaces and then proceed to see what could be done with them.

In [FK2] we showed that rich spaces exist, a result that requires nonstandard analysis, and presented the theory of neometric spaces within the most general possible nonstandard framework, which we called the huge neometric family. There we explained how the properties of internal sets in nonstandard hulls give rise to neocompact sets and how the saturation property of the nonstandard universe translates into

251

countable compactness for the neometric family.

A mathematician accustomed to working with nonstandard methods, and in particular within probability theory using liftings and standard parts, may be surprised by the way the results are presented in those papers. Our aim in this paper is to shed some light on the origins of our ideas. We are going to present "a nonstandard theory which explains the standard theory that came out of observing nonstandard practice." Moreover, the results here can be used as a translation tool between traditional nonstandard arguments and the new theory of neometric spaces.

The idea centers around a fundamental fact from nonstandard analysis: sequences indexed by $\mathbb{N}$, the natural numbers, can be extended to sequences indexed by $^*\mathbb{N}$, the hyperintegers. This is the reason for the name long sequence. This elementary procedure allows us to capture many important facts from nonstandard practice.

We shall refer to the survey paper [K6] in this volume for the definitions and basic facts concerning the general notion of a neometric family, and in particular the huge neometric family.

Long sequences are introduced in Section 2 of this paper, and the theory is developed further and applied to the huge neometric family in Section 3. Needless to say, we assume the reader is familiar with the superstructure approach to nonstandard analysis (see [L], [AFHL] and [C]). Acquaintance with [FK1] and [FK2] is highly desirable to get the complete picture of the subject.

This research was partially financed by Colciencias, the University of los Andes, the National Science Foundation, and the Vilas Trust Fund at the University of Wisconsin-Madison.

2. Long Sequences

One way to bring nonstandard analysis to bear in proofs by convergence is to use sequences indexed by the hyperintegers rather than the integers. We shall call such sequences **long sequences**. The paper [K2] made extensive use of long sequences, without using this name. In this section we develop some relations between long sequences and neometric spaces.

We fix an $\aleph_1$-saturated nonstandard universe and let $(\mathbf{H}, \mathcal{B}, \mathcal{C})$ be its huge neometric family as defined in [K6].

Definition 2.1. *A function $\langle x_n \rangle$ mapping $\mathbb{N}$ into a set S will be called a* **sequence** *in S, and a (possibly external) function $\langle x_J \rangle$ mapping $^*\mathbb{N}$ into S will be called a* **long sequence** *in S.*

As a warm-up, before establishing the connection between long sequences and neocompact spaces, we prove some basic facts about sets of hyperintegers and long sequences. We frequently use the following consequences of ω_1-saturation (e.g. see [SB]):

Lemma 2.2.
   (i) *The infinite hyperintegers have coinitiality ω_1, that is, every countable set of elements of $^*\mathbb{N} - \mathbb{N}$ has an infinite lower bound.*
   (ii) *For every internal set S and every sequence $\langle X_n \rangle$ in S, there exists an internal long sequence $\langle Y_J \rangle$ in S such that $Y_n = X_n$ for all $n \in \mathbb{N}$.* □

Definition 2.3. *We say that a statement $\phi(J)$ holds a.e., or that $\phi(J)$ holds for **all sufficiently small infinite** J, if there is an infinite hyperinteger K such that $\phi(J)$ is true for all infinite hyperintegers $J \leq K$.*

The following lemma is often used to verify that a statement holds a.e.

Lemma 2.4.
   (i) *(Overspill principle) If S is an internal subset of $^*\mathbb{N}$, then $J \in S$ a.e. if and only if $n \in S$ for all but finitely many $n \in \mathbb{N}$.*
   (ii) *(Countable completeness) The set of all $S \subset {}^*\mathbb{N}$ such that $J \in S$ a.e. is a countably complete filter.*

Proof. (i) is in any book on nonstandard analysis. (ii) If $J \in S$ a.e. and $S \subset T$, then obviously $J \in T$ a.e. Suppose $J \in S_n$ a.e. for all $n \in \mathbb{N}$, and let $S = \bigcap_n S_n$. Then for each $n \in \mathbb{N}$ there is an infinite hyperinteger K_n such that $J \in S_n$ for all infinite $J \leq K_n$. By ω_1-saturation there is an infinite hyperinteger K such that $K \leq K_n$ for all $n \in \mathbb{N}$. Then $J \in S$ for all infinite $J \leq K$, so $J \in S$ a.e. □

The overspill principle will often be used in the following form.

Corollary 2.5. *Let $\langle X_J \rangle$ be an internal long sequence in $^*\mathbb{R}$ and let $b \in {}^*\mathbb{R}$.*
   (i) $X_J \leq b$ *a.e. if and only if $X_n \leq b$ for all but finitely many $n \in \mathbb{N}$.*
   (ii) $X_J \approx 0$ *a.e. if and only if*
$$\lim_{n \to \infty} st(X_n) = 0.$$

Proof. (i) Apply the overspill principle to the internal set $S = \{J \in {}^*\mathbb{N} : X_J \leq b\}$.
(ii) follows from (i) and countable completeness. □

We now turn to long sequences in neometric spaces in the huge neometric family $(\mathbf{H}, \mathcal{B}, \mathcal{C})$. For the remainder of this paper, $\mathcal{M}$ and $\mathcal{N}$ will always belong to $\mathbf{H}$. The following is the key new concept we introduce in this paper.

Definition 2.6. *If $\langle x_J \rangle$ is a long sequence in $\mathcal{M}$ and $\langle X_J \rangle$ is an internal long sequence in $\bar{M}$ such that $x_J = {}^\circ X_J$ for all finite J and all sufficiently small infinite J, we say that $\langle X_J \rangle$ **lifts** $\langle x_J \rangle$. By an $\mathcal{M}$-**sequence** we shall mean a long sequence $\langle x_J \rangle$ in $\mathcal{M}$ which has a lifting. A (short) sequence $\langle x_n \rangle$ of elements of $\mathcal{M}$ will be said to be $\mathcal{M}$-**extendible** if it is the restriction to $^*\mathbb{N}$ of some $\mathcal{M}$-sequence $\langle x_J \rangle$, and $\langle x_J \rangle$ will be called an $\mathcal{M}$-**extension** of $\langle x_n \rangle$.*

By ω_1-saturation, for every sequence $\langle x_n \rangle$ in $\mathcal{M}$ there is an internal long sequence $\langle X_J \rangle$ such that X_n lifts x_n, and hence $X_n \in \text{monad}(\mathcal{M})$ for each $n \in \mathbb{N}$. If in addition we have $X_J \in \text{monad}(\mathcal{M})$ a.e., then the sequence $\langle x_n \rangle$ is $\mathcal{M}$-extendible and its $\mathcal{M}$-extension is the $\mathcal{M}$-sequence given by $x_J = {}^\circ X_J$ a.e. If $\langle x_n \rangle$ is an $\mathcal{M}$-extendible sequence, we use the convention that $\langle x_J \rangle$ denotes an $\mathcal{M}$-extension of $\langle x_n \rangle$.

The next proposition shows that the notion of an $\mathcal{M}$-extendible sequence is a generalization of the notion of a convergent sequence.

Proposition 2.7.
(i) If
$$\lim_{n \to \infty} x_n = b$$
in $\mathcal{M}$, then $\langle x_n \rangle$ is $\mathcal{M}$-extendible and $\langle x_J \rangle = b$ a.e.
(ii) If $\langle x_n \rangle$ is $\mathcal{M}$-extendible and $\langle y_n \rangle$ is a sequence in M such that
$$\lim_{n \to \infty} \rho(x_n, y_n) = 0,$$
then $\langle y_n \rangle$ is $\mathcal{M}$-extendible.

Proof. (i) Let X_n lift x_n, let Y lift b, and extend $\langle X_n \rangle$ to a long sequence $\langle X_J \rangle$. By overspill, $X_J \approx Y$ and hence ${}^\circ X_J = b \in \mathcal{M}$ a.e. Therefore $\langle x_J \rangle = \langle {}^\circ X_J \rangle$ is an $\mathcal{M}$-extension of $\langle x_n \rangle$ and $x_J = b$ a.e.

(ii) Let $\langle x_J \rangle$ be an $\mathcal{M}$-extension of $\langle x_n \rangle$ and let $\langle X_J \rangle$ lift $\langle x_J \rangle$. For each $n \in \mathbb{N}$, let Y_n lift y_n, and by ω_1-saturation let $\langle Y_J \rangle$ be an internal long sequence in $\bar{M}$ extending $\langle Y_n \rangle$. Then
$$\lim_{n \to \infty} {}^\circ \bar{\rho}(X_n, Y_n) = 0,$$
so by overspill, $X_J \approx Y_J$ a.e. It follows that $\langle Y_J \rangle$ lifts an $\mathcal{M}$-sequence $\langle y_J \rangle$ which extends $\langle y_n \rangle$, whence $\langle y_n \rangle$ is $\mathcal{M}$-extendible. □

Given a product $\mathcal{M} \times \mathcal{N}$ of two spaces $\mathcal{M}, \mathcal{N} \in \mathbf{H}$, $\langle z_J \rangle$ is an $(\mathcal{M} \times \mathcal{N})$-sequence if and only if there is an $\mathcal{M}$-sequence $\langle x_J \rangle$ and an $\mathcal{N}$-sequence $\langle y_J \rangle$ such that $z_J = (x_J, y_J)$ a.e. Thus if $\langle x_n \rangle$ is $\mathcal{M}$-extendible and $\langle y_n \rangle$ is $\mathcal{N}$-extendible, then the sequence of pairs $\langle z_n \rangle = \langle (x_n, y_n) \rangle$ is $\mathcal{M} \times \mathcal{N}$-extendible, and $z_J = (x_J, y_J)$ a.e.

The following shows that the $\mathcal{M}$-extension of a sequence is unique a.e.

Proposition 2.8.
(i) Let $\langle x_J \rangle$ and $\langle y_J \rangle$ be $\mathcal{M}$-sequences. Then
$$\lim_{n \to \infty} \rho(x_n, y_n) = 0$$
if and only if $x_J = y_J$ a.e.
(ii) (Uniqueness of the $\mathcal{M}$-extension) Let $\langle x_J \rangle$ and $\langle y_J \rangle$ be two $\mathcal{M}$-extensions of the same sequence $\langle x_n \rangle$ in M. Then $x_J = y_J$ a.e.

Proof. (i) Let $\langle X_J \rangle$ lift $\langle x_J \rangle$ and $\langle Y_J \rangle$ lift $\langle y_J \rangle$. By overspill, the following are equivalent:
$$\lim_{n \to \infty} \rho(x_n, y_n) = 0.$$
$$\lim_{n \to \infty} {}^\circ \bar{\rho}(X_n, Y_n) = 0.$$
$$\bar{\rho}(X_J, Y_J) \approx 0 \text{ a.e.}$$
$$x_J = y_J \text{ a.e.}$$

(ii) is a special case of (i). □

Proposition 2.9. *If $\langle x_n \rangle$ is $\mathcal{M}$-extendible, then for each $c \in \mathcal{M}$, the sequence $\langle \rho(x_n, c) \rangle$ is bounded in $\mathbb{R}$.*

Proof. Let $\langle X_J \rangle$ be a lifting of an $\mathcal{M}$ extension $\langle x_J \rangle$ of $\langle x_n \rangle$. Suppose $\rho(x_n, c)$ is not bounded. Let $\bar{c}$ lift c. Then for each $k \in \mathbb{N}$ there are arbitrarily large $n \in \mathbb{N}$ such that $\rho(x_n, c) > k$, and hence $\bar{\rho}(X_n, \bar{c}) \geq k$. By overspill, for each infinite $K \in {}^*\mathbb{N}$ there is an infinite $J \leq K$ such that $\bar{\rho}(X_J, \bar{c})$ is infinite. This contradicts the hypothesis that $\langle X_J \rangle$ is a lifting of $\langle x_J \rangle$. □

3. The Huge Neometric Family

We now give conditions for neocompactness, neoclosedness, neocontinuity, and neoseparability in the huge family **H** in terms of long sequences. The first proposition is crucial.

Proposition 3.1. *Let $\langle x_J \rangle$ be an $\mathcal{M}$-sequence. Then for all sufficiently small infinite K, the set $\{x_J : J \leq K\}$ is basic in $\mathcal{M}$ and the set*

$$\{x_J : J \leq K \text{ and } J \text{ is infinite }\}$$

is neocompact in $\mathcal{M}$.

Proof. Let $\langle X_J \rangle$ lift $\langle x_J \rangle$ with respect to $\mathcal{M}$. Then for all sufficiently small infinite K, ${}^\circ X_J = x_J$ for all $J \leq K$. For any such K, let

$$B = \{X_J : J \leq K\}, C = \{x_J : J \leq K\},$$

$$D = \{x_J : J \leq K \text{ and } J \text{ is infinite }\}.$$

Then B is internal and $C = {}^\circ B$, so C is basic. Moreover,

$$D = {}^\circ(\bigcap_n (B - \{X_m : m \leq n\})),$$

so D is neocompact. □

We need the notion of a countably determined set, which was introduced by Henson [He] and played an important role in [K2].

Definition 3.2. *A set $D \subset \bar{M}$ is **countably determined** if there is a countable sequence $\langle D_n \rangle$ of internal subsets of $\bar{M}$ such that D is an infinite Boolean combination of the D_n's. Equivalently, there is a countable sequence $\langle D_n \rangle$ of internal subsets of $\bar{M}$ and a set S of subsets of $\mathbb{N}$ such that*

(1) $$D = \bigcup_{F \in S} (\bigcap_{n \in F} D_n).$$

In fact, this representation can be chosen so that for any distinct $F, G \in S$, the intersections $\bigcap_{n \in F} D_n$ and $\bigcap_{n \in G} D_n$ are disjoint from each other.

Note that every internal, Π_1^0, and Σ_1^0 set is countably determined.

Theorem 3.3. *A set $C \subset \mathcal{M}$ is neocompact if and only if*
   (a) The monad of C is countably determined, and
   (b) Every (short) sequence $\langle x_n \rangle$ in C has an $\mathcal{M}$-extension to a long sequence $\langle x_J \rangle$ in C.

Proof. First assume that C is neocompact. By Basic Fact 2.3 in [K6], there exists a sequence of internal sets $\langle C_n \rangle$ such that
$$\text{monad}(C) = \bigcap_m ((C_m)^{1/m}).$$
So the monad of C is countably determined. Let $\langle x_n \rangle$ be a (short) sequence in C. $\langle x_n \rangle$ has a lifting $\langle X_J \rangle$. For each n and m, $X_n \in (C_m)^{1/m}$. By overspill and countable completeness,
$$X_J \in \bigcap_m ((C_m)^{1/m}) \text{ a.e.,}$$
so $\langle X_J \rangle$ lifts an $\mathcal{M}$-extension $\langle x_J \rangle$ in C.

Now assume that the monad of C is countably determined and that every (short) sequence $\langle x_n \rangle$ in C has an $\mathcal{M}$-extension to a long sequence $\langle x_J \rangle$ in C. Then the monad of C can be represented in the form (1) with any two distinct intersections being disjoint. We claim that
$$\text{monad}(C) = \bigcap \{B : \text{ for some finite } s \subset \mathbb{N}, B = (\bigcup_{n \in s} C_n) \supset \text{monad}(C)\}.$$
This will show that the monad of C is a Π^0_1 set, and hence that C is neocompact. Clearly monad(C) is included in the right side. Suppose X belongs to the right side. Let $G = \{n \in \mathbb{N} : X \notin C_n\}$. Since X belongs to the right side, for each n there exists
$$Y_n \in \text{monad}(C) - \bigcup\{C_k : n \geq k \in G\}.$$
Let $y_n = {}^\circ Y_n$. Then $\langle y_n \rangle$ is a (short) sequence in C, so by hypothesis it has an $\mathcal{M}$-extension $\langle Y_J \rangle$ in C. We have $Y_J \in \text{monad}(C)$ for all J. By overspill,
$$Y_J \notin \bigcup\{C_k : k \in G\} \text{ a.e.}$$
Take an infinite J with this property. Then for some $F \in S$, $Y_J \in \bigcap_{n \in F} C_n$. Moreover, for all $n \in F$, $Y_J \in C_n$ and hence $n \notin G$ and $X \in C_n$. Therefore
$$X \in \bigcap_{n \in F} C_n \subset \text{monad}(C).$$
This proves our claim and completes the proof. $\square$

Just as Basic Fact 2.3 in [K6] shows that the monad of a neocompact set is countably determined, Basic Fact 2.4 in [K6] shows that the monad of a neoseparable set is countably determined.

The following corollary is a good illustration of how our neometric theory is closely related to the classical theory of metric spaces. Notice what happens if you replace "$\mathcal{M}$-extendible" by "relatively compact".

Corollary 3.4. *A sequence $\langle x_n \rangle$ in $\mathcal{M}$ is $\mathcal{M}$-extendible if and only if there is a neocompact set $C \subset \mathcal{M}$ such that $x_n \in C$ for all $n \in \mathbb{N}$.*

Proof. If $\langle x_n \rangle$ has an $\mathcal{M}$-extension $\langle x_J \rangle$, then by Proposition 3.1 the set $C = \{x_J : J \leq K\}$ is neocompact for some infinite K, and $\{x_n : n \in \mathbb{N}\} \subset C \subset \mathcal{M}$. If C is neocompact and $\{x_n : n \in \mathbb{N}\} \subset C \subset \mathcal{M}$, then $\langle x_n \rangle$ is $\mathcal{M}$-extendible by Theorem 3.3. □

In applications of long sequences, it is important to know which sequences are $\mathcal{M}$-extendible. We can use Corollary 3.4 to characterize the $\mathcal{M}$-extendible sequences in various particular neometric spaces which have been studied in [K1], [FK1] and [FK2]. The next example characterizes the extendible sequences in a nonstandard hull.

Example 3.5. *A sequence $\langle x_n \rangle$ is $\mathcal{H}(\bar{M}, c)$-extendible if and only if $\rho(x_n, d)$ is bounded where $d \in \mathcal{H}(\bar{M}, c)$.*

Proof. By Proposition 2.9, for any $\mathcal{H}(\bar{M}, c)$-extendible sequence $\langle x_n \rangle$ and any $d \in \mathcal{H}(\bar{M}, c)$, the sequence $\langle \rho(x_n, d) \rangle$ is bounded. Suppose $\langle x_n \rangle$ is a sequence in $\mathcal{H}(\bar{M}, c)$ such that $\rho(x_n, d)$ has a finite bound b. Each x_n belongs to the closed ball $B = \{y \in \mathcal{H}(\bar{M}, c) : \rho(y, d) \leq b\}$. B is the standard part of an internal set and is therefore basic in $\mathcal{H}(\bar{M}, c)$. Thus by Corollary 3.4, $\langle x_n \rangle$ is $\mathcal{H}(\bar{M}, c)$-extendible. □

Let's now consider standard neometric spaces. We shall see that the only $\mathcal{M}$-extendible sequences on a standard neometric space $\mathcal{M} \in \mathbf{S}$ are the trivial ones, that is, the relatively compact sequences.

By definition, a sequence $\langle x_n \rangle$ in a complete metric space $\mathcal{M}$ is **relatively compact** if there is a compact set $C \subset \mathcal{M}$ which contains each x_n, or equivalently, every subsequence of $\langle x_n \rangle$ has a convergent subsequence. Thus a sequence in Euclidian space $\mathbb{R}^m$ is relatively compact if and only if it is bounded. By Corollary 3.4, in every neometric space $\mathcal{M}$, every relatively compact sequence is $\mathcal{M}$-extendible.

Example 3.6. *Let $\mathcal{M}$ be a standard neometric space. Then a sequence $\langle x_n \rangle$ in $\mathcal{M}$ is $\mathcal{M}$-extendible if and only if $\langle x_n \rangle$ is relatively compact.*

Proof. This follows from Corollary 3.4 and Basic Fact 2.5 in [K6]. □

For the following examples let Ω be a Loeb probability space. The paper [K1] gave characterizations of the $\mathcal{M}$-extendible sequences when M is a complete separable metric space and $\mathcal{M}$ is either the space $L^0(\Omega, M)$ of Loeb measurable functions with the metric of convergence of probability or the space $L^p(\Omega, M)$ where $p \in [1, \infty)$. In fact, these results were the original inspiration for the long sequences approach to neometric spaces.

If $x \in L^0(\Omega, M)$, the Borel probability measure on M induced by x is denoted by law(x). The space of all Borel probability measures on M with the Prohorov metric is denoted by Meas(M). (See, for example [EK]).

Example 3.7. *([K1], Theorem 3.2 and Lemmas 7.2 and 7.4). Let Ω be a Loeb probability space and M be a complete separable metric space.*

(i) A sequence $\langle x_n \rangle$ is $L^0(\Omega, M)$-extendible if and only if $\langle \text{law}(x_n) \rangle$ is relatively compact in Meas(M).

(ii) *Let $p \in [1, \infty)$. A sequence $\langle x_n \rangle$ is $L^p(\Omega, \mathcal{M})$-extendible if and only if $\langle law(x_n) \rangle$ is relatively compact in $Meas(\mathcal{M})$ and $(\rho(x_n(\cdot), a))^p$ is uniformly integrable for each $a \in \mathcal{M}$.*

The next result gives another characterization of neocompact sets in the case that $\mathcal{M}$ is neoseparable.

Proposition 3.8. *Suppose $\mathcal{M}$ is neoseparable. A set $C \subset \mathcal{M}$ is neocompact if and only if C is neoclosed in $\mathcal{M}$ and (b) of Theorem 3.3 holds, that is, every sequence $\langle x_n \rangle$ in C has an $\mathcal{M}$-extension $\langle x_J \rangle$ in C.*

Proof. Neocompactness implies neoclosed and (b) by Basic Fact 1.1 in [K6] and Theorem 3.3. Assume that (b) holds and that C is neoclosed but not neocompact. By Basic Fact 2.6 in [K6], C has a countable covering $\{O_n : n \in \mathbb{N}\}$ by neoopen sets in $\mathcal{M}$ which has no finite subcover. Let $C_n = C - (\bigcup_{k \leq n} O_k)$. Then $\langle C_n \rangle$ is a decreasing chain of nonempty neoclosed sets in $\mathcal{M}$, and $\bigcap_n C_n$ is empty. Choose $x_n \in C_n$. Then $x_n \in C$, and by (b) we can extend $\langle x_n \rangle$ to a long sequence $\langle x_J \rangle$ in C. By Proposition 3.1 we may choose an infinite K so that the set $S = \{x_J : J \leq K\}$ is basic. Then $\langle S \cap C_n \rangle$ is a decreasing chain of neocompact subsets of C. By countable compactness of the huge neometric family, the intersection

$$\bigcap_n (S \cap C_n) = S \cap \left(\bigcap_n C_n \right)$$

is nonempty, and this is a contradiction. □

Theorem 3.9. *Let $C \subset \mathcal{M}$. If C is neoclosed in $\mathcal{M}$ then*
(c) For every $\mathcal{M}$-sequence $\langle x_J \rangle$ such that $x_n \in C$ for all $n \in \mathbb{N}$, we have $x_J \in C$ a.e.

If the monad of C is countably determined, then C is neoclosed in $\mathcal{M}$ if and only if this condition holds.

Proof. Suppose first that C is neoclosed in $\mathcal{M}$. Let $\langle x_n \rangle$ be a sequence in C and $\langle x_J \rangle$ be an $\mathcal{M}$-extension of $\langle x_n \rangle$. By Proposition 3.1 there is an infinite K such that the set $D = \{x_J : J \leq K\}$ is basic in $\mathcal{M}$. Then $C \cap D$ is neocompact in $\mathcal{M}$, and $x_n \in C \cap D$ for all $n \in \mathbb{N}$. By Theorem 3.3, $\langle x_n \rangle$ has an $\mathcal{M}$-extension to a long sequence $\langle y_J \rangle$ in $C \cap D$. By uniqueness of the $\mathcal{M}$-extension, $y_J = x_J$ a.e. Then $x_J \in C$ a.e., and (c) is proved.

Now suppose that the monad of C is countably determined and (c) holds. Let D be neocompact in $\mathcal{M}$. Then the monad of D is countably determined. Since

$$\operatorname{monad}(C \cap D) = \operatorname{monad}(C) \cap \operatorname{monad}(D),$$

$\operatorname{monad}(C \cap D)$ is countably determined. Let $\langle x_n \rangle$ be a sequence in $C \cap D$. By Theorem 3.3, $\langle x_n \rangle$ has an $\mathcal{M}$-extension to a long sequence $\langle x_J \rangle$ in D. By condition (c), $x_J \in C$ a.e. Then $\langle x_J \rangle$ is a long sequence in $C \cap D$. We have shown that conditions (a) and (b) of Theorem 3.3 hold for $C \cap D$. By Theorem 3.3, $C \cap D$ is neocompact in $\mathcal{M}$, so C is neoclosed in $\mathcal{M}$. □

Corollary 3.10. *Suppose $C \subset \mathcal{M}$, C is neoclosed, and $\langle x_J \rangle$ is an $\mathcal{M}$-sequence such that*
$$\lim_{n \to \infty} \rho(x_n, C) = 0.$$
Then $x_J \in C$ a.e.

Proof. For each $n \in \mathbb{N}$ we may choose $y_n \in C$ such that $\rho(x_n, y_n) \leq 2\rho(x_n, C)$. Then
$$\lim_{n \to \infty} \rho(x_n, y_n) = 0,$$
so by Propositions 2.7 and 2.8, $\langle y_n \rangle$ is $\mathcal{M}$-extendible and $x_J = y_J$ a.e. Theorem 3.9 shows that $y_J \in C$ a.e., and therefore $x_J \in C$ a.e. $\square$

Observe that from the above theorem it follows right away that every neoclosed set in the huge neometric family is closed. Basic Fact 1.7 in [K6] says that this is true in all neometric families. Now, let's take a look at a characterization of neocontinuity in terms of $\mathcal{M}$-sequences.

Theorem 3.11. *Let $C \subset \mathcal{M}$ be neoclosed, and let $f : C \to \mathcal{N}$. If f is neocontinuous from $\mathcal{M}$ to $\mathcal{N}$, then*
(d) For any $\mathcal{M}$-sequence $\langle x_J \rangle$ in C, $\langle f(x_J) \rangle$ is an $\mathcal{N}$-sequence.
If the monad of the graph of f is countably determined, then f is neocontinuous from $\mathcal{M}$ to $\mathcal{N}$ if and only if this condition holds.

Proof. This generalizes a result from [K2]. Suppose first that f is neocontinuous from $\mathcal{M}$ to $\mathcal{N}$. Let $\langle x_J \rangle$ be an $\mathcal{M}$-sequence in C. By Proposition 3.1 there is an infinite K such that the set $D = \{x_J : J \leq K\}$ is basic in $\mathcal{M}$. By Basic Fact 2.7 in [K6] there is an internal function F such that $°F(X) = f(°X)$ for all $X \in \text{monad}(D)$. Since $\langle X_J \rangle$ and F are internal, $\langle F(X_J) \rangle$ is internal. For all $J \leq K$, $°F(X_J) = f(°X_J) = f(x_J)$. Therefore $\langle F(X_J) \rangle$ is a lifting of $\langle f(x_J) \rangle$ with respect to $\mathcal{N}$, so $\langle f(x_J) \rangle$ is an $\mathcal{N}$-sequence. This proves the first half.

Now suppose the monad of the graph of f is countably determined, and assume (d). Let $D \subset C$ be neocompact. Then $\text{monad}(D)$ is countably determined, so
$$\text{monad}(f|D) = \text{monad}(f) \cap (\text{monad}(D) \times \bar{N})$$
is countably determined. Let $\langle x_n, f(x_n) \rangle$ be a sequence in the graph of $f|D$. By Theorem 3.3, $\langle x_n \rangle$ has an $\mathcal{M}$-extension $\langle x_J \rangle$ in D. By hypothesis, $\langle f(x_J) \rangle$ is an $\mathcal{N}$-sequence. Then $\langle x_J, f(x_J) \rangle$ is an $(\mathcal{M} \times \mathcal{N})$-extension of $\langle x_n, f(x_n) \rangle$ in $f|D$. Therefore by Theorem 3.9, $f|D$ is neocompact in $\mathcal{M} \times \mathcal{N}$, so f is neocontinuous from $\mathcal{M}$ to $\mathcal{N}$. $\square$

Corollary 3.12. *Let $C \subset \mathcal{M}$, and let $f : C \to \mathcal{N}$ be neocontinuous from $\mathcal{M}$ to $\mathcal{N}$. If a sequence $\langle x_n \rangle$ in C is $\mathcal{M}$-extendible, then $\langle f(x_n) \rangle$ is $\mathcal{N}$-extendible to an $\mathcal{N}$-sequence $\langle y_J \rangle$, and $f(x_J) = y_J$ a.e.* $\square$

Finally, we give a necessary condition for neoseparability in terms of long sequences. An open question is whether this condition, together with the condition that the monad of the set is countably determined, is sufficient for neoseparability.

Proposition 3.13. *Let C be neoseparable in $\mathcal{M}$, and let $\langle x_J \rangle$ be an $\mathcal{M}$-sequence such that $x_J \in C$ a.e. Then for each $k \in \mathbb{N}$, $x_n \in C^{1/k}$ for all but finitely many $n \in \mathbb{N}$.*

Proof. Let $\mathrm{monad}(C) = \bigcap_n \bigcup_m (C_m)^{1/n}$. Suppose that there is a $k \in \mathbb{N}$ and an infinite subset $p \subset \mathbb{N}$ such that $x_n \notin C^{1/k}$ for all $n \in p$. By taking a subsequence, we may assume without loss of generality that $x_n \notin C^{1/k}$ for all $n \in \mathbb{N}$. Let $\langle X_J \rangle$ lift $\langle x_n \rangle$. Then $X_J \in \bigcap_n \bigcup_m (C_m)^{1/n}$ a.e., and hence $X_J \in \bigcup_m (C_m)^{1/k}$ a.e. However, for all $n \in \mathbb{N}$ we have $X_n \notin \bigcup_m (C_m)^{1/k}$ because $^\circ(\bigcup_m (C_m)^{1/k}) \subset C^{1/k}$. Then for each $n, m \in \mathbb{N}$ we have $X_n \notin (C_m)^{1/k}$, and by overspill, $X_J \notin (C_m)^{1/k}$ a.e. By countable completeness, $X_J \notin \bigcup_m (C_m)^{1/k}$ a.e., contrary to our previous assumption. $\square$

References

[AFHL] S. Albeverio, J. E. Fenstad, R. Hoegh-Krohn, and T. Lindstrøm. *Nonstandard Methods in Stochastic Analysis and Mathematical Physics*, Academic Press, New York (1986).

[C] N. Cutland. Editor. *Nonstandard Analysis and its Applications*, LMSST 10. Cambridge University Press. 1988.

[CK] N. Cutland and H. J. Keisler. *Applications of Neocompact Sets to Navier-Stokes Equations*, To appear.

[EK] S. Ethier and T. Kurtz. *Markov Processes*, Wiley (1986).

[FK1] S. Fajardo and H. J. Keisler. *Existence Theorems in Probability Theory*, To appear.

[FK2] S. Fajardo and H. J. Keisler. *Neometric Spaces*, To appear.

[FK3] S. Fajardo and H. J. Keisler. *Neometric Forcing*, To appear.

[H] C. W. Henson. *Analytic Sets, Baire Sets, and the Standard Part Map*, Canadian J. Math. 31 (1979), pp. 663-672.

[HK] C. W. Henson and H. J. Keisler. *On the strength of nonstandard analysis*, J. Symb. Logic, vol. 51 (1986), pp. 377-386.

[K1] H. J. Keisler. *From Discrete to Continuous Time*, Ann. Pure and Applied Logic 52 (1991), 99-141.

[K2] H. J. Keisler. *Quantifier Elimination for Neocompact Sets*, To appear.

[K3] H. J. Keisler. *Rich and Saturated Adapted Spaces*, To appear.

[K4] H. J. Keisler. *An Infinitesimal Approach to Stochastic Analysis*, Memoirs Amer. Math. Soc. 297 (1984).

[K5] H. J. Keisler. *Infinitesimals in Probability Theory*, In [C] above.

[K6] H. J. Keisler. *A Neometric Survey*, This Volume.

[Li] T. Lindstrøm. *An Invitation to Nonstandard Analysis*, pp 1105 in Nonstandard Analysis and its Applications, ed. by N. Cutland, London Math. Soc. (1988).

SERGIO FAJARDO
Universidad de Los Andes and Universidad Nacional
Bogota Colombia
E-mail sfajardo@cdcnet.uniandes.edu.co

H. JEROME KEISLER
University of Wisconsin
Madison WI USA
E-mail keisler@math.wisc.edu